ENVIRONMENTAL REMEDIATION TECHNOLOGIES, REGULATIONS AND SAFETY

ENVIRONMENTAL PROTECTION AGENCY (EPA)

ISSUES, CHALLENGES AND POLICIES

ENVIRONMENTAL REMEDIATION TECHNOLOGIES, REGULATIONS AND SAFETY

Additional books and e-books in this series can be found on Nova's website under the Series tab.

ENVIRONMENTAL REMEDIATION TECHNOLOGIES, REGULATIONS AND SAFETY

ENVIRONMENTAL PROTECTION AGENCY (EPA)

ISSUES, CHALLENGES AND POLICIES

COBY SCHAEFER
EDITOR

snova
New York

NOTICE TO THE READER

Library of Congress Cataloging-in-Publication Data

ISBN: 978-1-53614-933-3

Published by Nova Science Publishers, Inc. † New York

CONTENTS

Preface		**vii**
Chapter 1	EPA Regulations: Too Much, Too Little, or on Track? (Updated) *James E. McCarthy and Claudia Copeland*	**1**
Chapter 2	Federal Pollution Control Laws: How Are They Enforced? *Robert Esworthy*	**49**
Chapter 3	EPA Policies Concerning Integrated Planning and Affordability of Water Infrastructure *Jonathan L. Ramseur*	**105**
Chapter 4	EPA's Vessel General Permits: Background and Issues *Claudia Copeland*	**119**
Chapter 5	Implementing EPA's 2015 Ozone Air Quality Standards *James E. McCarthy*	**145**
Chapter 6	EPA's Mid-Term Evaluation of Vehicle Greenhouse Gas Emissions Standards *Richard K. Lattanzio*	**179**
Chapter 7	EPA's Methane Regulations: Legal Overview *Linda Tsang*	**183**
Chapter 8	EPA's and BLM's Methane Rules *Richard K. Lattanzio*	**209**
Chapter 9	EPA's Wood Stove/Wood Heater Regulations: Frequently Asked Questions *James E. McCarthy and Kate C. Shouse*	**213**

Chapter 10 EPA's Role in Emergency Planning and Notification at Chemical Facilities **227**
Richard K. Lattanzio and David M. Bearden

Index **237**

PREFACE

Since 2009, the U.S. Environmental Protection Agency (EPA) has proposed and promulgated numerous regulations to implement the pollution control statutes enacted by Congress. Critics have reacted strongly. Some, both within Congress and outside of it, have accused the agency of reaching beyond the authority given it by Congress and ignoring or underestimating the costs and economic impacts and overestimating the benefits of proposed and promulgated rules.

Chapter 1 - Since Barack Obama was sworn in as President in 2009, the U.S. Environmental Protection Agency (EPA) has proposed and promulgated numerous regulations to implement the pollution control statutes enacted by Congress. Critics have reacted strongly. Some, both within Congress and outside of it, have accused the agency of reaching beyond the authority given it by Congress and ignoring or underestimating the costs and economic impacts and overestimating the benefits of proposed and promulgated rules. The House conducted vigorous oversight of the agency in the 112th and 113th Congresses, and approved several bills that would overturn specific regulations or limit the agency's authority. Particular attention was paid to the Clean Air Act, but there also was congressional scrutiny of other environmental statutes and regulations implemented by EPA. With Republican majorities in both the House and Senate, the 114th Congress accelerated oversight of the Administration's initiatives and renewed efforts to limit EPA's regulatory activities. Environmental groups and other supporters of the agency disagree that EPA has overreached. Many of them believe that the agency has, in fact, moved in the right direction, including taking action on significant issues that had been long delayed or ignored in the past. In several cases, environment and public health advocates would have liked the regulatory actions to be stronger. EPA has stated that critics' focus on the cost of controls obscured the benefits of new regulations, which, it estimates, far exceed the costs. It maintains that pollution control is an important source of economic activity, exports, and American jobs. Further, the agency and its supporters have said that EPA is carrying out the mandates detailed by Congress in the federal environmental statutes. This chapter provides background information on EPA regulatory activity during the Obama Administration to help address these issues. It examines major or controversial regulatory actions taken by or under development at EPA from January 2009 to late 2016, providing details on the regulatory action itself, presenting an estimated timeline for completion of rules not yet promulgated (including identification of related court or statutory deadlines), and, in general, providing EPA's estimates of costs and benefits, where available. The chapter also discusses factors that affect the time frame in which regulations take effect, including statutory and judicial deadlines, public comment periods, judicial review, and permitting procedures, the net results of which are that existing facilities are likely to have several years before being required to comply with most of the regulatory actions under discussion. Unable to account for such factors, which will vary from case to case, timelines that show dates for proposal and promulgation of EPA regulations effectively underestimate the complexities of the regulatory process and overstate the near-term impact of many of the regulatory actions.

Chapter 2 - As a result of enforcement actions and settlements for noncompliance with federal pollution control requirements, the U.S. Environmental Protection Agency (EPA) reported that, during FY2013, regulated entities committed to invest an estimated $7.0 billion for judicially mandated actions and equipment to control pollution (injunctive relief), and $22.0 million for implementing mutually agreed-upon (supplemental) environmentally beneficial projects. EPA estimated that these compliance/enforcement efforts achieved commitments to reduce or eliminate 1.3 billion pounds of pollutants in the environment, primarily from air and water, and to treat, minimize, or properly dispose of 148 million pounds of hazardous waste. Noncompliance with federal pollution control laws remains a continuing concern. The overall effectiveness of the enforcement organizational framework, the balance between state autonomy and federal oversight, and the adequacy of funding are long-standing congressional concerns. This chapter provides an overview of the statutory framework, key players, infrastructure, resources, tools, and operations associated with enforcement and compliance of the major pollution control laws and regulations administered by EPA. It also outlines the roles of federal (including regional offices) and state regulators, as well as the regulated community. Understanding the many facets of how all federal pollution control laws are enforced, and the responsible parties involved, can be challenging. Enforcement of the considerable body of these laws involves a complex framework and organizational setting. The array of enforcement/compliance tools employed to achieve and maintain compliance includes monitoring, investigation, administrative and judicial (civil and criminal) actions and penalties, and compliance assistance and incentive approaches. Most compliance violations are resolved administratively by the states and EPA. EPA concluded 1,440 final administrative penalty orders in FY2013. Civil judicial actions, which may be filed by states or EPA, are the next most frequent enforcement action. EPA may refer civil cases to the U.S. Department of Justice (DOJ), referring 138 civil cases in FY2013. The U.S. Attorney General's Office and DOJ's Environmental Crimes Section, or the state attorneys general, in coordination with EPA criminal investigators and general counsel, may prosecute criminal violations against individuals or entities who knowingly disregard environmental laws or are criminally negligent. EPA reported the assessment of nearly $1.15 billion in civil penalties (administrative and judicial) and $4.5 billion in combined criminal fines, restitution, and court-ordered environmental projects during FY2013. Of the FY2013 totals, $1.0 billion in civil penalties and $4.0 billion in criminal fines, restitution, and court-ordered projects were associated with the *Deepwater Horizon* Gulf of Mexico judicial and criminal cases. Federal appropriations for environmental enforcement and compliance activities have remained relatively constant in recent fiscal years. Some contend that overall funding for enforcement activities has not kept pace with inflation or with the increasingly complex federal pollution control requirements. Congress appropriated $560.9 million for enforcement activities for FY2014, a 1.4% increase above the $553.1 million enacted for FY2013 (post-sequestration), but roughly 3.8% less than the $583.4 million enacted for FY2012. The President's FY2015 budget request included $583.0 million for EPA enforcement activities. On September 19, 2014, President Obama signed into law the Continuing Appropriations Resolution, 2015 (P.L. 113-164). The act provides FY2015 appropriations to federal agencies (including EPA) for continuing projects and activities generally at the rate and under the authority and conditions provided in the applicable conditions of the Consolidated Appropriations Act, 2014 (P.L. 113-76), less a 0.0554% rescission, until December 11, 2014, or until enactment of regular appropriations legislation.

Chapter 3 - For several years, the U.S. Environmental Protection Agency (EPA) has been working with states and cities to develop and implement new approaches that will achieve water quality goals cost-effectively and in a manner that "addresses the most pressing public health and environmental protection issues first." Two recent EPA initiatives are an integrated planning policy and a framework policy for assessing a community's financial capability to meet objectives and requirements of the Clean Water Act (CWA). Pressed by municipalities about the financial challenges that they face in addressing needs for wastewater and storm water control projects, in 2012 EPA issued an integrated permitting and planning policy. The intention of the policy is to provide communities with flexibility to prioritize and sequence needed water infrastructure investments so that limited public dollars can be invested in ways that each municipality finds most valuable. Water utilities and municipalities have welcomed the opportunity for flexibility under the integrated planning policy. But they have sought clarification of a number of issues, including EPA and state roles in developing integrated plans. A major point of contention between EPA and local government stakeholders has been the agency's reliance on administrative orders or judicially approved consent decrees to codify integrated pollution reduction plans, rather than through modification of CWA permits. City and town officials say that they would prefer that EPA allow compliance flexibility through permits, rather than subjecting cities and towns to legally binding consent decrees with penalties and fines for noncompliance. The agency takes the position that both enforcement and permits are necessary, depending on individual circumstances. EPA typically uses a consent decree after periods of permit noncompliance. While integrated planning may be helpful in identifying communities' relative priorities, a longstanding concern for local governments is EPA's process for evaluating how much communities can afford for CWA-mandated and other water infrastructure improvements. EPA has worked with communities to refine how the agency determines when a project is affordable for individual communities, because affordability considerations can influence schedules established for a community to meet CWA requirements. In 2014, EPA released a Financial Capability Assessment Framework that identifies a range of information related to a community's financial strength that may help provide a complete picture of cities' financial capabilities in relation to water infrastructure investments. State and local governments generally support EPA's efforts to encourage ways for communities to prioritize their CWA infrastructure investments and to consider a wide range of factors that affect affordability. Nevertheless, cities and states have continuing concerns with aspects of both policies. For example, some criticize EPA for relying in part on Median Household Income (MHI) as a measure of community affordability. Further, cities are critical that integrated plans that have been approved so far have been incorporated only into new or amended consent decrees, not CWA permits. EPA's integrated planning process and water infrastructure affordability are issues of interest to legislators. In recent years, Members have offered bills that would address these issues in different ways. Some bills have sought to codify EPA's integrated planning approach—as written in the 2012 framework—into the CWA. Other proposals would go beyond codifying the integrated planning approach by including other provisions that alter the CWA's existing framework to varying degrees (e.g., H.R. 465 in the 115th Congress). In general, some environmental groups have opposed such changes proposed in bills from prior Congresses, but municipalities have generally supported the additional flexibility.

Chapter 4 - In November 2011 the Environmental Protection Agency (EPA) proposed two Clean Water Act (CWA) permits to regulate certain types of vessel discharges into U.S. waters. The proposed permits would replace a single Vessel General Permit (VGP) issued in 2008 that was due to expire in December 2013. As proposed, the permits would apply to approximately 71,000 large domestic and foreign vessels and perhaps as many as 138,000 small vessels. This universe of regulated entities is diverse as well as large, consisting of tankers, freighters, barges, cruise ships and other passenger vessels, and commercial fishing vessels. Their discharges are similarly diverse, including among other pollutants aquatic nuisance species (ANS), nutrients, pathogens, oil and grease, metals, and toxic chemical compounds that can have a broad array of effects on aquatic species and human health, many of which can be harmful. EPA proposed two permits, one for large vessels to replace the 2008 VGP, and one for smaller vessels covered by a congressionally enacted temporary moratorium. Both were proposed well in advance of the VGP's expiration to provide ample time for the regulated community to prepare for new requirements. On March 28, 2013, EPA issued a final version of the VGP for large vessels. It took effect December 19, 2013. The permit for smaller vessels, the sVGP, was issued on September 10, 2014, and was scheduled to take effect on December 19, 2014. However, in December 2014, Congress passed legislation (S. 2444/P.L. 113-281) extending until December 18, 2017, the date when small vessels will need a CWA permit.

The CWA requires that all regulated discharges must meet effluent limitations representing applicable levels of technology-based control. The 2013 VGP largely retains the current permit's approach of relying on best management practices to control most discharges, because EPA concluded that it is infeasible to develop numeric effluent limits for most controlled discharges. However, the new VGP includes for the first time numeric ballast water discharge limits, which are consistent with standards in a 2012 Coast Guard rule and an international convention. The 2013 VGP raises two key issues. One concerns inclusion of specific numeric ballast water discharge limits in the permit. At issue had been whether EPA would propose more stringent numeric limits, as some environmental groups have favored and a few states have already adopted. A second issue concerns the role of states in regulating vessel discharges. Environmental groups and Canadian shippers challenged the 2013 permit in federal court. In October 2015, the court supported the environmentalists' challenge, ruling that the permit violated the CWA because it did not require the use of best available technology to control discharges of invasive species from ships' ballast water. The court remanded the permit to EPA, but did not vacate it; the permit remains in effect until EPA issues a new permit. Congressional interest in this topic has been evident for some time. In 2008 Congress enacted two bills to exempt certain vessels from a CWA permit requirement, thus restricting the population of vessels subject to the VGP. One was a permanent permit moratorium for recreational vessels of all sizes. The other act was a temporary permit moratorium for small commercial vessels and commercial fishing vessels, which was extended twice by Congress and would have expired December 18, 2014, had Congress not enacted an additional three-year extension in P.L. 113-281. In the 114th Congress, bills addressing the temporary permit moratorium for small vessels and regulation of ballast water discharges have been introduced (S. 373/H.R. 980 and S. 371). The Senate Commerce Committee approved S. 373 in February 2015; the committee also included provisions of this bill in S. 2829, which the Senate passed in June 2016, and S. 1611. Further, the House has passed H.R. 4909, which includes the text of H.R. 980 as one title of that bill.

Chapter 5 - Implementation of revised ozone standards by the U.S. Environmental Protection Agency (EPA) is now moving forward, after the agency designated 52 areas with just over 200 counties or partial counties and two tribal areas as "nonattainment" for the standards. The standards— formally known as National Ambient Air Quality Standards (NAAQS) for ground-level ozone— are standards for outdoor (ambient) air. In 2015, EPA tightened both the primary (health-based) and secondary (welfare-based) standards from 75 parts per billion (ppb) to 70 ppb after concluding that protecting public health and welfare requires lower concentrations of ozone than were previously judged to be safe. Ozone aggravates heart and lung diseases and may contribute to premature death; the primary standard addresses these concerns. Ozone can also have negative effects on forests and crop yields, which the secondary NAAQS is intended to protect. The designated nonattainment areas include counties in 22 states and the District of Columbia. Most of these areas have had previous experience as nonattainment for earlier versions of the NAAQS. Designation as nonattainment imposes more stringent permitting and pollution control requirement for new and modified stationary sources of emissions as compared with the requirements in areas that are in attainment of the NAAQS, and requires the development of State Implementation Plans demonstrating how emissions will be reduced sufficiently to reach attainment. EPA estimates the cost of meeting the 70 ppb standard in all states except California at $1.4 billion annually in 2025. Because most California areas would have until the 2030s to reach attainment, EPA provided separate cost estimates for California ($0.80 billion annually, post-2025). These cost estimates are substantially less than those from the National Association of Manufacturers and other industry sources, which have been widely cited. The benefits of reducing ozone concentrations were estimated by EPA at $2.9-$5.9 billion annually by 2025. The dollar value of avoided premature deaths accounts for 94% to 98% of this estimate. The agency projects that most areas will be able to reach attainment of the new standards by 2025 as a result of already promulgated regulations for gasoline, autos, power plants, and other emission sources. These regulations are being implemented independently of the 2015 NAAQS revision. Members of Congress have shown particular interest in whether the expected benefits of the standards justify their projected costs. There is controversy over the methods used to estimate both costs and benefits. As the Clean Air Act is currently written, however, the agency is prohibited from weighing costs against benefits in setting NAAQS standards. The statute simply directs EPA to set the primary standard at a level requisite to protect public health, allowing an adequate margin of safety. Various interest groups have lobbied against strengthening the standards. In the 115th Congress, the House has passed H.R. 806 to delay implementation of the 2015 NAAQS until the mid-2020s and to make changes to the process of future NAAQS revisions. The House Appropriations Committee reported a similar delay as a rider to EPA's 2018 appropriation (in Section 432 of H.R. 3354), but the final version of the appropriation (in H.R. 1625/P.L. 115-141) did not include the rider.

Chapter 6 - In 2009, the Obama Administration—through authorities provided to the National Highway Traffic Safety Administration (NHTSA) and the Environmental Protection Agency (EPA)—developed joint standards for fuel economy and greenhouse gas (GHG) emissions for new light-duty vehicles (defined generally as passenger cars and light trucks). The standards (referred to as the One National Program) were established in two phases: Phase 1 for vehicle model years (MY) 2012-2016, finalized on May 7, 2010; and Phase 2 for MY2017-2025, finalized on October 15, 2012. The agencies promulgated the joint rulemakings with the support of an array of stakeholders—including auto manufacturers, labor unions, the environmental community, the state of California, and other states—and intended them to provide industry with a single regulatory voice (to avoid conflicting regulations within the federal government and among the states) and a long-term regime (to provide greater certainty for product planning and engineering).

Chapter 7 - On March 28, 2017, President Trump signed Executive Order 13783, directing federal agencies to review existing regulations and policies that potentially burden the development or use of domestically produced energy resources. Acting pursuant to the order, the U.S. Environmental Protection Agency (EPA) is reviewing and reconsidering several regulations issued during the Obama Administration that address methane emissions from various industrial sectors. Methane is a short-lived greenhouse gas (GHG) with a Global Warming Potential of more than 25 times carbon dioxide that is emitted from various industrial activities. President Trump's executive order specifically requires EPA to review the revised emission standards for new, modified, and reconstructed equipment, processes, and activities of the oil and natural gas sector issued by the Obama Administration in June 2016. EPA issued these standards for methane and volatile organic compound (VOC) emissions pursuant to Section 111 of the Clean Air Act (CAA). Based on its review and in response to several administrative petitions for reconsideration, EPA is now reconsidering certain emission requirements from the June 2016 rule, which remain in effect unless EPA finalizes a proposed two-year stay of these requirements or otherwise repeals those requirements. In addition, EPA is reconsidering the emission standards and guidelines for new and existing municipal solid waste (MSW) landfills updated by the Obama Administration in August 2016. In those rules, EPA issued the updated and revised emission standards for MSW landfills built after 2014 to further reduce emissions, including methane emissions. The agency also revised emission guidelines established in 1996 for existing landfills operating prior to that date. At this time, the 2016 landfill rules are in effect. EPA has not formally proposed any revisions to the 2016 rules or initiated a public comment period for any issues under reconsideration. EPA's review of the oil and natural gas sector and landfill methane rules has influenced the pending judicial challenges to the various 2016 rules. The U.S. Court of Appeals for the District of Columbia Circuit granted EPA's requests to pause the judicial challenges of both rules to allow EPA to complete its review of them. In addition, stakeholders have successfully challenged in court EPA's attempts to stay the various requirements that the agency is currently reconsidering. Judicial review of EPA's attempts to stay rules in effect could more broadly impact the Trump Administration's efforts to similarly stay other rules that are under reconsideration. This chapter examines the statutory authority for issuing the methane regulations, legal challenges to the standards, and legal issues related to the reconsideration and stay of the regulations.

Chapter 8 - In 2016, the U.S. Environmental Protection Agency (EPA) and the Bureau of Land Management (BLM) promulgated separate regulations intended to control "methane emissions" at crude oil and natural gas production facilities. Some stakeholders have argued that the EPA and BLM rules are duplicative and outside of the agencies' statutory authorities. On January 30, 2017, a joint resolution of disapproval under the Congressional Review Act was introduced to overturn the BLM rule. H.J.Res. 36 would vacate the rule and preempt the agency from promulgating similar proposals in the future. The measure passed the House by a 221 to 191 vote and now moves to the Senate.

Chapter 9 - On March 7, 2018, the House passed H.R. 1917, a bill that would delay for three years the implementation of more stringent emission standards for new residential wood heaters. The emission standards were promulgated by the U.S. Environmental Protection Agency (EPA) in 2015, and are becoming effective through a two-step process. Step 1 standards took effect on May 15, 2015; unless delayed, more stringent Step 2 standards will become effective on May 15, 2020. EPA's action revises standards for wood stoves and pellet stoves that were set in 1988, and establishes standards for other types of wood heaters, principally forced air furnaces and hydronic heaters, for the first time. According to EPA, smoke from wood heaters and fireplaces contributes "hundreds of thousands of tons" of fine particles to the air each year, nationally accounting for nearly 25% of all area source air toxics cancer risks and 15% of non-cancer respiratory effects. In many areas, in wintertime, wood heaters are the largest source of particulate air pollution. Until the implementation of this rule, however, many heater types were not subject to any federal emission standard. The rule will only gradually reduce this pollution, because it only applies to new heaters (not those already in use) and it gives the industry a five-year grace period (until 2020) before its most stringent (Step 2) standards take effect. Nevertheless, EPA estimated that implementing the rule will eliminate 360 to 810 premature deaths annually in the 2015-2020 period as well as reduce hospital admissions and lost work days due to respiratory illness. EPA quantified these benefits at $3.4 billion to $7.6 billion per year during the 2015-2020 period, more than 70 times the agency's estimated annualized compliance cost to manufacturers, $46 million. Trade associations representing the affected industries and affected companies have mixed views of the 2015 standards. While generally supporting revision of the 1988 standards and the inclusion of additional heater types, some have expressed concern that the standards will impose too great a cost. Homeowners may continue to use current, highly polluting equipment, rather than buy more expensive replacements. In its analysis of the final rule, EPA estimated the vast majority of cost (88%) would be borne by the hydronic heater and forced air furnace segments of the industry. Specifically, EPA estimated compliance costs for hydronic heaters and forced-air furnaces will be about 17% of total sales revenues for the two product types. EPA did not estimate the effect of these costs on prices or output, but the agency's analysis does conclude that high unit compliance costs could lead to "potential nontrivial increases in market price to wood-burning appliance consumers and potential decreases in output." In addition to cost issues, many commenters on the *proposed* rule expressed concerns regarding the process to be used in certifying compliance and the short period of time in which units available at the time of the rule's promulgation could be tested and certified. EPA addressed many of these comments in the final rule, making changes intended to ease the burden of certifying compliance. Industry representatives continue to express concern that there will not be enough time to develop new models, certify compliance, and meet lead-time requirements of product retailers for the Step 2 (2020) compliance deadline. This chapter presents some of the most frequent questions raised concerning the rule, so as to provide basic background information about the final wood heater rule, its potential impacts, and stakeholder reactions to it.

Chapter 10 - Chemicals and the facilities that manufacture, store, distribute, and use them are essential to the U.S. economy. However, incidents occasioned by natural disasters, unintentional events, or security threats show that the handling and storage of chemicals are not without risk. Federal agencies implement a number of programs to help prevent chemical facility accidents, reduce risks of terrorist attacks on chemical facilities, protect chemical facility workers, collect and share relevant information with the public and decisionmakers, and prepare communities and local, tribal, and state first-responders to respond to potential large-scale accidents. This chapter reviews the U.S. Environmental Protection Agency's (EPA's) authorities regarding risk management, emergency planning, and release notification, among others, at chemical facilities. In doing so, it describes the statutory authorities—and makes note of some of the more prominent, subsequent regulations—as provided by the following:

- *Facility risk management planning requirements under Section 112(r)(7) of the Clean Air Act (CAA).* EPA's Risk Management Program (RMP) is aimed at reducing chemical risk at the local level. EPA regulations require owners and operators of a facility that manufactures, uses, stores, or otherwise handles certain listed flammable and toxic substances to develop a risk management program that includes hazard assessment (including an evaluation of worst-case and alternative accidental release scenarios), prevention mechanisms, and emergency response measures.
- *Emergency planning notification requirements under the Emergency Planning and Community Right-to-Know Act of 1986 (EPCRA).* The requirements are designed to promote emergency planning and preparedness at the state, local, and tribal levels. EPCRA helps ensure local communities and first responders have needed information on potential chemical hazards within their communities in order to develop community emergency response plans.
- *Emergency release notification requirements under the Comprehensive Environmental Response, Compensation, and Liability Act of 1980 (CERCLA).* CERCLA obligates a facility to report certain releases of hazardous substances to the National Response Center to inform decisions about federal involvement in responding to the incident to coordinate with

state and local officials. The requirements also establish liability for response costs and natural resource damages.

- *Duties of the Chemical Safety and Hazard Investigation Board, known as the Chemical Safety Board (CSB), under Section 112(r)(6) of the CAA.* The purpose of the CSB is to investigate accidents to determine the conditions and circumstances that led up to the event and to identify the cause or causes so that similar events might be prevented.
- *Toxic release inventory reporting requirements.* EPCRA authorizes EPA to establish and maintain a Toxic Release Inventory (TRI) of facilities that manufacture, import, process, or use certain types of toxic chemicals by providing public disclosure of the locations of such facilities.

In: Environmental Protection Agency (EPA) ISBN: 978-1-53614-933-3
Editor: Coby Schaefer

Chapter 1

EPA Regulations: Too Much, Too Little, or on Track? (Updated)*

James E. McCarthy and Claudia Copeland

Summary

Since Barack Obama was sworn in as President in 2009, the U.S. Environmental Protection Agency (EPA) has proposed and promulgated numerous regulations to implement the pollution control statutes enacted by Congress. Critics have reacted strongly. Some, both within Congress and outside of it, have accused the agency of reaching beyond the authority given it by Congress and ignoring or underestimating the costs and economic impacts and overestimating the benefits of proposed and promulgated rules. The House conducted vigorous oversight of the agency in the 112th and 113th Congresses, and approved several bills that would overturn specific regulations or limit the agency's authority. Particular attention was paid to the Clean Air Act, but there also was congressional scrutiny of other environmental statutes and regulations implemented by EPA. With Republican majorities in both the House and Senate, the 114th Congress accelerated oversight of the Administration's initiatives and renewed efforts to limit EPA's regulatory activities.

Environmental groups and other supporters of the agency disagree that EPA has overreached. Many of them believe that the agency has, in fact, moved in the right direction, including taking action on significant issues that had been long delayed or ignored in the past. In several cases, environment and public health advocates would have liked the regulatory actions to be stronger.

EPA has stated that critics' focus on the cost of controls obscured the benefits of new regulations, which, it estimates, far exceed the costs. It maintains that pollution control is an important source of economic activity, exports, and American jobs. Further, the agency and its supporters have said that EPA is carrying out the mandates detailed by Congress in the federal environmental statutes.

* This is an edited, reformatted and augmented version of Congressional Research Service Publication No. R41561, dated December 30, 2016.

This chapter provides background information on EPA regulatory activity during the Obama Administration to help address these issues. It examines major or controversial regulatory actions taken by or under development at EPA from January 2009 to late 2016, providing details on the regulatory action itself, presenting an estimated timeline for completion of rules not yet promulgated (including identification of related court or statutory deadlines), and, in general, providing EPA's estimates of costs and benefits, where available.

The chapter also discusses factors that affect the time frame in which regulations take effect, including statutory and judicial deadlines, public comment periods, judicial review, and permitting procedures, the net results of which are that existing facilities are likely to have several years before being required to comply with most of the regulatory actions under discussion. Unable to account for such factors, which will vary from case to case, timelines that show dates for proposal and promulgation of EPA regulations effectively underestimate the complexities of the regulatory process and overstate the near-term impact of many of the regulatory actions.

INTRODUCTION

Has the Obama Administration's EPA Overreached? Conflicting Views

Since Barack Obama was sworn in as President of the United States in 2009, the Environmental Protection Agency (EPA) has proposed and promulgated numerous regulations under many of the 11 pollution control statutes Congress has directed it to implement.[1] Most of these statutes were enacted decades ago and have not been amended for more than a decade, yet the agency is still addressing for the first time numerous directives given to it by Congress, while also addressing newly emerging pollution problems and issues. The statutes also mandate that EPA conduct periodic reviews of many of the standards and rules that it issues. The agency has conducted these reviews, as well, and revised some regulations in response to the reviews.

Although supporters have maintained that EPA is just doing its job, the agency's regulatory actions over the eight years have drawn attention for several reasons. In some cases, such as regulation of greenhouse gas emissions, they represented a new departure. Based on a 2007 Supreme Court ruling that greenhouse gas emissions are air pollutants under the Clean Air Act's definition of that term,[2] the agency undertook numerous regulatory actions setting emission standards or laying the framework for a future regulatory structure. In other cases, the agency revised emissions, effluent, and waste management regulatory decisions made during earlier Administrations and promulgated more stringent standards to address pollution that persists as long as 40 years after Congress directed the agency to take action. These actions were driven by statutory

---

[1] For a summary of each of the 11 statutes and their principal requirements, see CRS Report RL30798, *Environmental Laws: Summaries of Major Statutes Administered by the Environmental Protection Agency*, coordinated by David M. Bearden.

[2] See CRS Report R40984, *Legal Consequences of EPA's Endangerment Finding for New Motor Vehicle Greenhouse Gas Emissions*, by Robert Meltz.

requirements to reexamine regulations, by legal challenges and court decisions, or because of changing technologies or new and evolving scientific information.

EPA's actions, both individually and in sum, generated controversy. For example, the U.S. Chamber of Commerce described EPA's actions as "a series of one-sided, politically-charged regulations that are intended to take the place of legislation that cannot achieve a consensus in the Congress."[3] Affected parties, such as the National Petrochemical & Refiners Association (now known as the American Fuel and Petrochemical Manufacturers), labeled the agency's actions "overreaching government regulation" and "a clear distortion of current environmental law,"[4] while the National Mining Association said that "even at a time of great economic stress, EPA is poised to enact a series of back-door mandates that threaten to cost millions of American jobs, and increase the cost of their electricity while they're at it."[5]

Both Democrats and Republicans in Congress expressed concerns and criticism, through bipartisan letters commenting on proposed regulations, in statements at congressional hearings, and through introduced legislation that would delay, limit, or prevent certain EPA actions.[6] Committees have conducted vigorous oversight of the agency's actions; oversight and efforts to limit EPA's activities accelerated in the 114th Congress. Arguing in favor of riders to the FY2016 Interior, Environment, and Related Agencies appropriations bill intended to restrain the Administration's regulatory efforts, the Senate majority leader reportedly referred to Administration regulatory efforts as "an all-out assault on the American economy.... But in all the overreach, none has been more dramatic than at the EPA." In response to announcement of a final rule defining "waters of the United States" under the Clean Water Act, the chairman of the Senate Environment and Public Works Committee characterized the rule as "a measure that hugely expands the ability of this aggressive agency to disregard the conservation efforts of American states and interfere with the daily lives and property of the American people."[7]

EPA was not silent as the agency's actions came under attack. In a 2010 letter to the ranking Members of the Energy and Commerce Committee and its Subcommittee on Oversight and Investigations, then-EPA Administrator Lisa Jackson stated the following:

---

[3] U.S. Chamber of Commerce, "Regulatory Areas, Energy, and the Environment," http://www. uschamber.com/ regulations/areas, accessed June 15, 2015.

[4] NPRA, "NPRA Says Court Decision on GHGs Bad for Consumers," December 10, 2010, at http://www.npra.org/newsRoom/?fa=viewCmsItem&title=Latest%20News&articleID=5980

[5] National Mining Association, "EPA's Regulatory Train Wreck," 2011, http://www.nma. org/pdf/fact_sheets/epa_tw.pdf.

[6] For discussion of some of these congressional actions, see CRS Report R43851, *Clean Air Issues in the 114th Congress*, by James E. McCarthy; CRS Report R43943, *The "Waters of the United States" Rule: Legislative Options and 114th Congress Responses*, by Claudia Copeland; CRS Report R41698, *H.R. 1 Full-Year FY2011 Continuing Resolution: Overview of Environmental Protection Agency (EPA) Provisions*, by Robert Esworthy; and CRS Report R41979, *Environmental Protection Agency (EPA) FY2012 Appropriations: Overview of Provisions in H.R. 2584 as Reported*, by Robert Esworthy.

[7] Senator James Inhofe, "Why every property owner should fear EPA's 'waters of the United States' rule," *FoxNews.com*, June 2, 2015, http://www.foxnews.com/opinion/2015/ 06/02/why-every-property-owner-should-fearepas-waters-united-states-rule.html.

> The pace of EPA's Clean Air Act regulatory work under this administration is actually not faster than the pace under either of the two previous administrations. In fact, EPA has finalized or proposed fewer Clean Air Act rules (87) over the past 21 months than in the first two years of either President George W. Bush's administration (146) or President Clinton's administration (115).[8]

In congressional testimony and other fora, Administrator Gina McCarthy, head of the agency from July 2013 to January 2017, sought to rebut critics' challenges to EPA's actions and initiatives.

> When it comes to the American economy, cutting pollution doesn't dull our competitive edge, it sharpens it. Thanks to our fuel efficiency standards, the auto industry is once again a source of economic strength. The number of cars coming off American assembly lines, made by American workers, is the highest it's been in 12 years. From catalytic converters to smoke-stack scrubbers, America has a legacy of innovating the world's leading environmental technologies—accounting for more than 1.5 million jobs and $44 billion in exports in 2008 alone. That's more than other big sectors like plastics and rubber products. If you want to talk return on investment, in over four decades, we've cut air pollution by 70 percent, while our GDP has tripled.[9]

Environmental groups generally believe that the agency moved in the right direction, but in several cases they would have liked the regulatory actions to be stronger or to have occurred sooner. Many also feared that decisions to delay the issuance or implementation of several standards were bad omens. Commenting on EPA's 2010 request to delay the issuance of air emission standards for boilers, for example, Clean Air Watch stated that "there is an unfortunate appearance here that political pressure from Congress is affecting the situation. That EPA is running scared."[10]

While industry groups have been the most frequent critics of EPA regulation, industry sometimes complains of a lack of EPA action. In 2014, the CEO of the American Petroleum Institute complained that regulations dealing with renewable fuels were overdue, calling it "unacceptable," and "an example of government at its worst." The American Fuel and Petrochemical Manufacturers, seeking issuance of the same rule, called the agency's track record in failing to issue regulations "an egregious pattern of non-compliance."[11]

---

[8] Letter of Lisa P. Jackson, EPA Administrator, to Hon. Joe Barton and Hon. Michael C. Burgess, November 8, 2010, p. 1. According to the letter, "All three counts include all Clean Air Act rules that amend the Code of Federal Regulations and that require the EPA Administrator's signature." Administrator Jackson's letter was written in response to an October 14, 2010, letter from Reps. Barton and Burgess in which they expressed concern regarding the cumulative impacts of new regulations being proposed under the Clean Air Act.

[9] Gina McCarthy, EPA Administrator, Remarks at Resources for the Future, Washington, DC, September 25, 2014.

[10] Clean Air Watch, "EPA Seeks Big Delay in Final Toxic Rule for Boilers," December 7, 2010, at http://blogforcleanair.blogspot.com/2010/12/epa-seeks-big-delay-in-final-toxic-rule.html.

[11] See "EPA Won't Finalize Renewable Fuel Standard in 2014, Cites Lengthy Delays," *Daily Environment Report*, November 24, 2014.

It is not this chapter's purpose to render a verdict on whether EPA overreached, ran scared, or followed the directions and used the authorities given to it by Congress during the eight years of the Obama presidency. Statements characterizing EPA's actions, such as those cited above, depend on judgments as to whether the agency correctly determined the level of stringency needed to address an environmental problem, and whether the agency's actions are justified by the legislative mandates that Congress imposed and statutory authorities that Congress provided. Congress and the courts may render these judgments.

WHAT THIS CHAPTER DOES

This chapter provides a factual basis for discussion of these issues, which must ultimately be evaluated on a case-by-case basis. The chapter identifies and briefly characterizes major regulatory actions[12] promulgated, proposed, or under development by EPA from January 2009 to late 2016. The chapter uses data from EPA's Semiannual Regulatory Agendas[13] and the list of economically significant reviews completed by the Office of Management and Budget (OMB)[14] to compile a list of regulatory actions proposed, promulgated, or under development by the agency. The list includes most EPA rules considered "economically significant" by OMB since January 2009,[15] as well as some others that were not so designated but have been widely discussed.

Each entry in This chapter (1) gives the name or, where appropriate, the common name of the regulatory action (e.g., the Clean Power Plan or Mountaintop Mining in Appalachia); (2) explains what the action does; (3) states the current status of the rule or action (e.g., proposed September 20, 2013); (4) explains the significance of the action, providing information on estimated costs and benefits,[16] where available; (5) discusses

---

[12] This report uses the terms "regulatory action," "regulation," "rule," "standard," and "guidelines" for the actions it describes. There are slight differences among these terms, which are explained, if necessary to understand how the regulatory action will be implemented. In general, "regulatory action" is the broadest of the terms and includes each of the others.

[13] Executive Office of the President, Office of Management and Budget, Office of Information and Regulatory Affairs, *Semiannual Regulatory Agenda: Fall 2016,* November 17, 2016, at http://www.reginfo.gov /public/do/eAgendaMain.

[14] OMB, Office of Information and Regulatory Affairs (OIRA), *Historical Reports,* at http://www .reginfo.gov/public/do/eoHistReviewSearch.

[15] OIRA (the regulatory affairs staff within OMB) considers a rule to be "economically significant" if it is "likely to have an annual effect on the economy of $100 million or more or adversely affect in a material way the economy, a sector of the economy, productivity, competition, jobs, the environment, public health or safety, or State, local, or tribal governments or communities." OMB, *FAQs/Resources*, at http://www.reginfo. gov/public/jsp/Utilities/faq.jsp.

[16] Costs and benefits of EPA regulations are another issue frequently raised by critics of the agency, many of whom have questioned whether EPA's cost estimates are too low or estimates of benefits too high. There are recurring questions regarding the methodologies used to estimate both costs and benefits, including what to choose as the baseline against which to measure changes resulting from a regulation; how to monetize improvements in public health, such as the avoidance of premature death; whether to count both direct benefits and cobenefits (i.e., benefits achieved that were not the purpose of the regulation); how to account for benefits for which there is no accepted measurement or valuation methodology; whether to include reductions in the "social cost of carbon" as a benefit and, if so, how to measure those benefits; and whether certain benefits or

the timeline for implementation, and whether there is a nondiscretionary congressional deadline or a court order or remand driving its development; and (6) identifies a CRS analyst who would be the contact for further information. To simplify presentation, in some cases, we have summarized several separate, but related, regulations under one heading.

This is not a complete list of the regulations that EPA has proposed or promulgated during the Obama Administration. Rather, it is an attempt to identify the most significant and most controversial. A complete list would be substantially longer.

A Few Caveats Regarding Timing

Not all of the rules discussed here are Obama Administration initiatives. Many began development under the Bush Administration (or earlier, in some cases), including several that were promulgated under that Administration and subsequently were vacated or remanded to EPA by the courts. Within the Clean Air Act group, for example, most of the major rules, including the agency's boiler rules and two of the major rules affecting electric power plants (the Cross-State Air Pollution Rule and the Utility MACT/MATS rule) fit that description. Similarly, EPA's regulation of power plant cooling water intake structures under the Clean Water Act is governed by a 1995 consent decree and rules issued and proposed in several phases beginning in 2001. Other EPA actions, such as reconsideration of the ozone air quality standard, actually delayed for several years implementation of Bush Administration rules that strengthened existing standards. All of these are described in detail below.

Several other generalizations are worth underlining:

- Many proposed and "pre-proposal" rules linger for years without being promulgated. They may also be substantially altered before they become final and take effect, as a result of the proposal and public comment process, and/or judicial review.
- For proposed rules, we attempted to provide known deadlines, whether statutory or court-ordered. If there were no known deadlines, we attempted to provide EPA's estimate of the schedule for promulgation. Proposal or promulgation often took longer than estimated in cases that did not have a court-ordered deadline.
- Although they were the most likely deadlines to be met, even court-ordered dates for proposal or promulgation occasionally changed. It was not uncommon for EPA to request extensions of time, often due to the need to analyze extensive comments or reevaluate technical information.

---

costs are double-counted when simultaneous proposals address the same pollutant. These issues are not the subject of this report.

- Promulgation of standards is not the end of the road. Virtually all major EPA regulatory actions are subjected to court challenge, frequently delaying implementation for years. As noted earlier, many of the regulatory actions described here are the result of courts remanding and/or vacating rules promulgated by previous Administrations. EPA has also, in several cases, reconsidered rules after promulgation, changing what were announced as "final" standards, and, in some cases, granting additional time for compliance.
- In many cases, EPA rules must be adopted by states to which the program has been delegated before actual implementation occurs (e.g., establishing air quality plans or issuing permits). Moreover, many states require that the legislature review new regulations before the new rules would take effect.
- Federal standards for stationary sources under the air, water, and solid waste laws are generally implemented through permits, which would be individually issued by state and local permitting authorities after the federal standards take effect. When finalized, a permit would generally include a compliance schedule, typically giving the permittee several years for installation of required control equipment. Existing sources generally have several years following promulgation and effective dates of standards, therefore, to comply with any standards.

In short, the road to EPA regulation is rarely a straight path. There are numerous possible causes of delay. It would be unusual if the regulatory actions described here were all implemented on the anticipated schedule, and even if they were, existing facilities often have several years before being required to comply.

Congressional Activity

Since the 111th Congress, a number of EPA's regulatory actions have been the subject of legislative proposals, including stand-alone bills that would have delayed or prohibited EPA actions, resolutions of disapproval under the Congressional Review Act, and potential riders on EPA's appropriation. In the 112th and 113th Congresses, criticism of EPA actions increased, and several bills to prevent or delay EPA action passed the House but were not considered in the Senate. Some proposals were broad in nature, targeting all regulatory agencies or a lengthy list of specific regulations, while others focused more narrowly on individual rules or actions. In the 114th Congress, opponents of EPA's initiatives, who maintained that the agency was exceeding its authority, again considered various approaches to altering the agency's course, including riders on appropriations bills, stand-alone legislation, resolutions of disapproval under the Congressional Review Act, and amendments to the underlying environmental statutes.

Throughout this period, congressional oversight of EPA actions has been vigorous. With Republican majorities in both the House and the Senate, the 114th Congress accelerated oversight activities of the Administration's initiatives and legislative efforts to limit EPA's regulatory activities. The House passed several bills intended to halt or redirect EPA rules or policies. In addition, both the Senate and House passed three joint resolutions disapproving EPA rules under the Congressional Review Act. Each of the joint resolutions was vetoed by the President.

Beyond the criticism of specific regulations, there have also been calls for broad regulatory reforms of government—for example, to reinforce the role of economic considerations in agency decisionmaking, or to increase Congress's role in approving or disapproving regulatory decisions. Congress's interest in these types of regulatory reform proposals accelerated in the 114th Congress; many bills were introduced. Topics covered include reforms such as requiring agencies to adopt the least costly rule that meets relevant statutory objectives unless the benefits justify additional costs; establishing commissions or other mechanisms to review the effectiveness of federal programs and identify outdated or overly burdensome rules for repeal; requiring independent regulatory agencies to comply with regulatory analysis requirements; providing for judicial review of certain requirements and determinations for which judicial review is not currently available; altering judicial deference to agency interpretations of rules; and/or placing moratoria on the issuance of new regulations.

While many of these proposals were broad in nature and would have applied regulatory reforms to multiple federal agencies, some bills in the 114th Congress targeted EPA expressly and particularly. Examples include H.R. 1029/S. 543, which would alter qualifications for membership of the Science Advisory Board (SAB), which provides recommendations and scientific advice to the EPA Administrator, and revise public participation and other procedures of the SAB; and H.R. 1030/S. 544, which would bar EPA from proposing or promulgating rules based upon science that is not transparent or reproducible. Supporters of bills such as these say that they are intended to restore transparency and confidence in EPA's regulatory processes, while opponents contend that the legislation is intended to limit the quality of scientific advice provided to EPA and to impede EPA's ability to use science to protect public health and the environment. The Obama Administration opposed these bills.

CONCLUSION

This chapter has been updated frequently since the first version was released early in 2011. That original report and numerous updates to it tracked nearly four dozen rules. Many of the issues that were raised in 2011 regarding specific regulations have now been resolved—at least to the extent that proposed rules have been finalized. Still, the broader question of whether the Obama Administration's EPA "overreached" in its regulatory efforts has not gone away. Critics both in Congress and outside of it regularly accuse the agency of overreach. In 2013, in a case involving four of EPA's greenhouse gas regulatory actions, for example, a dozen states led by Texas asked the Supreme Court to "rein in a usurpatious agency and remind the President and his subordinates that they cannot rule by executive decree."[17] Legal challenges to the "waters of the United States" rule filed by multiple industries and more than half of the states similarly claimed overreach, beyond what Congress and the courts have allowed.

What is different since our first report is that there is now a more detailed record of EPA actions to be evaluated. Reviewing that record, we find the following:

- Most of the proposals that were controversial when our first report was released are now final.
- The final versions of many of the most controversial rules were made less stringent than what was proposed.
- The Clean Power Plan, which set state-by-state average greenhouse gas emission requirements for existing fossil-fueled power plants, is one exception to the stringency generalization: it underwent a number of changes following proposal, some of which made the rule more stringent and others less stringent.
- In revising proposed rules, EPA often relied on data submitted by industry and other stakeholders, acknowledging that it had inadequate or incomplete data when it proposed the rules.
- In several instances, the regulated community was given more time to comply than originally expected.
- Regardless of modifications in the final rules, many of the promulgated regulations have been challenged in court by a variety of groups—some seeking more stringent rules, others less stringent.
- Although some of these challenges remain to be heard by the courts, thus far the courts have upheld EPA decisions on the final regulations in most cases.[18]
- The pace of new proposed and final regulations has waxed and waned. The pace slowed considerably for several years after 2011. In part, this may have been because a backlog of rules that were remanded to the agency during the Bush Administration had been largely addressed by 2012; other rules were delayed until after the 2012 election, perhaps due to political considerations at the time. Most of these have now been promulgated.
- 2015 was a particularly contentious year. Major rules were promulgated on greenhouse gas emissions and discharges to water from electric power plants, hard on the heels of a December 2014 rule on coal combustion residuals and a

---

[17] Petition for certiorari filed by State of Texas et al. at 2, granted, 2013 WL 1743433 (Oct. 15, 2013), from Coalition for Responsible Regulation v. EPA, 682 F.3d 102 (D.C. Cir. 2012). Five other petitions for certiorari from the same decision were also granted; in June 2014, the Supreme Court issued a ruling that largely upheld EPA's authority to regulate greenhouse gases through the agency's New Source Review program (Utility Air Regulatory Group v. EPA, 134 S. Ct. 2427 (2014)).

[18] See, for example, CRS Legal Sidebar WSLG921, EPA on a Roll?: D.C. Circuit Upholds Three EPA Rulemakings in April, by Robert Meltz; and CRS Legal Sidebar WSLG1292, Three More EPA Wins in the D.C. Circuit for Greenhouse Gas Rules.

May 2015 rule to define waters subject to the regulatory jurisdiction of the Clean Water Act. Ambient air quality standards for ozone, promulgated in October 2015, were also contentious.

With a change in presidential administration taking place in January 2017, this will be the final version of this chapter. It thus provides an overview of major environmental regulations and regulatory issues that were prominent and controversial during the eight years of the Obama Administration. The 115th Congress and the new Administration seem likely to revisit at least some of these issues, but that is a topic for future reports.

One additional point to note here is that EPA has continuing and ongoing statutory responsibilities that will extend beyond the term of the Obama Administration. Some of these are nondiscretionary statutory duties, such as the requirement to review and, if appropriate, revise National Ambient Air Quality Standards every five years, or to review New Source Performance Standards under the Clean Air Act (CAA) every eight years. EPA must evaluate the statutory annual requirements of the Renewable Fuels Standards every year, revising the CAA's requirements if attainment of the statutory standards is not feasible. Numerous other CAA requirements are also nondiscretionary and recurring. Other statutes require review and, if appropriate, revision of existing standards at regular intervals, too: the Solid Waste Disposal Act, in Section 2002(b), requires that each regulation promulgated under the act "shall be reviewed and, where necessary, revised not less frequently than every three years"; and the Safe Drinking Water Act requires EPA to review each drinking water regulation every six years. Further, the Clean Water Act requires EPA to review sludge management rules "not less often than every 2 years" in order to identify additional toxic pollutants for regulation.

In its Fall 2016 Regulatory Agenda, the agency identified 130 rules that are in the pre-rule, proposed rule, or final rule stage, with timelines stretching into 2017 and later.[19] A few of the listed rules are expected to be "economically significant," such as a rule to establish financial responsibility requirements under the Comprehensive Environmental Response, Compensation, and Liability Act (CERCLA) for facilities in the hardrock mining industry, which EPA plans to promulgate in December 2017.

ORGANIZATION OF THE REMAINDER OF THIS CHAPTER

Major or controversial EPA rules developed under the Obama Administration are organized below under these headings: "Clean Air Act," "Clean Water Act," "Solid Waste (RCRA)," and "Toxic Substances Control Act (TSCA)."

---

19 See http://www.reginfo.gov/public/do/eAgendaMain?operation=OPERATION_GET_ AGENCY_RULE_LIST&currentPub=true&agencyCode=&showStage=active&agencyCd=2000&Image58.x=39&Image58.y=17.

Clean Air Act

Because so many of the rules discussed in This chapter were developed under the Clean Air Act, we have divided this section into five subsections: (1) “Climate Change”; (2) “Ambient Air Quality Standards”; (3) “Electric Generating Units”; (4) “Boilers and Incinerators”; and (5) “Other Clean Air Act Rules.”

Climate Change

Greenhouse Gas Reporting Rule

In October 2009, in response to a congressional mandate in EPA’s FY2008 appropriation (P.L. 110-161), EPA promulgated the Greenhouse Gas Reporting Rule.[20] The rule required 31 categories of sources to report their emissions of greenhouse gases to EPA annually, beginning in 2011, if the sources emit 25,000 tons or more of carbon dioxide or the equivalent amount of five other greenhouse gases (GHGs).[21] (Eleven other categories of sources have since been added to the rule.) By itself, the rule imposes little cost ($867 per facility, according to EPA’s estimate) because it only requires reporting; but the sources who are required to report are expected to be the focus of EPA efforts as the agency develops regulations to control emissions of GHGs. The original reporting deadline was March 31, 2011. As that date approached, EPA extended the deadline to September 30, 2011. The first data submitted under the rule were released January 11, 2012.

Greenhouse Gas Endangerment Finding

On December 15, 2009, EPA issued findings that six greenhouse gases cause or contribute to air pollution that endangers public health and welfare.[22] The action was taken in response to an April 2007 Supreme Court decision (*Massachusetts v. EPA*) that required the agency to decide the issue or to conclude that climate change science is so uncertain as to preclude making such findings. These findings do not themselves impose any requirements on industry or other entities. However, the action was a prerequisite to finalizing EPA’s greenhouse gas emission standards for cars and light-duty trucks, which were jointly promulgated by EPA with fuel economy standards from the Department of Transportation, in May 2010. These, in turn, triggered permit requirements for stationary sources of GHGs, beginning January 2, 2011. On December 10, 2010, the U.S. Court of

---

[20] U.S. Environmental Protection Agency, “Mandatory Reporting of Greenhouse Gases; Final Rule,” 74 *Federal Register* 56260, October 30, 2009.

[21] GHG emissions consist of carbon dioxide (CO2), methane, nitrous oxide (N2O), sulfur hexafluoride (SF6), and two categories of gases—hydrofluorocarbons (HFCs) and perfluorocarbons (PFCs). Since each of these substances has a different global warming potential, the emissions of each are converted to the equivalent amount of CO2 emissions, based on how potent the substance is as compared to CO2, giving rise to the term “CO2-equivalent.”

[22] .S. Environmental Protection Agency, “Endangerment and Cause or Contribute Findings for Greenhouse Gases Under Section 202(a) of the Clean Air Act,” 74 *Federal Register* 66496, December 15, 2009.

Appeals for the D.C. Circuit denied industry and state motions to stay the endangerment finding and related regulations, and on June 26, 2012, the court upheld the regulations. The court's decision applied to 84 cases filed by a variety of industry groups and states (*Coalition for Responsible Regulation v. EPA*).

Light Duty Motor Vehicle Greenhouse Gas Emission Standards, Model Years 2012-2016

On May 7, 2010, EPA and the National Highway Traffic Safety Administration (NHTSA) promulgated integrated GHG emission standards and corporate average fuel economy (CAFE) standards for new cars and light trucks, a category that includes SUVs and minivans, as well as most pickup trucks.[23] NHTSA was required by the Energy Independence and Security Act of 2007 (EISA, P.L. 110-140) to promulgate CAFE standards so that by 2020, new cars and light trucks reach a combined average fuel economy of 35 miles per gallon (mpg). EPA simultaneously issued vehicle greenhouse gas standards in response to directives from the Supreme Court in *Massachusetts v. EPA*. The EPA regulations required a reduction in emissions to an estimated combined emission level of 250 grams of CO_2 per mile by model year 2016, about a 21% reduction in emissions when fully implemented. The Administration estimated that complying with the regulations would add $1,100 to the cost of an average vehicle, although this additional purchase cost was expected to be paid back through lifetime fuel savings. The new standards were phased in beginning with the 2012 model year. EPA estimates that the additional lifetime cost of 2012-2016 model year vehicles under the regulations will be about $52 billion; benefits are expected to be approximately $240 billion. This rule was also upheld by the D.C. Circuit in the June 26, 2012, *Coalition for Responsible Regulation* decision.

Light Duty Motor Vehicle Greenhouse Gas Emission Standards, Model Years 2017-2025

Using the same authority described in the item above, EPA and NHTSA promulgated joint GHG/fuel economy rules for 2017-2025 model year vehicles, on October 15, 2012.[24] Under these standards, GHG emissions from new cars and light trucks will be reduced about 50% by 2025 compared to 2010 levels, to an expected fleet average of 163 grams per mile; average fuel economy will rise to nearly 50 miles per gallon. The agencies estimated that the new technology to comply with the standards will cost

---

[23] U.S. Environmental Protection Agency, U.S. Department of Transportation, "Light-Duty Vehicle Greenhouse Gas Emission Standards and Corporate Average Fuel Economy Standards; Final Rule," 75 *Federal Register* 25324-25728, May 7, 2010. Information on benefits and costs are summarized in an April 2010 EPA Fact Sheet, at http://www.epa.gov/oms/climate/regulations/420f10014.pdf.

[24] U.S. Environmental Protection Agency, U.S. Department of Transportation, "2017 and Later Model Year Light-Duty Vehicle Greenhouse Gas Emissions and Corporate Average Fuel Economy Standards; Final Rule," 77 *Federal Register* 62624, October 15, 2012. A link to the rule and other information, including a Fact Sheet, is available at http://www.epa.gov/otaq/climate/regs-light-duty.htm#new1.

roughly $1,800 per vehicle in 2025, although lifetime fuel savings would total roughly $5,700 to $7,400.

In the final rule, EPA committed to a Midterm Evaluation of the standards for model years 2022- 2025. The evaluation is scheduled to be completed by April 2018, along with a decision about whether to retain the standards as promulgated, make them less stringent, or make them more stringent. According to its website, EPA is examining a wide range of factors in the evaluation, such as developments in powertrain technology, vehicle electrification, light-weighting and vehicle safety impacts, the penetration of fuel efficient technologies in the marketplace, consumer acceptance of fuel efficient technologies, trends in fuel prices and the vehicle fleet, employment impacts, and many others. As a first step, a draft Technical Assessment Report was made available for public comment in July 2016. Auto manufacturers have expressed concerns about a lack of harmonization between EPA, NHTSA, and California standards, and consumer acceptance of zero emission technology, particularly given the current relatively low price of gasoline.

Greenhouse Gas Tailoring Rule

On June 3, 2010, EPA promulgated a rule that defined which stationary sources would be required to obtain Clean Air Act permits for GHG emissions and how the requirements would be phased in, a rule the agency referred to as the "Tailoring Rule."[25] Under the Clean Air Act, stationary sources of pollution (power plants, refineries, steel mills, etc.) are required to obtain preconstruction and operating permits if they emit more than a threshold amount of any air pollutant. The statutory threshold (generally 100 tons per year of any air pollutant, but in some cases 250 tons) is not well-suited to carbon dioxide (CO_2), the principal greenhouse gas: requiring permits of facilities that emit more than 100 tons of CO_2 would lead to the "absurd result"[26] that as many as 6 million sources of GHGs would need to apply for permits.

Out of administrative necessity, therefore, the agency set higher annual thresholds. The threshold set by the Tailoring Rule (annual emissions of 75,000-100,000 tons of carbon dioxide equivalents) limited which facilities were required to obtain permits for their GHG emissions: from 2011 through 2016, the nation's largest GHG emitters, including power plants, refineries, cement production facilities, and about two dozen other categories of sources (an estimated 17,000 facilities annually) were to be the only sources required to obtain permits. Of these, most would have faced only an administrative requirement to provide an estimate of their GHG emissions. A few (EPA

[25] U.S. Environmental Protection Agency, "Prevention of Significant Deterioration and Title V Greenhouse Gas Tailoring Rule; Final Rule," 75 *Federal Register* 31514, June 3, 2010.

[26] In the preamble to the Tailoring Rule, EPA refers to three legal doctrines that guided its interpretation of the statutory requirements of the permit sections of the Clean Air Act, the first of which was "what we will call (1) The 'absurd results' doctrine, which authorizes agencies to apply statutory requirements differently than a literal reading would indicate, as necessary to effectuate congressional intent and avoid absurd results.... " 75 *Federal Register* 31516, June 3, 2010.

estimated 1,600) new or modified facilities would need to address whether they have the best available control technology for limiting emissions.[27]

The Tailoring Rule was effectively overturned by the Supreme Court on June 23, 2014. The Court agreed that EPA could require permits for GHG emissions from sources that were treated as major sources because of their emissions of conventional pollutants, and thus had to obtain CAA permits anyway. But the Court ruled that EPA could not require sources to obtain permits based solely on their GHG emissions. There is little practical effect from the Court's decision, since most of the sources that the Tailoring Rule would have required to obtain GHG permits are also in the "anyway" category.

Medium- and Heavy-Duty Vehicle Greenhouse Gas Rule

On September 15, 2011, EPA and NHTSA promulgated integrated GHG emission standards and fuel economy standards for medium- and heavy-duty vehicles.[28] Medium- and heavy-duty trucks are trucks with a gross vehicle weight of 8,500 pounds or more. EPA's endangerment finding (see above) specifically referenced medium- and heavy-duty trucks as among the sources that contribute to the GHG emissions for which it found endangerment. In addition, NHTSA was required by Section 102 of the Energy Independence and Security Act of 2007 (EISA, P.L. 110-140) to promulgate fuel economy standards for medium- and heavy-duty trucks, reflecting the "maximum feasible improvement" in fuel efficiency. The standards are being phased in between 2014 and 2018. When fully implemented, they will require an average per vehicle reduction in GHG emissions of 17% for diesel trucks and 12% for gasoline-powered trucks. The expected cost increase for the 2014-2018 vehicles affected by the rule is $8.1 billion. EPA projects benefits of $57 billion over the trucks' lifetimes, including $50 billion in fuel savings. In the President's June 25, 2013, Climate Action Plan, he committed to a second round of fuel efficiency and GHG emission standards for post-2018 heavy-duty vehicles.

Phase 2 Medium- and Heavy-Duty Truck Greenhouse Gas Emission Standards

On October 25, 2016, EPA and NHTSA promulgated a second round of greenhouse gas emission and fuel economy standards for medium- and heavy-duty trucks. The first round of standards (Phase 1), promulgated in 2011, covered model year 2014-2018 trucks. The Phase 2 standards will affect trucks beginning with the 2021 model year, and will set standards for trailers beginning in 2018. According to EPA, "The Phase 2 standards are expected to lower CO_2 emissions by approximately 1.1 billion metric tons,

---

[27] EPA reported, however, that in the first 11 months of the program, only 68 permit applications were received. See U.S. Environmental Protection Agency, "Prevention of Significant Deterioration and Title V Greenhouse Gas Tailoring Rule Step 3, GHG Plantwide Applicability Limitations and GHG Synthetic Minor Limitations, Proposed Rule," 77 *Federal Register* 14233, March 8, 2012.

[28] U.S. Environmental Protection Agency, U.S. Department of Transportation, "Greenhouse Gas Emissions Standards and Fuel Efficiency Standards for Medium- and Heavy-Duty Engines and Vehicles; Final Rules," 76 *Federal Register* 57106, September 15, 2011.

save vehicle owners fuel costs of about $170 billion, and reduce oil consumption by up to 2 billion barrels over the lifetime of the vehicles sold under the program."[29]

EPA projects the total cost of the rule at $29-31 billion over the lifetime of model year 2018-2029 trucks. The standards will increase the cost of a long-haul tractor-trailer by as much as $13,500 in model year 2027, according to the agency, but the buyer would recoup the investment in fuel-efficient technology in less than two years through fuel savings. In EPA's analysis, fuel consumption of 2027 model tractor-trailers will decline by 34% as a result of the rule.[30] As with the earlier rule, EPA and NHTSA worked on the Phase 2 standards with stakeholders and with the California Air Resources Board (CARB) with the goal of ensuring that manufacturers can continue to build a single national fleet.

Carbon Pollution Standards for New, Modified, and Reconstructed Power Plants

Electric generating units are the largest U.S. source of greenhouse gas emissions, accounting for about 30% of all anthropogenic GHG emissions. In a settlement agreement with 11 states and other parties, EPA agreed to propose New Source Performance Standards (NSPS) for power plant GHG emissions by July 26, 2011, and take final action on the proposal by May 26, 2012. This schedule encountered delays: NSPS were not proposed until April 13, 2012,[31] and although EPA faced a statutory deadline of one year after the date of proposal (i.e., April 13, 2013) for promulgation of final standards, it did not meet that deadline. The agency received nearly 2.7 million comments on the proposed rule—until that time, the most it had received on any rule in its 40-year history.

On June 25, 2013, the President directed EPA to re-propose the rule by September 20, 2013. EPA met that deadline, releasing a modified proposal on that date. The re-proposal appeared in the *Federal Register* on January 8, 2014.[32] A four-month public comment period ensued. Modified and reconstructed units are also subject to NSPS: EPA proposed separate standards for these units on June 2, 2014. EPA combined these rulemakings, finalizing a rule that covers new, reconstructed, and modified units on August 3, 2015; it was published in the *Federal Register* on October 23, 2015.[33]

---

29 U.S. Environmental Protection Agency, Office of Transportation and Air Quality, "EPA and NHTSA Adopt Standards to Reduce Greenhouse Gas Emissions and Improve Fuel Efficiency of Medium- and Heavy-Duty Vehicles for Model Year 2018 and Beyond," Regulatory Announcement, August 2016, at https://nepis.epa.gov/Exe/ZyPDF.cgi/ P100P7NL.PDF?Dockey=P100P7NL.PDF.

30 U.S. Environmental Protection Agency, U.S. Department of Transportation, "Greenhouse Gas Emissions and Fuel Efficiency Standards for Medium- and Heavy-Duty Engines and Vehicles – Phase 2; Final Rule," 81 *Federal Register* 73478, 73482, October 25, 2016. Payback period information is found in the EPA regulatory announcement cited above.

31 The standards appeared in the *Federal Register* on April 13, 2012. The standards and supporting materials are available at http://www.epa.gov/carbonpollutionstandard/ actions.html.

32 U.S. Environmental Protection Agency, "Standards of Performance for Greenhouse Gas Emissions from New Stationary Sources: Electric Utility Generating Units, Proposed Rule," 79 *Federal Register* 1430, January 8, 2014.

33 U.S. Environmental Protection Agency, "Standards of Performance for Greenhouse Gas Emissions from New, Modified, and Reconstructed Stationary Sources: Electric Utility Generating Units, Final Rule," 80 *Federal Register* 64510, October 23, 2015.

The GHG emission standards for new units were set at levels achievable by most natural-gas-fired units without added pollution controls; new coal-fired units, on the other hand, would need to use carbon capture and storage (CCS) technology to capture about 20% of their uncontrolled emissions. Although the components of CCS technology have been demonstrated, no operating power plant in the United States combined them all in an operating unit as of the date of the rule's promulgation. The electric power industry has generally concluded that a CCS requirement would effectively prohibit the construction of new coal-fired plants, other than those already permitted. EPA maintains otherwise, but it also says that, because of low natural gas prices and abundant existing generation capacity, it believes no new coal-fired units subject to the proposed standards will be constructed between 2015 and 2020.

The final standards for modified and reconstructed units[34] are less stringent. These units would have to meet emission standards equal to their best yearly performance from 2002 to the year of modification, or, if reconstructed, the emissions of the best demonstrated generating technology for the type of unit. Neither would be subject to CCS requirements. EPA says "we anticipate few covered units will trigger the reconstruction or modification provisions in the period of analysis [through 2025]. As a result, we anticipate negligible costs or benefits associated with those standards."[35]

Even if the rule's costs are negligible, the New Source standards play an important role. Under Section 111 of the Clean Air Act, EPA can only promulgate standards for emissions from *existing* sources if it has set standards for emissions from new sources in the same category. Existing source standards were promulgated at the same time as the NSPS (see next item).

Clean Power Plan (Carbon Pollution Standards for Existing Power Plants)

In the settlement agreement that EPA reached in 2010 (see previous item), the agency also agreed to develop guidelines for GHG emissions from existing fossil-fueled power plants. The President directed EPA to propose these guidelines by June 1, 2014; finalize them by June 1, 2015; and require the states to submit implementation plans by June 30, 2016. EPA promulgated this rule— which it calls the Clean Power Plan—at the same time as the NSPS standards for power plants.[36] The Clean Power Plan sets state-specific goals for CO_2 emissions from fossil-fueled power plants. EPA set these goals based on

---

[34] A modification is any physical or operational change to an existing source that increases the source's maximum achievable hourly rate of air pollutant emissions. A reconstructed source is a unit that replaces components to such an extent that the capital cost of the new components exceeds 50% of the capital cost of an entirely new comparable facility. U.S. Environmental Protection Agency, "Carbon Pollution Standards for Modified and Reconstructed Stationary Sources: Electric Utility Generating Units, Proposed Rule," 79 *Federal Register* 34960, June 18, 2014.

[35] U.S. Environmental Protection Agency, Regulatory Impact Analysis for the Proposed Carbon Pollution Guidelines for Existing Power Plants and Emission Standards for Modified and Reconstructed Power Plants, June 2014, p. ES-26, at http://www2.epa.gov/sites/production/ files/2014-06/documents/20140602ria-clean-power-plan.pdf.

[36] U.S. Environmental Protection Agency, "Carbon Pollution Emission Guidelines for Existing Stationary Sources: Electric Utility Generating Units, Final Rule," 80 *Federal Register* 64661, October 23, 2015.

three "building blocks": improved efficiency at coal-fired power plants; substitution of natural gas combined cycle generation for coal-fired power; and zero-emission power generation (from increased renewable or nuclear power). Two sets of goals were proposed: an interim set, which would apply to the average emissions rate in a state in the 2022- 2029 time period; and a final set for the years 2030 and beyond. EPA estimates the cost of the rule at $5.1 billion to $8.4 billion annually in 2030, with annual benefits of $32 billion to $54 billion in that year.

In addition to targeting emission rates for the states, the Clean Power Plan set emission and emission-rate targets for three areas of Indian country. Emission rates for Alaska and Hawaii and two U.S. territories (Guam and Puerto Rico) were not finalized due to incomplete information. EPA stated it will determine how to address the requirements of Section 111(d) with respect to these jurisdictions at a later time. Legal challenges to the Clean Power Plan were filed by more than 100 parties following its promulgation in October 2015. On February 9, 2016, the Supreme Court stayed the rule pending adjudication of the legal challenges.

Federal Implementation Plan for Existing Power Plants Under the Clean Power Plan

As described above, the Clean Power Plan, which sets greenhouse gas emission standards for existing power plants, would set *average* emission rates for fossil-fueled power plants separately for each state. The rule would be implemented through state regulations, with each state submitting an implementation plan to EPA outlining how it would achieve its state goal. This presents a potential problem for EPA: what to do if a state does not submit a plan—if it "just says no." Under the Tenth Amendment to the Constitution, EPA cannot require states to undertake sovereign acts, such as implementing federal regulatory requirements or writing the federal regulations into state law. If a state does not submit an implementation plan, EPA's only recourse under the Constitution and the Clean Air Act would be to promulgate requirements for the state through a Federal Implementation Plan. EPA proposed a template for such plans and model emissions trading rules that could be used by the states, at the same time that it finalized the Clean Power Plan, on August 3, 2015. The agency sent a final version of the model trading rules to the Office of Management and Budget for interagency review on November 3, 2016.

Methane Emission Standards for New and Modified Sources in the Oil and Gas Industry

On June 3, 2016, EPA promulgated a suite of New Source Performance Standards (NSPS) under Section 111 of the Clean Air Act to set controls for the first time on methane emissions from sources in the crude oil and natural gas production sector and the natural gas transmission and storage sector.[37] The rule builds on the agency's 2012 NSPS for volatile organic compound (VOC) emissions[38] and would extend controls for methane and VOC emissions beyond the existing requirements to include new or modified hydraulically fractured oil wells; pneumatic pumps; compressor stations; and leak detection and repair at well sites, gathering and boosting stations, and processing plants. The Administration stated that the rule is a key component under the President's Climate Action Plan, and that plan's *Strategy to Reduce Methane Emissions*,[39] needed to set the Administration on track to achieve its goal to cut methane emissions from the oil and gas sector by 40%-45% from 2012 levels by 2025, and to reduce all domestic greenhouse gas emissions by 26%-28% from 2005 levels by 2025.

Methane—the key constituent of natural gas—is a potent greenhouse gas with a global warming potential (GWP) more than 25 times greater than that of carbon dioxide. According to EPA's *Inventory of U.S. Greenhouse Gas Emissions and Sinks*, methane is the second-most-prevalent greenhouse gas emitted in the United States from human activities, and nearly 30% of those emissions come from oil production and the production, transmission, and distribution of natural gas.[40]

EPA projects that the standards for new, reconstructed, and modified sources will reduce methane emissions by 510,000 short tons in 2025, the equivalent of reducing 11 million metric tons of carbon dioxide. In conjunction with the proposal, EPA conducted a regulatory impact analysis (RIA) that looked at the illustrative benefits and costs of the proposed NSPS: in 2025, EPA estimated the rule will have costs of $530 million and climate benefits of $690 million (in constant 2012 dollars). The rule will also reduce emissions of VOCs and hazardous air pollutants. EPA was not able to quantify the benefits of these reductions.

GHG Emission Standards for Commercial Aircraft

On August 15, 2016, EPA issued a finding that greenhouse gas emissions (including CO_2 emissions) from commercial aircraft contribute to the pollution that causes climate

---

[37] .S. Environmental Protection Agency, "Oil and Natural Gas Sector: Emission Standards for New, Reconstructed, and Modified Sources; Final Rule," 81 *Federal Register* 35824, June 3, 2016.

[38] U.S. Environmental Protection Agency, "Oil and Natural Gas Sector: New Source Performance Standards and National Emission Standards for Hazardous Air Pollutants Reviews, Final Rule," 77 *Federal Register* 49489, August 16, 2012.

[39] There are other regulatory and nonregulatory components to the Administration's *Strategy to Reduce Methane Emissions*, including a proposal for Control Techniques Guidelines for VOC reduction requirements at existing oil and gas sources in ozone nonattainment areas and states in the Ozone Transport Region, a proposal to clarify the definitions for "major source" categories in the oil and natural gas sector for the purpose of permitting, and a proposal to expand voluntary efforts under the Natural Gas STAR program and to provide several new mechanisms through which oil and gas companies could make and track commitments to reduce methane emissions. For more information, see Executive Office of the President (EOP), *The President's Climate Action Plan*, June 2013; EOP, *Climate Action Plan: Strategy to Reduce Methane Emissions*, March 2014; EOP, "Fact Sheet: Administration Takes Steps Forward on Climate Action Plan by Announcing Actions to Cut Methane Emissions," January 14, 2015; and CRS Report R43860, *Methane: An Introduction to Emission Sources and Reduction Strategies*.

[40] GWP as calculated over 100 years. U.S. Environmental Protection Agency, *Inventory of U.S. Greenhouse Gas Emissions and Sinks: 1990-2013*, Washington, DC, EPA 430-R-15-004, April 15, 2015.

change and endangers the health and welfare of Americans.[41] The finding, under Section 231 of the Clean Air Act, is the precondition for GHG emission standards for commercial aircraft. U.S. aircraft emit roughly 11% of GHG emissions from the U.S. transportation sector and 29% of GHG emissions from all aircraft globally.

In February 2016, the International Civil Aviation Organization (ICAO) agreed on international CO_2 standards for aircraft, beginning in 2020. ICAO is a specialized body of the United Nations with 191 member states. For the past five years, ICAO had been working with the aviation industry and other stakeholders to develop international CO_2 emission standards for aircraft. EPA and the Federal Aviation Administration, representing the United States, have participated in ICAO's process. In October 2016, ICAO's governing council also agreed on a system for offsetting future carbon emissions from aviation.

EPA's endangerment finding lays the necessary foundation for the adoption and implementation of domestic aircraft CO_2 standards, in accordance with the Clean Air Act and the ICAO agreement. The emissions offset system agreed to in the ICAO process is voluntary for the next decade and has been agreed to by the U.S. industry.

Ambient Air Quality Standards

Ozone Ambient Air Quality Standards

On October 1, 2015, EPA finalized a revision of the National Ambient Air Quality Standards (NAAQS) for ozone.[42] The Clean Air Act requires EPA to review the NAAQS every five years. The previous review was completed in March 2008, so the agency was behind the statutory schedule for completion of its review, and was under a court order to sign a final rule by October 1, 2015.

NAAQS are the cornerstone of the Clean Air Act, in effect defining what EPA considers to be clean air for six widespread categories of air pollution. Ozone is the most widespread of the six pollutants. Exposure to it in concentrations above the standard has been linked to respiratory illnesses, heart attacks, and premature death. Ozone also has negative effects on forests and crop yields, which the NAAQS are also supposed to protect.

In setting standards for ambient air quality, NAAQS do not directly limit emissions, but they set in motion a process under which "nonattainment areas" are identified and states and EPA develop plans and regulations to reduce pollution in those areas. Nonattainment designations may also trigger statutory requirements, including that new major sources offset certain emissions by reducing emissions from existing sources.

---

[41] U.S. Environmental Protection Agency, "Finding That Greenhouse Gas Emissions from Aircraft Cause or Contribute to Air Pollution That May Reasonably Be Anticipated to Endanger Public Health and Welfare," 81 *Federal Register* 54422, August 15, 2016.

[42] The rule appeared in the *Federal Register* on October 26, 2015: U.S. Environmental Protection Agency, "National Ambient Air Quality Standards for Ozone, Final Rule," 80 *Federal Register* 65292.

The ozone NAAQS promulgated in 2015 were based on a review of 2,300 scientific studies. The NAAQS set both primary (health-based) and secondary (welfare-based) standards of 70 parts per billion (ppb)—as opposed to the previous standards of 75 ppb—for ozone concentrations in ambient air. EPA has identified 241 counties that would violate the NAAQS's proposed range, if the most recent three years of data available in 2015 were used to determine attainment. For comparison, 155 counties had monitors showing violation of the currently implemented (2008) standard. Actual designation of nonattainment areas will be made using later data than that available in 2015, however—most likely data for 2014-2016. Given emission trends, a revised standard may affect fewer counties than the current projection might suggest. (Note: From late 2009 to September 2011, EPA conducted a reconsideration of the 2008 ozone NAAQS, and proposed changes. At the President's request, however, the proposal was withdrawn without final action. The reconsideration is discussed below.)

EPA estimates that the costs of implementing the proposal would be $1.4 billion annually in 2025 in states other than California. The monetized value of the health benefits associated with changing the standard would range from $2.9 billion to $5.9 billion annually in 2025, according to the agency. Separate cost and benefit estimates were prepared for California; under the statute, most areas in California will have until the 2030s to attain the standard. Relying on their own estimates, which make different assumptions regarding the cost of emission controls, industry sources (notably the National Association of Manufacturers) have stated that the ozone proposal could be the most expensive rule in EPA history, and have supported efforts in Congress to delay promulgation of a final standard.

2010-2011 Ozone Ambient Air Quality Standards Review

On January 19, 2010, EPA proposed a revision of the National Ambient Air Quality Standard (NAAQS) for ozone.[43] At the President's request, on September 2, 2011, this proposal was withdrawn, leaving EPA to implement previously promulgated (2008) ozone standards. As noted above, the Clean Air Act requires EPA to review NAAQS every five years. EPA had completed a review of the ozone NAAQS in 2008, and made the standard more stringent, but the Obama Administration's EPA suspended implementation of the 2008 standard in 2009 in order to consider further strengthening it.

The reconsidered ozone NAAQS that was proposed in January 2010 was among the most controversial standards under consideration at EPA, because of its wide reach and potential cost. In the 2010 proposal, EPA identified at least 515 counties that would violate the proposed NAAQS if the most recent three years of data available at the time of proposal were used to determine attainment (compared to 85 counties that violated the 1997 standard in effect at that time). The agency estimated that the costs of implementing

---

[43] U.S. Environmental Protection Agency, "National Ambient Air Quality Standards for Ozone; Proposed Rule," 75 *Federal Register* 2938, January 19, 2010.

the reconsidered ozone NAAQS, as proposed, would range from $19 billion to $25 billion annually in 2020, with benefits of roughly the same amount.

On September 2, 2011, the White House announced that the President had requested that EPA Administrator Jackson withdraw the draft ozone standards, since work was already under way to update a review of the science that would result in the reconsideration of the ozone standard in 2013.[44] That review led to the 2015 revision of the standards described above.

Particulate Matter (Including "Farm Dust") NAAQS

EPA considers particulate matter (PM) to be among the most serious air pollutants, responsible for tens of thousands of premature deaths annually. The current NAAQS sets standards for both "fine" particulates ($PM_{2.5}$) and larger, "coarse" particles (PM_{10}). The $PM_{2.5}$ standards affect far more people and far more counties than the standard for PM_{10}, and both sets of standards have affected mostly industrial, urban areas.

EPA completed a review of the PM NAAQS in 2006. The agency is required by the Clean Air Act to review NAAQS at five-year intervals, so another review was due in 2011. As the review process was getting under way, in February 2009, the D.C. Circuit Court of Appeals remanded the 2006 standard for $PM_{2.5}$ to EPA, saying that the standard was "contrary to law and unsupported by adequately reasoned decisionmaking."[45] As a result, EPA combined the statutory five-year review of the standard and its response to the D.C. Circuit decision, completing a review of the PM standard that served both purposes in January 2013.[46] The review left the standard for coarse particles unchanged, as well as the standard for 24-hour exposures to $PM_{2.5}$. But it lowered the standard for annual exposures to $PM_{2.5}$, as suggested by the agency's outside scientific advisers, from 15 micrograms per cubic meter to 12.

Although this appears to be a significant strengthening of a standard that potentially affects a wide array of mobile and stationary sources, EPA projects the incremental cost of the revision at a relatively modest $53 million to $350 million annually. The cost of compliance with the PM NAAQS is moderated by the fact that other EPA standards (for various emission sources) are reducing exposures to $PM_{2.5}$ even without a strengthening of the ambient standard. Annual benefits of the more stringent NAAQS were estimated to range from $4.0 billion to $9.1 billion.

In the 112th Congress, attention to PM issues focused on the larger, coarse particles, PM_{10}, even though EPA did not propose to change them. Members of the House and Senators discussed the need to prevent a supposed EPA plan to use the revision of the PM_{10} standard to impose controls on "farm dust." The House passed legislation to prevent

---

[44] The White House, Office of the Press Secretary, "Statement by the President on the Ozone National Ambient Air Quality Standards," September 2, 2011.

[45] American Farm Bureau Fed'n v. EPA, 559 F.3d 512 (D.C. Cir. 2009).

[46] U.S. Environmental Protection Agency, "National Ambient Air Quality Standards for Particulate Matter; Final Rule," 78 *Federal Register* 3086, January 15, 2013. A link to the standards and other supporting materials can be found at http://www.epa.gov/ pm/actions.html.

EPA from tightening standards for PM10 for one year and to permanently limit EPA's authority to regulate dust in rural areas. EPA stated early in the PM review process that it did not intend to change the PM10 standard, and the final revision made no change.

Sulfur Dioxide NAAQS

Three other NAAQS reviews (for sulfur dioxide,[47] nitrogen dioxide,[48] and carbon monoxide) were completed in 2010 and 2011. Of these, only the sulfur dioxide (SO2) NAAQS was considered an economically significant rule.[49] EPA estimated the cost of the more stringent SO_2 NAAQS at $1.5 billion annually, with benefits 9-24 times that amount.

Electric Generating Units[50]

Cross-State Air Pollution (Clean Air Transport) Rule

Under the Clean Air Act's "good neighbor" provisions (in Section 110(a) of the act), states are required to prohibit emissions that contribute significantly to nonattainment (or interfere with maintenance) of ambient air quality standards by downwind states. If a state fails to meet these requirements, EPA must issue federal regulations addressing the problem. Under the Bush Administration, EPA found that 28 eastern states had failed to adequately address this requirement, and issued the Clean Air Interstate Rule (CAIR) to address the good neighbor provisions. The rule was remanded to the agency by the D.C. Circuit Court of Appeals in 2008. EPA promulgated a replacement, the Cross-State Air Pollution Rule (CSAPR, pronounced "Casper"), on August 8, 2011.[51] The original CAIR rule established cap-and-trade programs for sulfur dioxide and nitrogen oxide emissions from coal-fired electric power plants in 28 eastern states, at an estimated annual cost of $3.6 billion in 2015. The replacement rule also applies to 28 states; it allows unlimited intrastate allowance trading, but limits interstate trading in response to the D.C. Circuit decision. Its annual compliance cost was estimated at $800 million annually in 2014, on top of $1.6 billion already being spent to comply with CAIR. EPA estimates the benefits of CSAPR at $120 billion to $280 billion annually, chiefly the avoidance of 13,000 to 34,000 annual premature deaths.

---

[47] U.S. Environmental Protection Agency, "Primary National Ambient Air Quality Standard for Sulfur Dioxide; Final Rule," 75 *Federal Register* 35520, June 22, 2010.

[48] U.S. Environmental Protection Agency, "Primary National Ambient Air Quality Standards for Nitrogen Dioxide; Final Rule," 75 *Federal Register* 6473, February 9, 2010.

[49] The agency concluded that the nitrogen dioxide NAAQS, even though it was strengthened, would have no costs or benefits, since the agency projected no areas to be nonattainment for the revised standard. The agency decided not to change the carbon monoxide NAAQS, so there were no costs or benefits associated with that review, either.

[50] See also Items 8, 9, and 10 under the subhead "Climate Change."

[51] U.S. Environmental Protection Agency, "Federal Implementation Plans: Interstate Transport of Fine Particulate Matter and Ozone and Correction of SIP Approvals," 76 *Federal Register* 48208, August 8, 2011. Explanatory material can be found at http://www.epa.gov/ crossstaterule/actions.html. The rule was generally referred to as the Clean Air Transport Rule prior to being finalized.

Numerous parties petitioned the D.C. Circuit for review of CSAPR, and the court stayed its implementation pending the completion of the court's proceedings. On August 21, 2012, the court vacated the standards and remanded them to EPA. In June 2014, however, the Supreme Court overturned the D.C. Circuit decision and remanded the case to the D.C. Circuit for further action. In light of the Supreme Court decision, the stay on implementation of CSAPR was lifted: the first phase of the rule was implemented in 2015, with a second phase scheduled for 2017. Because of the earlier CAIR requirements, which remained in effect pending their replacement and, more recently, because power companies have replaced substantial amounts of coal-fired generation with cheaper (and cleaner) natural-gas-fired units, electric generators had already (in 2012) achieved more than two-thirds of the pollution reductions necessary to comply with the 2017 standards.

Cross-State Air Pollution Rule Update

On October 26, 2016, EPA promulgated updates to the Cross-State Air Pollution Rule (CSAPR).[52] The CSAPR rule was first promulgated in August 2011, and took effect in January 2015. As explained above, CSAPR is designed to address what are called the "good neighbor" provisions of the Clean Air Act by requiring states to limit emissions of pollution that cause air quality problems in other, downwind states. The rule established cap-and-trade programs for sulfur dioxide (SO_2) and nitrogen oxide (NOx) emissions from coal-fired electric power plants in 28 eastern states.

On July 28, 2015, the D.C. Circuit Court of Appeals remanded CSAPR's ozone-season NOx emission budgets for 11 states and the annual SO_2 budgets of four states to EPA for modification. The updated rule, in part, addresses the court's remand of the ozone-season NOx emission budgets. The updated rule also sets new requirements for 22 states to limit NOx pollution that is contributing to nonattainment of the 2008 ambient ozone standard in downwind states; the 2011 CSAPR rule was designed to achieve an earlier, less stringent version of the ozone standard.[53]

The updated rule will take effect in May 2017. EPA estimates the annual cost of the rule at $68 million, with benefits of $880 million annually. The agency expects that the rule's requirements can be met quickly and at low cost by optimizing operation of existing pollution control technology, turning on existing pollution controls that are currently idled, upgrading to state-of-the-art low-NOx combustion controls, and shifting generation to lower-emitting power plants.

NOx emissions from power plants in the eastern United States will be lowered by 20% in 2017 under the update and "other changes already underway in the power sector," with three-fourths of the reductions occurring in Pennsylvania, Indiana, West Virginia, Ohio, Alabama, and Kentucky.

---

[52] U.S. EPA, "Cross-State Air Pollution Rule Update for the 2008 Ozone NAAQS, Final Rule," 81 *Federal Register* 74504, October 26, 2016.

[53] The court's remand of the SO2 budgets was addressed separately in a memorandum that proposed two options for future state compliance with SO2 requirements. The memorandum is at https://www3.epa.gov/airtransport/CSAPR/ pdfs/CSAPR_SO2_Remand_Memo.pdf.

Mercury and Air Toxics Standards/MACT for Electric Generating Units ("Utility MACT" or MATS Rule)

In 2005, EPA promulgated regulations establishing a cap-and-trade system to limit emissions of mercury from coal-fired power plants. The rules were challenged, and the D.C. Circuit Court of Appeals vacated them in 2008. Rather than appeal the ruling to the Supreme Court, EPA agreed to propose and promulgate Maximum Achievable Control Technology (MACT) standards by the end of 2011.

EPA stated that the standards for existing units, promulgated February 16, 2012,[54] could be met by 56% of coal- and oil-fired electric generating units using pollution control equipment already installed; the other 44% would be required to install technology that would reduce uncontrolled mercury and acid gas emissions by about 90%, at an annual cost of $9.6 billion. Standards for new facilities are more stringent, and many (including the industry that manufactures pollution control and monitoring equipment), doubted whether compliance with the mercury portion of these standards could be measured. In response to industry petitions, EPA reconsidered the mercury limit for new facilities, and announced changes to the standards for new facilities on March 29, 2013.

EPA estimated that the annual benefits of the Utility MACT, including the avoidance of up to 11,000 premature deaths annually, will be between $37 billion and $90 billion. Virtually all of the avoided deaths and monetized benefits come from the rule's effect on emissions of particulates, rather than from identified effects of reducing mercury and air toxics exposure. Numerous parties petitioned the courts for review of the rule, contending in part that EPA had failed to conduct a cost-benefit analysis in its initial determination that control of air toxics from electric power plants was "appropriate and necessary," and that the agency's later cost-benefit analysis indicated that the rule failed this test. In June 2015, the Supreme Court agreed with the petitioners, remanding the rule to the D.C. Circuit for further consideration.

As of this writing, the effect of the Supreme Court decision remains unclear. Existing power plants had until April 2015, with a possible one-year extension, to meet the standards. The majority of the affected plants had already complied with the rule before the Court's decision. In response to the Court's decision, EPA prepared a supplemental "appropriate and necessary" finding that, after taking public comment, it finalized in the April 25, 2016, *Federal Register*. As of late 2016, the MATS rule remains in effect while the circuit court considers whether EPA's action in response to the Supreme Court decision has properly addressed the Court's concerns.

---

[54] U.S. Environmental Protection Agency, "National Emission Standards for Hazardous Air Pollutants from Coal and Oil-Fired Electric Utility Steam Generating Units and Standards of Performance for Fossil-Fuel-Fired Electric Utility, Industrial-Commercial-Institutional, and Small Industrial-Commercial-Institutional Steam Generating Units," 77 *Federal Register* 9304, February 16, 2012.

Boilers and Incinerators

MACT and Area Source Standards for Boilers.

EPA proposed Maximum Achievable Control Technology standards to control emissions of toxic air pollutants from commercial and industrial boilers in June 2010. A final rule was issued February 21, 2011, under a court order by the Federal District Court for the District of Columbia.[55] Because of voluminous comments and new information received from industry during a public comment period, EPA had asked the court to extend the deadline for promulgating final standards to April 2012. Having been denied that extension, the agency initiated a reconsideration after it released the final rule, and it promulgated changes to the rule on January 31, 2013.[56] In addition to adjusting the rule's emission standards, the January 2013 rule reset the clock for compliance, effectively giving industry almost two additional years to install control equipment.

Boilers are used to provide heat and/or power throughout industry and in many commercial and institutional facilities. The D.C. Circuit vacated EPA's previous MACT rule for this category in 2007, saying EPA had wrongly excluded many industrial boilers from the definition of solid waste incinerators, which have more stringent emissions requirements under the Clean Air Act. The vacated rule had estimated annual costs of $837 million, with a benefit-cost ratio of about 20- to-1. The January 2013 rule set more stringent standards. It affects about 14,000 boilers, according to the agency, with annual costs estimated at $1.2 billion and benefits of $25 billion to $67 billion annually, including the avoidance of 3,100 to 7,900 premature deaths.

EPA also promulgated what are called "area source" standards for smaller boilers at the same time as the MACT.[57] The area source standards affect 183,000 boilers, most of which do not need to install pollution controls. Their sole requirement is to perform a tune-up every two to five years to comply with the regulations. EPA estimated the net cost of the area source rule to be $490 million annually, with partial benefits ranging from $210 million to $520 million annually.

Commercial and Industrial Solid Waste Incinerator (CISWI) Standards

A third regulation promulgated and reconsidered at the same time as the boiler MACT and area source boiler rules sets standards for emissions from commercial and industrial solid waste incinerators.[58] These standards are related to the D.C. Circuit's

[55] The final rule appeared in the *Federal Register* on March 21, 2011. U.S. Environmental Protection Agency, "National Emission Standards for Hazardous Air Pollutants for Major Sources: Industrial, Commercial, and Institutional Boilers and Process Heaters; Final Rule," 76 *Federal Register* 15608, March 21, 2011.

[56] U.S. Environmental Protection Agency, "National Emission Standards for Hazardous Air Pollutants for Area Sources: Industrial, Commercial, and Institutional Boilers; Final Rule; Notice of Final Action on Reconsideration," 78 *Federal Register* 7138, January 31, 2013.

[57] .S. Environmental Protection Agency, "National Emission Standards for Hazardous Air Pollutants for Area Sources: Industrial, Commercial, and Institutional Boilers; Final Rule," 76 *Federal Register* 15554, March 21, 2011. Final action on reconsideration is at 78 *Federal Register* 7488, February 1, 2013.

[58] The 2011 standards were U.S. Environmental Protection Agency, "Standards of Performance for New Stationary Sources and Emission Guidelines for Existing Sources: Commercial and Industrial Solid Waste Incineration

remand of the boiler rules in 2007; they also faced a judicial deadline of February 21, 2011, and after being promulgated, were also reconsidered in early 2013. As reconsidered, the rules expanded the number of existing facilities subject to the more stringent CISWI standards from 20 to 106, with annual costs of $271 million, according to EPA, and benefits of $380 million-$1 billion annually.

Other Clean Air Act Rules

Tier 3 Emissions Standards for Passenger Cars and Light Trucks and Gasoline Standards

In February 2011, EPA began to scope out new emissions standards for conventional pollutants (i.e., nongreenhouse gases) from passenger cars and light trucks. In a May 2010 memorandum from the White House to the EPA Administrator, President Obama had directed EPA to review the adequacy of the current "Tier 2" emissions standards for these vehicles, which EPA finalized in February 2000, and were phased in between model years (MYs) 2004 and 2009.[59] EPA proposed "Tier 3" standards April 13, 2013, and released the final standards March 3, 2014. As with the Tier 2 standards, the proposed Tier 3 standards include changes to both vehicle emission limits and fuel formulation rules, lowering allowable sulfur content to reduce the fouling of catalytic converters on existing vehicles and facilitate the use of new technology. The proposal would lower allowable sulfur from 30 parts per million to a maximum of 10, and would require reductions in vehicle emissions of 60%-80%. In letters to the EPA Administrator, several Senators asked EPA to delay its rulemaking over concerns that the new fuel standards would raise the price of gasoline,[60] but EPA maintained that the rule as proposed would add less than a penny a gallon to the price of gasoline. The rules will be phased in, beginning in 2017. EPA estimates the cost of the rules at $1.1 billion to $1.5 billion annually, with annual benefits ranging from $7 billion to $19 billion.

Wood Stove Emission Standards

On February 3, 2015, EPA released final emission standards for new residential wood heaters, the most common of which are wood stoves, pellet stoves, hydronic heaters, and forced air furnaces. The regulations, which revise New Source Performance Standards (NSPS) set in 1988 for wood stoves and pellet stoves and establish standards

---

Units; Final Rule," 76 *Federal Register* 15704, March 21, 2011. The notice of final action on reconsideration is at 78 *Federal Register* 9112, February 7, 2013.

[59] For more information on the Tier 3 standards, see U.S. EPA, "Tier 3 Vehicle Emission and Fuel Standards Program," at http://www.epa.gov/otaq/tier3.htm.

[60] Jeremy P. Jacobs, "Bipartisan Senate Group Seeks Delay in EPA Tailpipe Rules," *E&E News PM*, January 12, 2012. Also, "Four Democratic Senators Urge EPA to Issue Advance Notice on Proposed Rule for Tier 3," *Daily Environment Report*, March 22, 2013.

for other types of wood heaters for the first time, appeared in the *Federal Register* on March 16, 2015.[61]

According to EPA, smoke from wood heaters contributes "hundreds of thousands of tons" of fine particles to the air throughout the country each year, accounting for nearly 25% of all area source air toxics cancer risks and 15% of noncancer respiratory effects. In many areas, in wintertime, wood heaters are the largest source of particulate air pollution; yet, many heater types have not been subject to any federal emission standard prior to promulgation of this rule. The rule would only gradually reduce this pollution, because it would apply only to new heaters (not those already in use) and it would give the industry a five-year grace period before its most stringent standards would take effect. Wood heaters can last for 40 years or more, so it will be decades before the full health benefits of the rule will be attained. Nevertheless, the rule would eliminate an estimated 360 to 810 premature deaths annually in the 2015-2020 period, according to EPA, as well as reduce hospital admissions and lost work days due to respiratory illness. EPA quantifies these benefits at $3.4 billion to $7.6 billion per year during the 2015-2020 period, more than 70 times the agency's estimate of the annualized cost to manufacturers, $46 million. Trade associations representing the affected industries and companies in the industry supported revision of the standards and the inclusion of additional heater types, but they expressed concern that the standards will impose too great a cost. Facing higher costs for new units, homeowners will continue to use current, highly polluting equipment, rather than replace it, the industry maintains. Many have also expressed concerns regarding the process to be used in certifying compliance and the short period of time in which currently available units could be tested and certified.

Ethanol Waiver ("Blend Wall")

Section 211(f) of the Clean Air Act effectively limits the amount of oxygen in gasoline unless EPA issues a waiver. Since ethanol contains oxygen, an increase in the ethanol content of gasoline offered for sale can occur only if EPA issues such a waiver. EPA may issue a waiver if the agency determines that the fuel or fuel additive will not cause or contribute to the failure of any emission control device or system used by vehicle manufacturers to achieve compliance with emission standards under the act.

On March 6, 2009, Growth Energy (on behalf of 52 U.S. ethanol producers) applied to EPA for a waiver from the then-current regulation limiting the ethanol content in gasoline to a maximum of 10% (E10). The application requested an increase in the maximum concentration to 15% (E15). A complete waiver would allow the use of significantly more ethanol in gasoline than has been permitted under the Clean Air Act. Limiting ethanol content to 10% leads to an upper bound of roughly 15 billion gallons of ethanol in all U.S. gasoline. This "blend wall" could limit the fuel industry's ability to

[61] U.S. EPA, "Standards of Performance for New Residential Wood Heaters, New Residential Hydronic Heaters and Forced-Air Furnaces; Final Rule," 80 *Federal Register* 13672, March 16, 2015.

meet the Energy Independence and Security Act's requirement to use increasing amounts of renewable fuels (including ethanol) in transportation.

On November 4, 2010, EPA granted a partial waiver allowing the use of E15 in MY2007 vehicles and newer.[62] The agency delayed a decision on MY2001-MY2006 vehicles until the Department of Energy completed testing of those vehicles. On January 21, 2011, EPA announced that the waiver would be expanded to include MY2001-MY2006 vehicles.[63] EPA determined that data were insufficient to address concerns that had been raised over emissions from MY2000 and older vehicles, as well as heavy-duty vehicles, motorcycles, and nonroad applications, and thus a waiver for these vehicles/engines was denied.

EPA has noted that granting the waiver eliminates only one impediment to the use of E15—other factors, including retail and blending infrastructure, state and local laws and regulations, and manufacturers' warranties, would still need to be addressed. Because of concerns over potential damage by E15 to equipment not designed for its use, this partial waiver has been challenged in court by a group of vehicle and engine manufacturers, although that case was dismissed because none of the petitioners had been injured *in fact*.[64] On June 23, 2011, EPA issued final rules, including new labeling requirements, to prevent the accidental use of E15 in vehicles and engines not approved for its use.

Because of various factors, expansion of E15 supply has been slow. The Department of Energy reports E15 is sold at approximately 100 stations, mostly in the Midwest.[65] Further, one renewable fuel organization reports that eight auto companies approve of the use of E15 in all of their vehicle models for MY2017.[66]

Expanded Renewable Fuel Standard (RFS2)

The RFS is a mandate that requires the nation's transportation fuel supply contain increasing amounts of renewable fuel on an annual basis (Section 211[o] of the Clean Air Act, as amended by the Energy Independence and Security Act of 2007 [EISA, P.L. 110-140]).[67] The total renewable fuel volume under the RFS consists of both conventional biofuel (e.g., corn starch ethanol) and advanced biofuel (e.g., cellulosic biofuel, biomass-based diesel). EPA is responsible for administering the RFS. One of EPA's annual responsibilities is to release a final rule announcing the standard for each year. Another responsibility is to reduce the statutory volumes given certain conditions.

---

[62] U.S. Environmental Protection Agency, "Partial Grant and Partial Denial of Clean Air Act Waiver Application Submitted by Growth Energy to Increase the Allowable Ethanol Content of Gasoline to 15 Percent; Decision of the Administrator; Notice," 75 *Federal Register* 68094-68150, November 4, 2010.

[63] U.S. Environmental Protection Agency, "Partial Grant of Clean Air Act Waiver Application Submitted by Growth Energy to Increase the Allowable Ethanol Content of Gasoline to 15 Percent; Decision of the Administrator," 76 *Federal Register* 4662, January 26, 2011.

[64] Grocery Mfrs. Ass'n v. EPA, 693 F.3d 169 (D.C. Cir. 2012).

[65] U.S. Department of Energy, *Alternative Fuels Data Center, E15*, March 30, 2016.

[66] Renewable Fuels Association, E15 Approval Status for Conventional (Non-FFV) Automobiles, 2016.

[67] For more information, see CRS Report R43325, *The Renewable Fuel Standard (RFS): In Brief*, by Kelsi Bracmort.

On November 23, 2016, EPA announced the final 2017 renewable fuel standards setting the total renewable fuel volume at 19.28 billion gallons, which was below the 24.0 billion gallon statutory requirement.[68] EPA set the conventional biofuel mandate at 15.0 billion gallons, in line with the statutory requirement of 15.0 billion gallons for 2017. EPA set the 2017 advanced biofuel mandate at 4.28 billion gallons, which was below the statutory requirement of 9.0 billion gallons. EPA used the cellulosic biofuel waiver authority to reduce the cellulosic biofuel and the total renewable fuel standards. EPA asserts that "real-world constraints, such as the slower than expected development of the cellulosic biofuel industry and constraints in the marketplace related to supply of certain biofuels to consumers, have made the timeline laid out by Congress for the growth in renewable fuel use (other than for biomass-based diesel) impossible to achieve."[69] For the 2017 final rule, EPA followed the trend from its previous final rule for the RFS—which covered 2014, 2015, and 2016—by lowering the total renewable fuel volume required.[70]

There have been numerous difficulties with implementing the RFS. It is a complicated, technical policy, subject to multiple factors that determine its outcome. Also, the RFS was established and expanded at a time when gasoline consumption was expected to be higher than what is currently observed, allowing for more blending of biofuels into gasoline. Moreover, it was thought that advanced biofuel production would occur at a faster pace, and there was no forewarning about an economic downturn and the slower-than-anticipated investment in advanced biofuel projects. Lastly, EPA has had to implement the RFS using their resources, stakeholder input, and guidance from the courts, all of which have contributed to the uncertainty and delay of certain aspects of the program (e.g., issuance of the annual standards, approval of eligible fuel pathways).

One distinct issue of concern with the RFS is the "reset" provision. The reset provision requires that the EPA Administrator modify the applicable volumes of the RFS in future years starting in 2016 if certain conditions are met.[71] The latest final rule has triggered the use of this provision for both the advanced biofuel and cellulosic biofuel categories. This is primarily a result of the lack of cellulosic biofuel production. Cellulosic biofuel production has not reached the statutory requirements for various reasons, including technological challenges, financial challenges, and more.[72] EPA has repeatedly reduced the annual cellulosic biofuel requirement each year from 2010 to 2017. This lack of production has a domino effect on the advanced biofuel category, particularly when there is not enough other advanced biofuel production to backfill the cellulosic biofuel shortfall.

---

[68] U.S. Environmental Protection Agency, "Renewable Fuel Standard Program: Standards for 2017 and Biomass-Based Diesel Volume for 2018; Final Rule," November 23, 2016; EISA, P.L. 110-140.

[69] Ibid.

[70] U.S. Environmental Protection Agency, "Renewable Fuel Standard Program: Standards for 2014, 2015, and 2016 and Biomass-Based Diesel Volume for 2017; Final Rule," 80 *Federal Register* 239, December 14, 2015.; EISA, P.L. 110-140.

[71] For more information, see CRS Report R44045, *The Renewable Fuel Standard (RFS): Waiver Authority and Modification of Volumes.*

[72] For more information, see CRS Report R41106, *The Renewable Fuel Standard (RFS): Cellulosic Biofuels.*

Legislation introduced in the 114th Congress would amend the Clean Air Act to modify and/or repeal the RFS. For instance, proposed legislation would have modified the way EPA projects yearly cellulosic biofuel mandates. Various other bills would also amend the RFS, including legislation to eliminate portions of the RFS, lower the annual mandated levels, or eliminate the program entirely.[73]

Oil and Natural Gas Air Pollution Standards

In February 2010, EPA signed a consent agreement under which it was to promulgate revisions of the New Source Performance Standards and Hazardous Air Pollutant standards for oil and gas production by November 30, 2011. The agency promulgated these rules on August 16, 2012.[74] Under the CAA, EPA is required to review New Source Performance Standards every eight years; the revisions update NSPS rules for VOCs and SO_2 that were promulgated in 1985. Similarly, EPA had a statutory obligation to review hazardous air pollutant standards for oil and natural gas production, which were issued in 1999, by 2007. Additionally, the 2012 rules are the first regulations to address emissions from natural gas wells that use hydraulic fracturing ("fracking"). The new standards, which were to be fully implemented by 2015, require companies to capture natural gas and volatile organic compounds (VOCs) that escape when hydraulically fractured gas wells are prepared for production. The rules affect production, processing, transmission, and storage, but not distribution to customers. EPA estimated that the rules would result in the capture of 95% of the VOCs otherwise emitted. Although there are costs associated with the use of equipment to capture the emissions, EPA estimated that the rules will produce a net annual savings of $11 million to $19 million for the industry, because the captured gas and condensate can be sold. Some states already required similar measures, and EPA estimates that about half of fracked natural gas wells already met the standards at the time of promulgation. In 2013, EPA promulgated updates to the storage tank portions of the rules in response to petitions for reconsideration. The updates provided additional time for compliance and an alternative emissions limit.[75] Industry groups have filed lawsuits challenging both the 2012 standards and the 2013 updates.

Portland Cement Manufacturing

On September 9, 2010, EPA promulgated New Source Performance Standards (NSPS) for conventional pollutants from new cement kilns and Maximum Achievable Control Technology (MACT) standards for hazardous air pollutants from both existing

---

[73] See, for example, S. 1584, S. 577, S. 934, H.R. 434, H.R. 703, H.R. 704, and H.R. 3228.

[74] U.S. Environmental Protection Agency, "Oil and Natural Gas Sector: New Source Performance Standards and National Emission Standards for Hazardous Air Pollutants Reviews; Final Rule," 77 *Federal Register* 49490, August 16, 2012. For information, see http://www.epa.gov/airquality/oilandgas/actions.html.

[75] U.S. Environmental Protection Agency, "Oil and Natural Gas Sector: Reconsideration of Certain Provisions of New Source Performance Standards," 78 *Federal Register* 58416-58448, September 23, 2013.

and new cement kilns.[76] When fully implemented in late 2013, the standards would have required a 92% reduction in emissions of both particulate matter and mercury and a 97% reduction in emissions of acid gases, according to EPA, as well as controlling other pollutants. EPA had previously issued emission standards for this industry in 1999, but the standards were challenged in court and remanded to the agency by the D.C. Circuit Court of Appeals. The new rules reflect EPA's reconsideration of the 1999 standards.

The agency estimated that it would cost the industry $350 million annually to comply with the 2010 standards, but that benefits (including the avoidance of 960 to 2,500 premature deaths in people with heart disease) would be worth $6.7 billion to $18 billion annually. The trade association representing the industry said the standards would cause some facilities to close. In 2011, the D.C. Circuit Court of Appeals remanded the 2010 standards to EPA for the agency to reconsider emission standards for kilns that use solid waste as fuel. The court did not stay implementation of the 2010 standards, but EPA, in proposing changes to the particulate portion of the standards in 2012, announced its intention to give the industry an additional two years to comply, with a third year available if needed. The changes were estimated to reduce industry costs by $52 million annually, compared to the 2010 rule. EPA finalized these changes February 12, 2013.[77]

Further regulation of this industry, which is the third-highest stationary U.S. source of carbon dioxide emissions, was at one point under consideration: when EPA promulgated the rule in September 2010, it stated in the preamble to the rule that it is "working towards a proposal for GHG standards" for these plants.[78] But action on these standards was not mentioned in the President's June 2013 Climate Action Plan, and has not taken place.

Stationary Internal Combustion Engines

EPA set standards for both compression ignition[79] (generally diesel) and spark ignition (generally gasoline) stationary engines[80] in 2010. The agency subsequently amended the rules on January 30, 2013.[81] The regulations would affect stationary engines, such as emergency power generators used by hospitals and other sources and electric power generators used for compressors and pumps by a wide array of industrial, agricultural, and oil and gas industry sources. The rules are referred to as the RICE

---

[76] U.S. Environmental Protection Agency, "National Emission Standards for Hazardous Air Pollutants from the Portland Cement Manufacturing Industry and Standards of Performance for Portland Cement Plants; Final Rule," 75 *Federal Register* 54970, September 9, 2010.

[77] 78 *Federal Register* 10006, February 12, 2013.

[78] Ibid., p. 54997.

[79] U.S. Environmental Protection Agency, "National Emission Standards for Hazardous Air Pollutants: Reciprocating Internal Combustion Engines; Final Rule," 75 *Federal Register* 9648, March 3, 2010.

[80] U.S. Environmental Protection Agency, "National Emission Standards for Hazardous Air Pollutants for Reciprocating Internal Combustion Engines; Final Rule," 75 *Federal Register* 51570, August 20, 2010.

[81] 78 *Federal Register* 6674, January 30, 2013.

(Reciprocating Internal Combustion Engine) rules. They apply to engines that meet specific siting, age, and size criteria (generally engines of 500 horsepower or less). EPA estimates that more than 1.2 million engines will be affected by the regulations. Depending on the type of engine, owners were required to install pollution control equipment or follow certain work practice standards, such as burning low sulfur fuel or performing oil changes and inspections. EPA estimated that the health benefits of the two rules would be between $1.45 billion and $3.5 billion annually in 2013. Annualized costs for the rules were estimated to be $626 million in 2013. EPA stated that the 2013 amendments will reduce the annualized costs by $139 million (to $487 million). The amendments were issued in response to a suit by the Engine Manufacturers Association. The most controversial of the amendments allows backup generators to operate for up to 100 hours per year during emergency or peak power use periods without being subject to emission limits, although beginning in 2015 they needed to use low-sulfur fuel. In May 2015, the D.C. Circuit Court of Appeals found that EPA acted arbitrarily and capriciously in allowing backup generators to operate without emission controls for up to 100 hours per year as part of an emergency demand-response program. The court remanded that portion of the rules to EPA.

Ocean-Going Ships

EPA took two steps to control emissions from ocean-going ships in 2009 and 2010. It promulgated emission standards for new marine engines[82] and it proposed the establishment of Emission Control Areas (ECAs) extending 200 nautical miles off most U.S. shores.[83] In the ECAs, which received final approval in March 2010, both U.S. and foreign ships were required to use low sulfur fuel, beginning in 2012. In both cases, the actions reflect international standards that the United States and other maritime nations had agreed to under the International Convention for the Prevention of Pollution from Ships (MARPOL). EPA estimated the cost of these two initiatives at over $3 billion annually by 2030, mostly attributable to the cleaner fuel requirement. The agency also estimated that monetized benefits of the requirements will exceed costs by more than 30 to 1. The ECAs and the new standards were supported by both industry and environmental groups, and were extended to cover the U.S. Caribbean, beginning in 2014. In July 2012, controversy arose over the requirement that ships in Alaskan waters use low-sulfur fuel, with the state of Alaska filing suit to block implementation of the fuel requirement. In September 2013, a Federal District Court dismissed the state's lawsuit.

---

[82] U.S. Environmental Protection Agency, "Control of Emissions from New Marine Compression-Ignition Engines at or Above 30 Liters per Cylinder; Final Rule," 75 *Federal Register* 22896, April 30, 2010.

[83] International Maritime Organization, Marine Environmental Protection Committee, "Proposal to Designate an Emission Control Area for Nitrogen Oxides, Sulphur Oxides and Particulate Matter, Submitted by the United States and Canada," April 2, 2009, at http://www.epa.gov/oms/regs/nonroad/marine/ci/mepc-59-eca-proposal.pdf.

Flares and Process Heaters at Petroleum Refineries

On September 12, 2012, EPA promulgated amendments to New Source Performance Standards for flares and process heaters at petroleum refineries.[84] The amendments are the result of the agency's reconsideration of standards it promulgated on June 24, 2008. The agency estimated that the reconsidered rules will have capital costs of $460 million, but will result in savings to the industry of $79 million per year, while resulting in monetized benefits of $240 million to $580 million annually, principally from the avoided health impacts caused by reduced emissions of sulfur dioxide and nitrogen oxides.

Brick and Clay Product NESHAP

EPA promulgated National Emission Standards for Hazardous Air Pollutants (NESHAPs) for manufacturers of bricks, structural clay products, and clay ceramics in 2003, but the standards were vacated by the D.C. Circuit Court of Appeals in 2007. The agency was sued by the Sierra Club for its failure to take further action. Under a November 2012 consent decree, the agency agreed to sign proposed standards to replace the vacated rule by November 20, 2014, and to sign a final rule by September 24, 2015. In compliance with this schedule, EPA finalized national emission standards for mercury, particulate matter, acid gases, dioxins, and furans on September 24, 2015.[85] EPA estimated the cost of the rule at $25 million annually, with monetized cobenefits three to eight times the cost. The Brick Industry Association called the proposal "a much more reasonable rule than the one EPA first envisioned several years ago," but they and others have continued to express concerns regarding the cost and achievability of the standards. Environmental groups and the association of state air pollution officials are concerned for different reasons: in their view, EPA improperly set standards under a section of the act that allows an alternative to the Maximum Achievable Control Technology requirement that generally applies to hazardous air pollutant standards. Both industry and environmental groups have filed petitions for review of the standards with the D.C. Circuit Court of Appeals.

Clean Water Act

Mountaintop Mining in Appalachia

Since 2009, EPA and other federal agencies (the Office of Surface Mining and Reclamation [OSM], in the Department of the Interior, and the U.S. Army Corps of Engineers) have been working on a series of actions and regulatory proposals to reduce the harmful environmental and health impacts of surface coal mining, including

---

[84] U.S. Environmental Protection Agency, "Standards of Performance for Petroleum Refineries; Performance Standards for Petroleum Refineries for Which Construction, Reconstruction, or Modification Began After May 14, 2007," 77 *Federal Register*, 56422, September 12, 2012.

[85] U.S. EPA, "NESHAP for Brick and Structural Clay Products Manufacturing; and NESHAP for Clay Ceramics Manufacturing; Final Rule," 80 *Federal Register* 65470, October 26, 2015.

mountaintop removal mining, in Appalachia. The actions are intended to improve regulation and strengthen environmental reviews of permit requirements under the CWA and the Surface Mining Control and Reclamation Act (SMCRA). Viewed broadly, the Administration's combined actions on mountaintop mining displease both industry and environmental advocates. The additional scrutiny of permits and more stringent requirements have angered the coal industry and many of its supporters. At the same time, while environmental groups support EPA's steps to restrict the practice, many favor tougher requirements or even total rejection of mountaintop mining in Appalachia. Many of the actions have been highly controversial in Congress. One action—not directly involving EPA—that is related to the Obama Administration's focus on mountaintop mining remains under way. It is issuance of a set of nationwide permits by the Army Corps of Engineers that apply to coal mining operations and a number of other activities that discharge to U.S. waters. The existing nationwide permits, which were issued in 2012, will expire in March 2017, and the Army Corps is expected to reissue the permits before then.[86]

Another regulatory action involves a rule issued by OSM on December 19, 2016, to revise SMCRA regulations on surface coal mining operations.[87] The Obama Administration identified a 2008 rule, called the stream buffer zone rule, which exempts so-called valley fills and other mining waste disposal activities from requirements to protect a 100-foot buffer zone around streams, for revision as part of the series of actions concerning surface coal mining in Appalachia. The 2016 final revised SMCRA rule is intended to avoid or minimize adverse impacts of coal mining on surface water, groundwater, fish, wildlife, and other natural resources by limiting the mining of coal in or through streams, restricting placement of waste in streams, and limiting the generation of mining waste. Most of the regulations being replaced by the 2016 rule were promulgated in 1983. OSM asserts that updated rules, which have been under development since 2009, are needed to reflect current science, technology, and modern mining practices. The revised rule will apply nationwide, not just in Appalachia. Potential changes to the 1983 rules drew wide controversy and criticism. Mining industry groups are very critical of the costs of the rule, while environmental groups that have generally supported strengthening SMCRA regulations contend that the rule should be stronger to provide more protection to streams. A number of states say that the rule would undermine state authority to administer SMCRA and that OSM failed to consult adequately with states during development of the rule. Legislation in the 114th Congress proposed to prevent OSM from releasing the rule pending a study by the National Research Council on economic and scientific data in the rule (H.R. 1644/S. 1458). Legislation to overturn

[86] For information see CRS Report 97-223, The Army Corps of Engineers' Nationwide Permits Program: Issues and Regulatory Developments, by Claudia Copeland and Jonathan L. Ramseur.

[87] For information, see CRS Insight IN10627, OSM Finalizes the Stream Protection Rule, by Claudia Copeland; and CRS Report R44150, The Office of Surface Mining's Proposed Stream Protection Rule: An Overview, by Claudia Copeland.

the 2016 final rule under procedures of the Congressional Review Act also was introduced (H.J.Res. 107).

Controversy also was generated by EPA's 2011 veto of a CWA permit that had been issued by the Corps for a surface coal mining project in West Virginia. EPA's veto was overturned by a federal court, but that ruling was reversed on appeal, and the Supreme Court declined to review the case. Legislation to restrict EPA's veto authority has been introduced on several occasions, including in the 114th Congress (H.R. 896, H.R. 1203, and S. 54/S. 234).

Pesticide Application General Permit

EPA has developed a Clean Water Act (CWA) general permit to control pesticides that are applied to waters of the United States, such as aerial application of insecticide to control mosquitoes. The general permit was issued on October 31, 2011, in response to a 2009 federal court decision that invalidated a 2006 EPA rule, which had codified the agency's long-standing view that pesticide applications that comply with federal pesticides law do not require CWA permits.[88] The estimated universe of affected activities is approximately 5.6 million applications annually, which are performed by 365,000 applicators, in four use patterns: mosquito and other flying insect pest control, aquatic weed and algae control, aquatic nuisance animal control, and forest canopy pest control. The permit requires all operators covered by the permit to minimize pesticide discharges to waters by practices such as using the lowest amount of pesticide product that is optimal for controlling the target pest. It also requires operators to prepare plans to document their pest management practices. Under OMB's criteria, the permit is not a significant rule, but is "economically significant."[89] EPA reissued the 2011 permit without significant change in October 2016. Meanwhile, in the 114th Congress, legislation intended to overturn the court's 2009 ruling by exempting aerial pesticide application activities from clean water permit requirements has been introduced (H.R. 897, S. 1500, and a provision of S. 659). The House passed a modified version of H.R. 897, retitled the Zika Vector Control Act, on May 24, 2016.[90]

Florida Nutrient Water Quality Standards

The CWA directs states to adopt water quality standards for their waters and authorizes EPA to promulgate new or revised standards if a state's actions fail to meet

---

88 U.S. Environmental Protection Agency, "Final National Pollutant Discharge Elimination System (NPDES) General Permit for Point Source Discharges from the Application of Pesticides; Notice of Final Permit," 76 *Federal Register* 68750-68756, November 7, 2011.

89 "Significant" rules are a broader OMB category that includes not only the economically significant (i.e., primarily those with an annual effect on the economy of $100 million or more), but also rules that "create a serious inconsistency or otherwise interfere with an action taken or planned by another agency"; "materially alter the budgetary impact of entitlements, grants, user fees, or loan programs or the rights and obligations of recipients thereof"; or "raise novel legal or policy issues arising out of legal mandates, the President's priorities, or the principles set forth" in Executive Order 12866.

90 For additional information, see CRS Report RL32884, *Pesticide Use and Water Quality: Are the Laws Complementary or in Conflict?*, by Claudia Copeland.

CWA requirements. Water quality standards consist of designated uses, criteria to protect the designated uses, and an antidegradation statement. They serve as the framework for pollution control measures specified for individual sources. Because of severe water quality impairment of Florida waters by nutrients (nitrogen and phosphorus) from diverse sources including agriculture and livestock, municipal and industrial wastewater discharges, and urban stormwater runoff, EPA determined in 2009 that Florida's existing *narrative* water quality standards for nutrients must be revised in the form of *numeric* criteria that will enable Florida to better control nutrient pollution. In 2009 EPA entered into a consent decree with environmental litigants requiring the agency to promulgate numeric nutrient water quality standards for Florida. To meet the legal deadline, EPA promulgated the first phase of these standards, called the "inland waters rule," in December 2010. Water quality standards do not have the force of law until the state translates them into permit limits or otherwise imposes pollution control requirements on dischargers. The rule would not establish any requirements directly applicable to regulated entities or other sources of nutrient pollution.

The 2010 rule did not go into effect, because, in response to criticism of the standards, EPA delayed the effective date of the rule to allow local governments, businesses, and the state of Florida time to review the standards and develop implementation strategies. While few dispute the need to reduce nutrients in Florida's waters, EPA's 2010 rule has been controversial, involving disputes about the data underlying the proposal, potential costs of complying with numeric standards when they are incorporated by the state into discharge permit limitations, and disputes over administrative flexibility. EPA has said all along that it prefers that Florida implement its own numeric nutrient water quality criteria, and in 2012 the state submitted revised standards with numeric nutrient criteria. In response, EPA indicated to the state that the agency likely would approve the standards, at which time the agency would initiate administrative action to repeal the 2010 federal rule. EPA's deadline for issuing the second phase of standards, for estuaries, coastal waters, and flowing waters in the South Florida Region, also was extended several times to allow the state to develop its own standards.

In 2013, EPA and the state reached agreement in principle on steps that will put the state in charge of determining numeric limits on nutrient pollution in Florida waterways. Under that agreement, Florida pledged to move forward with rulemaking and legislation to complete the job of setting numeric nutrient criteria for Florida waterways. In response to the state's actions, EPA approved the state's implementation plan for controlling nutrient pollution in Florida waters and petitioned the federal court in Florida to allow it to approve the state's water quality standards, although they lack numeric criteria for all waters. In January 2014, the U.S. District Court for the Northern District of Florida agreed to amend the 2009 consent decree in light of the adoption of new nutrient criteria, thus lifting the requirement for EPA to issue numeric nutrient standards under the second phase of rulemaking, and in September 2014, EPA finalized a rule withdrawing the overlapping federally promulgated water quality standards to allow Florida to implement its standards to address nutrient pollution.[91] Environmental groups have criticized EPA's approval of the standards that Florida adopted, saying that EPA's actions are inconsistent with its 2009 determination that numeric criteria are necessary to protect Florida's waters, but their legal challenge to the plan was rejected by the Florida federal district court's January 2014 ruling.

Chesapeake Bay TMDL

Pursuant to a court-ordered schedule, EPA has developed a plan, called a Total Maximum Daily Limit (TMDL), to restore nutrient-impaired waters of the Chesapeake Bay. The TMDL is required because jurisdictions in the Chesapeake Bay watershed have failed to meet deadlines to attain water quality goals for the bay, thus triggering Clean Water Act requirements that the federal government must develop a plan to do so. The TMDL is not a regulation. A TMDL represents the maximum amount of a pollutant that a body of water may receive and still meet its water quality standards.[92] Individual actions needed to meet the overall pollutant limits specified in the TMDL, such as discharge permit limits or other controls, are to be developed by the Chesapeake Bay jurisdictions in Watershed Implementation Plans (WIPs). The Chesapeake Bay TMDL is the largest ever developed by EPA or any state, since it will apply to all impaired waters of the 64,000 square miles of the six states in the bay watershed. On December 29, 2010, EPA issued the TMDL.[93] Pursuant to the schedule of steps in the TMDL, jurisdictions are now developing specific WIPs, which outline the types of controls and best management practices that will be used to reduce pollution in the bay. The TMDL has been controversial with agricultural and other groups that are concerned about the likely mandatory nature of many of EPA's and states' upcoming actions. In 2013, a federal district court upheld the TMDL, in a lawsuit filed by the American Farm Bureau Federation that had challenged EPA's authority to set pollution limits in the multistate plan. The district court's ruling was upheld on appeal in 2015.

Airport Deicing Effluent Limitations Guidelines and New Source Performance Standards

In 2012, EPA promulgated regulations under the CWA to limit water pollution from aircraft and airport runway deicing operations.[94] The rule is intended to limit runoff of deicing fluid, because it contains urea and other contaminants that contribute to low oxygen levels in streams, which can cause fish kills, algal blooms, and contamination of

---

[91] For additional information, see http://water.epa.gov/lawsregs/rulesregs/florida_index.cfm.

[92] For additional information, see CRS Report R42752, *Clean Water Act and Pollutant Total Maximum Daily Loads (TMDLs)*, by Claudia Copeland.

[93] Notice of the TMDL appeared in the *Federal Register* January 5, 2011. U.S. Environmental Protection Agency, "Clean Water Act Section 303(d): Notice for the Establishment of the Total Maximum Daily Load (TMDL) for the Chesapeake Bay," 76 *Federal Register* 549-550, January 5, 2011.

[94] U.S. Environmental Protection Agency, "Effluent Limitations Guidelines and New Source Performance Standards for the Airport Deicing Category; Final Rule," 77 *Federal Register* 29168-29205, May 16, 2012.

surface water or groundwater. The rule, which had been under development for several years and was proposed in 2009, is part of ongoing EPA activities under the CWA to regulate wastewater discharges from categories of industries through new and revised effluent limitations guidelines. EPA estimated that the final rule will reduce the volume of deicing-related pollutants by 16.4 million pounds at a cost of $3.5 million annually. Those estimates are substantially less than the 44.6 million pounds of pollutants estimated in the proposed rule, which was projected to cost the industry $91.3 million annually. EPA estimated that the final rule will apply to 198 existing airports.

Construction Site Effluent Limitations Guidelines

In 2009, EPA promulgated regulations under the Clean Water Act (CWA), called effluent limitations guidelines (ELGs), to limit pollution from stormwater runoff at construction sites.[95] The rule is called the Construction and Development, or C&D, ELG. OMB determined that it is an economically significant rule. It requires construction sites that disturb one or more acres of land to use erosion and sediment control best management practices to ensure that soil disturbed during construction activity does not pollute nearby waterbodies. For construction sites disturbing 10 acres or more, the rule established, for the first time, enforceable numeric limits on stormwater runoff pollution. EPA issued the rule in response to a 2004 lawsuit filed by an environmental group; in 2006, a federal court ordered EPA to issue a final rule by December 1, 2009. The rule affects about 82,000 firms nationwide involved in residential, commercial, highway, street, and bridge construction. EPA has issued effluent guidelines for 56 industries that include many types of discharges, such as manufacturing and service industries. These guidelines are implemented in discharge permits issued by states and EPA. Several industry groups challenged the C&D ELG. In response, EPA examined the data set underlying a portion of the rule and concluded that it improperly interpreted the data. In August 2010, a federal appeals court granted EPA's request for remand of a portion of the rule to conduct a rulemaking to correct the numeric effluent limitation. In November 2010, EPA promulgated a direct final rule to stay the effectiveness of the numeric turbidity limit in the 2009 rule; other portions of the rule remain in effect.[96] To resolve industry challenges to the 2009 rule, in 2014, EPA finalized modifications of the 2009 rule, including withdrawal of the numeric turbidity effluent limitations in the 2009 rule, which had been controversial, and changes specific to the nonnumeric portions of the rule. The effluent limits in the 2014 rule emphasize best practices to manage erosion and stabilize soils during construction. If more data on numeric discharge standards for construction sites become available, EPA could initiate a new rulemaking in the future.[97]

---

[95] U.S. Environmental Protection Agency, "Effluent Limitation Guidelines for the Construction and Development Point Source Category," 74 *Federal Register* 62996-63058, December 1, 2009.

[96] U.S. Environmental Protection Agency, "Direct Final Rule Staying Numeric Limitation for the Construction and Development Point Source Category," 75 *Federal Register* 68215-68217, November 5, 2010.

[97] U.S. Environmental Protection Agency, "Effluent Limitations Guidelines and Standards for the Construction and Development Point Source Category, Final Rule," 79 *Federal Register* 12661-12667, March 6, 2014.

"Post-Construction" Stormwater Rule

For some time, EPA has explored options to strengthen the existing regulatory program for managing stormwater, which is a significant source of water quality impairments nationwide. Under the current program, large cities and most industry sources are subject to CWA rules issued in 1990; smaller cities, other industrial sources, and construction sites are covered by rules issued in 1999. Among the options that EPA considered to strengthen stormwater regulations was establishing rules containing post-construction requirements for stormwater discharges from new development and redevelopment, which currently are not regulated, focusing on stormwater discharges from developed or post-construction sites such as subdivisions, roadways, industrial facilities and commercial buildings, or shopping centers. Under a consent agreement with environmental groups, EPA was expected to propose a rule in 2013, and to issue a final rule by December 2014. EPA's efforts to develop the stormwater rule were controversial and, apparently, technically challenging, and the agency missed the 2013 deadline to propose a rule. In March 2014, EPA announced that it is deferring action on the stormwater rule and instead will provide incentives, technical assistance, and other approaches for cities and towns to address stormwater runoff themselves.[98]

Revised Cooling Water Intake Rule

Thermoelectric generating plants and manufacturing facilities withdraw large volumes of water for production and, especially, to absorb heat from their industrial processes. Water withdrawals by power producers and manufacturers represent more than one-half of water withdrawn daily for various uses in the United States. Although water withdrawal is a necessity for these facilities, it also presents special problems for aquatic resources. In particular, the process of drawing surface water into the plant through cooling water intake structures (CWIS) can simultaneously pull in fish, shellfish, and tiny organisms, injuring or killing them. In 2014, EPA promulgated a CWA rule to protect fish from entrainment by cooling water intake structures at existing power plants and certain other industrial facilities. The final rule applies to approximately 1,065 existing electric generating and manufacturing plants.[99]

The proposed CWIS rule was highly controversial. Many in industry had feared, while environmental groups had hoped, that EPA would require installation of technology that most effectively minimizes impacts of cooling water intake structures, but also is the most costly option. The EPA proposal declined to mandate such technology universally and instead favored a less costly, more flexible regulatory option. In the 2014 final rule, EPA again declined to mandate closed-cycle cooling as a uniform requirement and provided several compliance options that are more flexible and less

---

98 For additional information, see CRS Report 97-290, *Stormwater Permits: Status of EPA's Regulatory Program*, by Claudia Copeland.

99 U.S. Environmental Protection Agency, "National Pollutant Discharge Elimination System—Final Regulations to Establish Requirements for Cooling Water Intake Structures at Existing Facilities and Amend Requirements at Phase I Facilities; Final Rule," 79 *Federal Register* 48300-48439, August 15, 2014.

costly than the proposal.[100] Legal challenges to the final rule have been filed by industry and environmental groups.

Revised Steam Electric Effluent Limitations Guidelines

Under authority of CWA Section 304, EPA establishes national technology-based regulations, called effluent limitations guidelines (ELGs), to reduce pollutant discharges from industries directly to waters of the United States and indirectly to municipal wastewater treatment plants based on Best Available Technology. These requirements are incorporated into discharge permits issued by EPA and states. In a 2009 study, EPA found that then-existing regulations do not adequately address the pollutants being discharged and have not kept pace with changes that have occurred in the electric power industry over the last three decades. Pollutants of concern include metals (e.g., mercury, arsenic, and selenium), nutrients, and total dissolved solids.

In 2015, EPA promulgated revised effluent limitations for the steam electric industry to replace rules that were issued in 1982. The new rule, which was effective on January 4, 2016, applies to 1,079 steam electric power plants nationwide, but EPA believes that only 133 plants are likely to incur compliance costs, because a large portion of the industry has already implemented processes or technologies that are required by the rule. The 2015 rule regulates six pollutant wastestreams from steam electric power plants. The estimated annual compliance cost of the rule would be $340 million (after-tax compliance cost). The rule also would reduce pollutant discharges by 385 million pounds annually and reduce water use by 57 billion gallons per year. Many in industry are concerned that the 2015 rule will impose new requirements at the same time that power plants are implementing other EPA rules. Environmental advocates are generally satisfied with the final rule, but many do have concerns with issues such as compliance deadlines in the rule. Both industry groups and environmental groups have challenged the rule in federal court.[101]

Oil Spill Prevention, Control, and Countermeasure Requirements, Including Deadline Extension for Farms and Exemption for Milk Storage

To prevent the discharge of oil from onshore and offshore facilities, EPA issued CWA regulations for spill prevention control and countermeasure (SPCC) plans in 1973.[102] SPCC plans apply to owners or operators of certain nontransportation-related facilities. In general, SPCC plans focus on oil spill prevention, requiring, for example, secondary containment (e.g., dikes or berms) for oil-storage equipment.

Following the passage of the Oil Pollution Act of 1990, the agency proposed substantial changes and clarifications that were not made final until July 2002.[103]

---

[100] For more information, see CRS Report R41786, *Cooling Water Intake Structures: Summary of the EPA Rule*, by Claudia Copeland.

[101] For information, see CRS Report R43169, Regulation of Power Plant Wastewater Discharges: Summary of the EPA Final Rule, by Claudia Copeland.

[102] 38 Federal Register 34164, December 11, 1973.

[103] 67 Federal Register 47041, July 17, 2002.

However, EPA has both extended the 2002 rule's compliance date (on multiple occasions) and made further amendments to the 2002 rule. On one occasion, amendments offered by the Bush Administration's EPA in 2008 were eliminated by the Obama Administration's EPA the following year.[104]

For most types of facilities subject to SPCC requirements, the deadline for complying with the changes made in 2002 was November 10, 2011.[105] However, in a November 2011 rulemaking, EPA extended the compliance date for farms to May 10, 2013.[106]

Pursuant to the CWA definition of oil, the SPCC requirements apply to petroleum-based and nonpetroleum-based oil.[107] In a 1975 *Federal Register* notice, EPA clarified that its 1973 SPCC regulations apply to oils from animal and vegetable sources.[108] EPA subsequently stated that "milk typically contains a percentage of animal fat, which is a non-petroleum oil" and is thus potentially subject to SPCC provisions.[109] However, in January 2009, EPA proposed a conditional exemption from SPCC requirements for milk storage units.[110] EPA issued a final rule April 18, 2011, exempting all milk and milk product containers and associated piping from the SPCC requirements. EPA's rationale for the exemption is that these units are subject to industry standards for sanitation and may be regulated by other agencies, including the U.S. Department of Agriculture.[111] In addition, the final rule states that exempted milk storage units are not included in a facility's overall oil storage volume, a primary factor for SPCC applicability.

"Waters of the United States" Rule

From the earliest days, Congress has grappled with where to set the line between federal and state authority over the nation's waterways. Typically, this debate occurred in the context of federal legislation restricting uses of waterways that could impair navigation and commerce. The phrase Congress often used to specify waterways over which the federal government had authority was "navigable waters of the United States." However, in the legislation that became the CWA of 1972, Congress felt that the term was too constricted to define the reach of a law whose purpose was not maintaining navigability, as in the past, but rather preventing pollution. Accordingly, in the CWA Congress retained the traditional term "navigable waters," but defined it broadly to mean "waters of the United States." That phrase is important in the context of Section 404 of

---

[104] A November 13, 2009, rule (74 Federal Register 58784) eliminated specific exclusions/exemptions made by a December 5, 2008, rulemaking (73 Federal Register 74236).

[105] U.S. Environmental Protection Agency, "Oil Pollution Prevention; Spill Prevention, Control, and Countermeasure Rule Compliance Date Amendment," 75 Federal Register 63093, October 14, 2010.

[106] U.S. Environmental Protection Agency, "Oil Pollution Prevention: Spill Prevention, Control, and Countermeasure Rule—Compliance Date Amendment for Farms," 76 Federal Register 72120, November 22, 2011.

[107] See CWA Section 311(a) (33 U.S.C. 1321(a)).

[108] 40 *Federal Register* 28849, July 9, 1975.

[109] 74 *Federal Register* 2461, January 15, 2009.

[110] U.S. Environmental Protection Agency, "Oil Pollution Prevention: Spill Prevention, Control, and Countermeasure Rule Requirements—Amendments," 74 *Federal Register* 2461, January 15, 2009.

[111] 76 *Federal Register* 21652, April 18, 2011.

the law, a permit program jointly administered by EPA and the Army Corps of Engineers that regulates discharges of dredged and fill material to U.S. waters, including wetlands. The same phrase also defines the geographic extent of the other parts of the CWA, including state-established water quality standards, the discharge permit program in Section 402, oil spill liability, and enforcement. Consequently, how broadly or narrowly "waters of the United States" is defined has been a central question of CWA law and policy for nearly 40 years.

Controversies increased following two Supreme Court rulings, one in 2001 and one in 2006, on how "waters of the United States" are defined for purposes of the 404/wetlands permit program. Those two rulings interpreted the regulatory scope of the CWA more narrowly than previously, and created uncertainty about the appropriate scope of waters that are protected by the CWA. To resolve uncertainties about which waters are protected by the CWA, on May 27, 2015, EPA and the Army Corps of Engineers finalized regulations that define "waters of the United States." The agencies' intention in proposing it was to clarify questions of CWA jurisdiction, in view of the Supreme Court's rulings while reflecting the agencies' scientific and technical expertise. However, the rule as proposed in 2014 and finalized in 2015 has been and remains very controversial. Industries that are the primary applicants for CWA permits and agriculture groups (although farms are exempt from most permitting) raised numerous objections over how broadly they fear that the rule will be interpreted. Many states and state environmental agencies expressed support for a rule to clarify the scope of CWA jurisdiction, but there was no state consensus on the Corps-EPA proposed or final rule. Environmental groups defend the agencies' efforts to protect U.S. waters and reduce frustration resulting from unclear jurisdiction of the CWA. Still, some of them argue that the final rule should be stronger, such as by providing categorical protection for more waters.[112]

Legal challenges to the final rule were filed by more than 30 states and a variety of industries seeking to have the rule invalidated. Environmental groups also challenged parts of the rule that they believe are not stringent enough. Although the rule became effective August 28, 2015, a federal court imposed a nationwide stay in October 2015, pending further developments, which are ongoing. Congressional interest in the rule has been strong, with hearings held by House and Senate committees and multiple bills introduced to halt or redirect the rule. The Senate and House passed a resolution of disapproval of the rule (S.J.Res. 22), under procedures in the Congressional Review Act, but President Obama vetoed the joint resolution on January 19, 2016, and on January 21, the Senate failed to invoke cloture on a motion to proceed to override the veto.[113]

---

[112] For additional information on the rule, see CRS Report R43455, *EPA and the Army Corps' Rule to Define "Waters of the United States"*, by Claudia Copeland.

[113] For information, see CRS Report R43943, The "Waters of the United States" Rule: Legislative Options and 114th Congress Responses, by Claudia Copeland.

Solid Waste (RCRA)

Coal Combustion Residuals

In 2015, coal-fired power plants accounted for 33% of U.S. electric power, resulting in approximately 117 million tons of coal combustion residuals (CCR, commonly referred to as "coal ash"). On December 22, 2008, national attention was turned to risks associated with managing CCR when a breach in a surface impoundment pond at the Tennessee Valley Authority's Kingston, TN, plant released 1.1 billion gallons of coal ash slurry, covering hundreds of acres and damaging or destroying homes and property. In addition to the risk of a sudden, catastrophic release such as that at Kingston, EPA determined that CCR disposal in unlined landfills and surface impoundments presents substantial risks to human health and the environment from releases of toxic constituents (particularly arsenic and selenium) into surface and groundwater. To address those risks, on April 17, 2015, EPA issued a final rule to regulate solid waste disposal units that receive CCR.[114] EPA did so in accordance with its existing authorities to regulate solid waste under Subtitle D of the Resource Conservation and Recovery Act (RCRA). The regulations, which went into effect on October 19, 2015, establish national standards for landfills and surface impoundments that receive CCR.[115]

Under Subtitle D, the EPA standards were enforceable primarily by citizen suits. However, provisions in the Water Infrastructure Improvements for the Nation Act (WIIN Act, P.L. 114-322, signed into law on December 16, 2016) amended Subtitle D to expand EPA's potential authority to implement those standards. Specifically, Section 2301 amends Subtitle D to establish a framework for EPA to approve state permit programs to regulate CCR disposal units, and, in effect, authorize EPA to enforce the federal standards in states that choose not to adopt a permit program.[116]

In its Regulatory Impact Analysis, EPA estimated the average annualized regulatory costs to be $587 million a year under Subtitle D. The average annual benefits were estimated at $289 million per year, but the monetized benefits do not include a number of important benefits that EPA was not able to quantify. There could be additional costs or benefits depending on how the rule affects the recycling of coal ash.

Identification of Nonhazardous Materials That Are Solid Wastes When Burned

In conjunction with emission standards for boilers and solid waste incinerators (see discussion of "Boilers and Incinerators" section above), in February 2011, EPA finalized regulations intended to clarify when certain materials burned as fuel in a combustion unit

---

[114] U.S. Environmental Protection Agency's, "Hazardous and Solid Waste Management System; Disposal of Coal Combustion Residuals from Electric Utilities, Final Rule," 80 Federal Register 21302-21501, April 17, 2015. The final regulations are codified at 40 C.F.R. Part 257, Subpart D.

[115] For more information, see CRS Insight IN10583, *Overview of EPA Standards for "Coal Ash" Disposal*, by Linda Luther.

[116] For more information, see CRS Insight IN10585, *State Programs for "Coal Ash" Disposal in the WIIN Act*, by Linda Luther.

would be considered a "solid waste."[117] The definition of solid waste plays an important role in implementing the emission standards for both boilers and solid waste incinerators. The 2007 D.C. Circuit Court of Appeals decision that vacated EPA's previous emission standards for boilers also vacated EPA's definition of terms under its "CISWI Definitions Rule."[118] The D.C. Circuit concluded that EPA erred in defining "commercial and industrial solid waste" to exclude solid waste that is burned at a facility in a combustion unit whose design provides for energy recovery or which operates with energy recovery. Instead, the D.C. Circuit stated that the Clean Air Act "requires any unit that combusts 'any solid waste material at all'—regardless of whether the material is being burned for energy recovery—to be regulated as a 'solid waste incineration unit.'"[119]

The 2011 final rule addresses issues brought up by the D.C. Circuit and, in doing so, significantly narrows the current universe of nonhazardous secondary materials that could be burned in boilers.[120] EPA anticipates that boiler operators that burn materials newly identified as a solid waste would switch to a nonwaste fuel, rather than being subject to the more stringent emission standards applicable to solid waste incinerators. The final rule also addresses a host of concerns raised by various stakeholders during the public comment period for the proposed rule, including those of several Members of Congress. In particular, the final rule clarifies that the definition of solid waste would not affect current used oil recycling regulations (which allows burning used oil in space heaters, under certain conditions) and explicitly excludes from the definition of solid waste "scrap tires used in a combustion unit that are ... managed under the oversight of established tire collection programs." EPA states that this regulatory action would not directly invoke any costs or benefits. Instead, any costs or benefits would be related to the Boiler MACT and CISWI Standards (discussed above). On February 7, 2013, EPA amended the 2011 rule to clarify specific elements of the regulations. The amendments were jointly promulgated with EPA's reconsideration of the CISWI proposed rule (discussed above).

Underground Storage Tanks

On June 19, 2015, the EPA Administrator signed final regulations revising the agency's 1988 Underground Storage Tank (UST) technical regulations and state program approval regulations promulgated under Subtitle I of the Solid Waste Disposal Act (SWDA).[121] The agency proposed the revisions in 2011.[122] The final regulations address

---

[117] U.S. Environmental Protection Agency, "Identification of Non-Hazardous Secondary Materials That Are Solid Waste, Final Rule," 76 *Federal Register* 15455, March 21, 2011.

[118] U.S. Environmental Protection Agency, "Standards of Performance for New Stationary Sources and Emission Guidelines for Existing Sources: Commercial and Industrial Solid Waste Incineration Units, Final Rule," 70 *Federal Register* 55568, September 22, 2005.

[119] This and related court finding are discussed in the final rule at 76 *Federal Register* 15461.

[120] See EPA's web page "Identification of Non-Hazardous Materials That Are Solid Waste: Final Rulemaking" at http://www.epa.gov/epawaste/nonhaz/define/index.htm.

[121] U.S. Environmental Protection Agency, "Revising Underground Storage Tank Regulations—Revisions to Existing Requirements and New Requirements for Secondary Containment and Operator Training," 80

amendments to Subtitle I made in the Energy Policy Act of 2005 (P.L. 109-58)[123] and also update UST leak prevention and detection technologies and requirements.[124] The revisions are intended to improve leak prevention and detection of releases from USTs, which EPA and states report as a leading source of groundwater contamination. The final revisions to UST technical regulations include (1) secondary containment requirements for new and replaced tanks and piping, (2) training requirements for UST owners and operators, (3) new operation and maintenance requirements, (4) new release prevention and detection technologies, (5) requirements to ensure UST systems are compatible with certain biofuel blends before storing them, and (6) updated codes of practice. The regulations also revise state program approval requirements to incorporate the new requirements, and phase out past deferrals for several types of tanks (emergency generator tanks, airport hydrant systems, and field-constructed tanks).

The Energy Policy Act of 2005 (EPAct) amended the SWDA to require states that receive federal funding under Subtitle I to meet certain requirements (such as operator training and secondary containment requirements). The final rule expands on EPAct and further applies these requirements in Indian country and in states that do not receive Subtitle I funds. EPA's stated goal is to make UST requirements similar in all states and in Indian country. Additionally, the final rule expands the scope of certain requirements beyond those established in law. For example, EPAct requires states receiving Subtitle I money to require either that (1) new or replaced tanks installed within 1,000 feet of community water systems or drinking water wells be equipped with secondary containment, or (2) UST manufacturers and installers maintain evidence of financial responsibility to cover the costs of corrective actions. In the final rule, EPA requires secondary containment for new or replaced tanks in all locations. The agency noted that states are implementing or plan to implement this approach, and most tanks are within that distance of a water system (including water distribution lines) or potable well.

Among the updates to the 1988 UST rules, the rule modifies the requirement that UST systems must be compatible with stored substances, by adding options for owners and operators to demonstrate that UST systems are compatible with fuel containing more than 10% ethanol (E10) or 20% biodiesel (B20).[125]

Federal Register 41566, July 15, 2015. The Solid Waste Disposal Act (SWDA) is commonly referred to as the Resource Conversation and Recovery Act.

[122] U.S. Environmental Protection Agency, "Revising Underground Storage Tank Regulations-Revisions to Existing Requirements for Secondary Containment and Operator Training," 76 *Federal Register* 71708, November 18, 2011.

[123] P.L. 109-58, Title XV, Subtitle B, comprises the Underground Storage Tank Compliance Act (USTCA), which broadly amended the UST leak prevention and provisions of SWDA Subtitle I (42 U.S.C. §6991-6991m).

[124] Technical regulations are codified at 40 C.F.R. §280. EPA also revised state program approval requirements in 40 C.F.R. §281 to incorporate changes to the technical regulations.

[125] The new regulations and support documents (including a comparison of the key differences between the 1988 regulations and the 2015 revised UST regulations) are available at http://www.epa.gov/oust/fedlaws/revregs.html.

Toxic Substances Control Act (TSCA)

Lead: Renovation, Repair, and Painting (RRP) Rule Amendments

Stakeholders have differed on what federal requirements are appropriate to protect children from lead-based paint hazards that may occur from renovation and remodeling of older residential, public, or commercial buildings. Certain public interest and environmental organizations have sought to expand the scope of buildings and circumstances in which federal training, certification, and work practice requirements would apply, while renovators have argued that such requirements may not be warranted in terms of costs to the renovator and property owner depending on the level of risk and other circumstances. In 2008, pursuant to Section 402(c)(3) of the Toxic Substances Control Act (TSCA), EPA promulgated the Lead Renovation, Repair, and Painting (RRP) rule that included certain provisions for which there has been disagreement.[126] Shortly after promulgation of the rule, trade associations who represent renovators and certain public interest and environmental organizations challenged it.[127] In 2009, EPA signed a settlement agreement with the public interest and environmental organization petitioners (but not the trade association), committing to propose certain amendments to the RRP rule.[128] In response to the settlement agreement, EPA promulgated two amendments to the rule and continues to consider a third rulemaking. Among other provisions, the RRP rule amendments

- eliminated the "opt-out" provision that had exempted renovators from training and work practice requirements if they obtained certification from the property owner that no child under age six or pregnant woman resided in the building and no children spent significant amounts of time there; and
- allowed certified renovators the option to collect paint samples from components to be affected by a renovation for analysis by a nationally accredited laboratory instead of using EPA-recognized test kits.[129]

EPA did not amend the RRP rule as the agency originally proposed to require dust wipe testing and clearance requirements after the completion of a renovation activity.

---

[126] 15 U.S.C. §2582(c)(3). This provision of TSCA was enacted by the Lead-Based Paint Exposure Reduction Act (Subtitle B) as part of the Residential Lead-Based Paint Hazard Reduction Act of 1992 (P.L. 102-550, Title X). The RRP rule is codified at 40 C.F.R. Part 745, Subpart E. Additional information is available on EPA's website "Lead Renovation, Repair, and Painting Program Rules," April 19, 2016, https://www.epa.gov /lead/lead-renovation-repairand-painting-program-rules.

[127] See *National Association of Home Builders v. EPA*, D.C. Cir., No. 08-1193, consolidated with *New York City Coalition to End Lead Poisoning et al. v. EPA,* D.C. Cir., No. 08-1235 and *Sierra Club et al. v. EPA,* D.C. Cir., No. 08- 1258.

[128] The 2009 settlement agreement and certain amendments to the settlement agreement are available at http://www.regulations.gov/#!documentDetail;D=EPA-HQ-OPPT-2010-0173-0089.

[129] U.S. Environmental Protection Agency, "Amendment to the Opt-out and Recordkeeping Provisions in the Renovation, Repair, and Painting Program: Lead; Final Rule," 75 *Federal Register* 24802-24819, May 6, 2010, and"Lead; Clearance and Clearance Testing Requirements for the Renovation, Repair, and Painting Program, Final Rule," 76 *Federal Register* 47918-47946, August 5, 2011.

EPA continues to consider potential RRP requirements for public and commercial buildings that children do not frequent.[130] On June 9, 2016, EPA announced that the agency would review the entire RRP rule as required by Section 610 of the Regulatory Flexibility Act.[131] The agency anticipates completing the review by January 2017. In the 114th Congress, H.R. 2328 and S. 1987 would have amended TSCA, among other things, to limit the circumstances in which the RRP rule would apply.

---

[130] Regulatory Information Service Center, "View Rule: Lead; Renovation, Repair, and Painting Program for Public and Commercial Buildings," Fall 2016, http://www.reginfo.gov/ public/do/eAgendaViewRule?pubId=201610&RIN= 2070-AJ56.

[131] 5 U.S.C. §610. Section 610 of the Regulatory Flexibility Act requires federal agencies to review, within 10 years of promulgation, each rule that has or will have a "significant economic impact on a substantial number of small entities."

In: Environmental Protection Agency (EPA)
Editor: Coby Schaefer
ISBN: 978-1-53614-933-3

Chapter 2

FEDERAL POLLUTION CONTROL LAWS: HOW ARE THEY ENFORCED?*

Robert Esworthy

SUMMARY

As a result of enforcement actions and settlements for noncompliance with federal pollution control requirements, the U.S. Environmental Protection Agency (EPA) reported that, during FY2013, regulated entities committed to invest an estimated $7.0 billion for judicially mandated actions and equipment to control pollution (injunctive relief), and $22.0 million for implementing mutually agreed-upon (supplemental) environmentally beneficial projects. EPA estimated that these compliance/enforcement efforts achieved commitments to reduce or eliminate 1.3 billion pounds of pollutants in the environment, primarily from air and water, and to treat, minimize, or properly dispose of 148 million pounds of hazardous waste. Noncompliance with federal pollution control laws remains a continuing concern. The overall effectiveness of the enforcement organizational framework, the balance between state autonomy and federal oversight, and the adequacy of funding are long-standing congressional concerns.

This chapter provides an overview of the statutory framework, key players, infrastructure, resources, tools, and operations associated with enforcement and compliance of the major pollution control laws and regulations administered by EPA. It also outlines the roles of federal (including regional offices) and state regulators, as well as the regulated community. Understanding the many facets of how all federal pollution control laws are enforced, and the responsible parties involved, can be challenging. Enforcement of the considerable body of these laws involves a complex framework and organizational setting.

The array of enforcement/compliance tools employed to achieve and maintain compliance includes monitoring, investigation, administrative and judicial (civil and criminal) actions and penalties, and compliance assistance and incentive approaches. Most compliance violations are resolved administratively by the states and EPA. EPA

---

* This is an edited, reformatted and augmented version of Congressional Research Service Publication No. RL34384, dated October 7, 2014.

concluded 1,440 final administrative penalty orders in FY2013. Civil judicial actions, which may be filed by states or EPA, are the next most frequent enforcement action. EPA may refer civil cases to the U.S. Department of Justice (DOJ), referring 138 civil cases in FY2013. The U.S. Attorney General's Office and DOJ's Environmental Crimes Section, or the state attorneys general, in coordination with EPA criminal investigators and general counsel, may prosecute criminal violations against individuals or entities who knowingly disregard environmental laws or are criminally negligent. EPA reported the assessment of nearly $1.15 billion in civil penalties (administrative and judicial) and $4.5 billion in combined criminal fines, restitution, and court-ordered environmental projects during FY2013. Of the FY2013 totals, $1.0 billion in civil penalties and $4.0 billion in criminal fines, restitution, and court-ordered projects were associated with the *Deepwater Horizon* Gulf of Mexico judicial and criminal cases.

Federal appropriations for environmental enforcement and compliance activities have remained relatively constant in recent fiscal years. Some contend that overall funding for enforcement activities has not kept pace with inflation or with the increasingly complex federal pollution control requirements. Congress appropriated $560.9 million for enforcement activities for FY2014, a 1.4% increase above the $553.1 million enacted for FY2013 (post-sequestration), but roughly 3.8% less than the $583.4 million enacted for FY2012. The President's FY2015 budget request included $583.0 million for EPA enforcement activities. On September 19, 2014, President Obama signed into law the Continuing Appropriations Resolution, 2015 (P.L. 113-164). The act provides FY2015 appropriations to federal agencies (including EPA) for continuing projects and activities generally at the rate and under the authority and conditions provided in the applicable conditions of the Consolidated Appropriations Act, 2014 (P.L. 113-76), less a 0.0554% rescission, until December 11, 2014, or until enactment of regular appropriations legislation.

INTRODUCTION

Congress has enacted laws requiring individuals and facilities to take measures to protect environmental quality and public health by limiting potentially harmful emissions and discharges, and remediating damage. Enforcement of federal pollution control laws in the United States occurs within a highly diverse, complex, and dynamic statutory framework and organizational setting. Multiple statutes address a number of environmental pollution issues, such as those associated with air emissions, water discharges, hazardous wastes, and toxic substances in commerce. Regulators and citizens take action to enforce regulatory requirements in a variety of ways to bring violators into compliance, to deter sources from violating the requirements, or to clean up contamination (which may have occurred prior to passage of the statutes). Implementation and enforcement provisions vary substantially from statute to statute, and are often driven by specific circumstances associated with a particular pollution concern. Given these many factors, it is difficult to generalize about environmental enforcement.

This chapter focuses on enforcement of federal environmental pollution control requirements under the Clean Air Act (CAA); the Clean Water Act (CWA); the Comprehensive Environmental Response, Compensation, and Liability Act (CERCLA or

Superfund); and other statutes for which EPA is the primary federal implementing agency.[1] The chapter provides a brief synopsis of the statutory framework that serves as the basis for pollution control enforcement, including an overview of the key players responsible for correcting violations and maintaining compliance. Implementation and enforcement of pollution control laws are interdependent and carried out by a wide range of actors including federal, state, tribal, and local governments; the regulated entities themselves; the courts; interest groups; and the general public. *Figure 1*, below, presents the array of local, state, tribal, and federal entities that constitutes the environmental pollution control enforcement/compliance framework and organizational setting.

Source: Diagram prepared by the Congressional Research Service (CRS).

Figure 1. Key Players in Enforcement of Pollution Control Laws.

A diverse set of regulatory approaches and enforcement tools is applied to a sizeable universe of regulated entities by these multiple regulating authorities to ensure compliance. A general discussion of enforcement monitoring and response tools is included in this chapter, followed by a summary of recent fiscal year federal funding levels for enforcement activities. Discussion of available enforcement data sources, as well as tables illustrating examples of trends in enforcement activities, is presented in the two appendixes.

While this chapter touches on many aspects of environmental enforcement, it does not describe every aspect and statute in detail. Rather, the chapter is intended to provide a broad perspective of environmental enforcement by highlighting key elements, and a general context for the range of related issues frequently debated. Information included in this chapter is derived from a variety of sources. These sources, including relevant subject-matter reports providing in-depth discussion of specific topics and laws, are referenced throughout.

---

[1] See CRS Report RL30798, *Environmental Laws: Summaries of Major Statutes Administered* by the Environmental Protection Agency.

Several themes reflecting congressional concerns over time since EPA was established in 1970 are reflected throughout the major sections of this chapter. Congress has conducted oversight, primarily in the form of hearings, on various aspects of the organizational infrastructure and operations designed to enforce pollution control statutes.[2] These aspects of enforcement have also been the topic of investigations by the Government Accountability Office (GAO)[3] and EPA's Office of Inspector General (EPA-OIG).[4] The federal government's oversight of and coordination with states in implementing and enforcing federal pollution control laws have been of particular interest to Congress.[5] The following sections briefly discuss some of the key issue areas.

Federal and State Government Interaction

Since many, but not all, of the federal pollution control statutes authorize a substantial role for states, state autonomy versus the extent of federal oversight is often at the center of debate with regard to environmental enforcement. Not unexpectedly, given the "cooperative federalism"[6] that is often used to characterize the federal, state, and

---

[2] See, for example: House Committee on Judiciary Subcommittee on Regulatory Reform, Commercial and Antitrust Law hearing "H.R. 1493, *the Sunshine for Regulatory Decrees and Settlements Act*," June 5, 2013, http://judiciary.house.gov/hearings/ 113th/hear_06052013.html; House Committee on Energy and Commerce Subcommittee on Energy and Power hearing, "*EPA Enforcement Priorities and Practices*," June 6, 2012, http://energycommerce.house.gov/hearings/hearingdetail. aspx?NewsID=9552; Senate Committee on Environment and Public Works, "*Oversight Hearing on EPA Regional Inconsistencies*," June 28, 2006, http://epw.senate.gov/ hearing_statements.cfm?id=257928.

[3] Government Accountability Office (GAO): *Environmental Enforcement: EPA Needs to Improve the Accuracy and Transparency of Measures Used to Report on Program Effectiveness*, GAO-08-1111R, September 18, 2008; *Environmental Compliance and Enforcement: EPA's Effort to Improve and Make More Consistent Its Compliance and Enforcement Activities*, GAO-06-840T, June 28, 2006. All available at http://www.gao.gov.

[4] EPA's Office of Inspector General (EPA-OIG): *Response to Congressional Request on EPA Enforcement*, Rpt. No. 13-P0168, February 28, 2013; *EPA Needs to Improve Its Recording and Reporting of Fines and Penalties*, Rpt. No. 10- P-0077, March 9, 2010; *EPA Has Initiated Strategic Planning for Priority Enforcement Areas, but Key Elements Still Needed*, Rpt. No. 08-P-0278, September 25, 2008; *Overcoming Obstacles to Measuring Compliance: Practices in Selected Federal Agencies*, Rpt. No. 2007-P-00027, June 20, 2007, *Enforcement—Compliance with Enforcement Instruments*, Rpt. No. 2001-P-00006, March 29, 2001. All available at http://www.epa.gov/oig/.

[5] Ibid. footnote 3; see also GAO: *Drinking Water: Unreliable State Data Limit EPA's Ability to Target Enforcement Priorities and Communicate Water Systems' Performance*, GAO-11-381, June 17, 2011, *Environmental Protection Agency: Major Management Challenges*, GAO-11-422T, March 2, 2011, *Clean Water Act: Longstanding Issues Impact EPA's and States' Enforcement Efforts,* GAO-10-165T, October 15 2009, *Environmental, Protection: Collaborative EPA-State Effort Needed to Improve Performance Partnership System*, GAO/T-RCED-00-163, May 2, 2000, and *Environmental Protection: Overcoming Obstacles to Innovative State Regulatory Programs*, GAO-02-268, January 31, 2002. See also EPA-OIG (http://www.epa.gov/oig/): *EPA Must Improve Oversight of State Enforcement*, Report No. 12-P-0113, December 9, 2011; *EPA Needs to More Actively Promote State Self-Assessment of Environmental Programs*, Report No. 2003-P-00004, December 27, 2002.

[6] Many references discuss "cooperative federalism" in the context of environmental policy; these include Robert L. Fischman, Cooperative Federalism and Natural Resources Law, *New York University Envtl. L. J.* 179, vol. XIV 2006, Issue 1; Mark Agrast et al., How to Protect Environmental Protections?, *Envtl. Law Reporter*, vol. 35, 2005 (10413- 10417), the Environmental Law Institute; Philip J. Weiser, Towards a Constitutional Architecture for Cooperative Federalism, *North Carolina L. Rev.*, vol. 79, 2001 (663, 671), University of North Carolina; Vickie L. Patton, A Balanced Partnership, *The Envtl. Law Forum*, vol. 13, no. 3, May/June 1996; and, Robert V. Percival, Environmental Federalism: Historical Roots and Contemporary Models, *Maryland Law Rev.*, vol. 54, 1995 (1141).

tribal governments in the joint implementation and enforcement of pollution control requirements, relationships and interactions among these key enforcement players often have been less than harmonious.

Disagreements involving environmental priorities and strategic approaches, and balancing the relative roles of compliance assistance with enforcement, contribute to the complexity and friction that come with enforcing national pollution control laws. Other contributing factors include the increasing number of statutory and related regulatory pollution control requirements (some with conflicting mandates) and the adequacy of the resources available for their implementation.

The effects of variability among statutes, coupled with variability in federal and state interpretations and regulations, are often central to the debate. Some argue that this variability leads to too much inconsistency in enforcement actions from state to state, region to region, or between federal versus state actions. Others counter that this represents the flexibility and discretion intended by the statutes to address specific circumstances and pollution problems.

A July 2007 GAO report found that progress had been made regarding federal oversight of state environmental enforcement programs, and that there had been improvements with regard to cooperative federal-state planning and priority setting. However, the GAO concluded that a greater effort was needed to achieve more consistency and effectiveness, and that these issues continue to need improvements.[7] In a December 2011 report, the EPA OIG found that although "OECA had made efforts to improve state performance and consistency ... state performance remains inconsistent across the country, providing unequal environmental benefits to the public and an unlevel playing field for regulated industries."[8]

Federal Funding and Staffing for Enforcement Activities

The level of federal funding allocated to states and tribes to support effective enforcement of federal pollution control laws has also been a long-standing congressional concern.[9] In March 2012, the Environmental Council of the States (ECOS) reported concerns among state environmental agencies with regard to the extent of reductions in federal funding for state environmental protection activities.[10] In a 2008 study, ECOS[11]

---

[7] GAO, *Environmental Protection: EPA-State Enforcement Partnership Has Improved, but EPA's Oversight Needs Further Enhancement*. GAO-07-883, July 31, 2007, and *Environmental Protection Agency: Major Management Challenges*, GAO-11-422T, March 2, 2011.

[8] *EPA Must Improve Oversight of State Enforcement*, Report No. 12-P-0113, December 9, 2011, http://www.epa.gov/ oig/reports/2012/20111209-12-P-0113.pdf.

[9] For example, see EPA's Office of Inspector General, *Congressional Request on EPA Enforcement Resources and Accomplishments*, October 10, 2003, Report 2004-S-00001, http://www.epa.gov/oig/. GAO, *EPA's Execution of Its Fiscal Year 2007 New Budget Authority for Enforcement and Compliance Assurance Programs*. GAO-08-1109R, September 26, 2008.

[10] ECOS press release, "Prospects for Massive Cuts in Federal Funding Alarm State Environmental Agencies," March 26, 2012, http://www.ecos.org/. See also ECOS April 10, 2014, testimony before the House Committee

reported that during 2005-2008 states expected spending to implement federal environmental laws to double while federal appropriations declined.[12] Subsequently, ECOS reported that although federal funding for enforcement allocated to states increased marginally from FY2009 to FY2010, overall, reductions in state budget revenue are impacting their ability to maintain viable environmental enforcement programs.[13] In a September 2012 ECOS report, budget environmental agency data collected from 49 states indicated that total funding from the federal government decreased between FY2011 and FY2012, and further decreases were expected for FY2013.[14] In 2007, GAO reported that, although funding overall for enforcement activities had increased somewhat, it generally had not kept pace with the increasing number of mandates and regulations, or with inflation.[15]

The federal enforcement funding and personnel, primarily within EPA and the Department of Justice (DOJ), to ensure effective enforcement of environmental statutes has also been a concern of both appropriations and authorizing committees in Congress. Recently, in addition to funding priorities among the various EPA programs and activities (including enforcement), several promulgated and pending EPA regulatory actions[16] have been central to debates on EPA's appropriations. Some Members expressed concerns related to these actions during hearings of EPA's appropriations, and authorizing committees continued to address EPA regulatory actions through hearings and consideration of legislation during the 113th Congress. During the FY2011 and FY2012 appropriations deliberations, several provisions were proposed, and a subset adopted, that restricted the use of funding for the development, implementation, and enforcement of certain regulatory actions that cut across the various environmental pollution control statutes' programs and initiatives.[17] Although no separate FY2013[18] Interior,

---

on Appropriations Subcommittee on Interior, Environment, and Related Agencies, regarding funding for EPA included in the President's FY2015 Budget Request, see http://www.ecos.org/section/policy/testimony.

[11] The Environmental Council of the States (ECOS) is a national nonprofit (501(c)(6)), nonpartisan association of state and territorial environmental commissioners, established in December 1993. http://www.ecos.org.

[12] ECOS, *March 2008 Green Report: State Environmental Expenditures 2005-2008*, March 12, 2008, available at http://www.ecos.org/section/green_reports/. See also *The Funding Gap*, The Journal of the Environmental Council of the States, Winter 2004, http://www.ecos.org/section/publications.

[13] ECOS, *August 2010 Green Report: Status of State Environmental Agency Budgets, 2009-2011*, August 2010, and *Impacts of Reductions in FY 2010 on State Environmental Agency Budgets*, March 2010; *Funding Environmental Protection: State Budget Shortfalls and Ideas for Mitigating Them*, June 2009, available at http://www.ecos.org/ section/green_reports/.

[14] ECOS, *September 2012 Green Report: Status of State Environmental Agency Budgets, 2011-2013*, September 2012, TABLE 1, http://www.ecos.org/section/green_reports/.

[15] See footnote 7. See also GAO Testimony: *Management Challenges and Budget Observations*, before the Subcommittee on Oversight and Investigations, Committee on Energy and Commerce, House of Representatives, GAO-12-149T, Oct 12, 2011, http://www.gao.gov/assets/590/585707.pdf.

[16] See CRS Report R41561, *EPA Regulations: Too Much, Too Little, or On Track?*, by James E. McCarthy and Claudia Copeland, for a discussion of selected EPA regulatory actions.

[17] See CRS Report R41979, *Environmental Protection Agency (EPA) FY2012 Appropriations: Overview of Provisions in H.R. 2584 as Reported*, by Robert Esworthy. For an overview of proposed provisions contained in House-passed H.R. 1 and S.Amdt. 149, see CRS Report R41698, *H.R. 1 Full-Year FY2011 Continuing Resolution: Overview of Environmental Protection Agency (EPA) Provisions*, by Robert Esworthy.

[18] See CRS Report R42520, *Environmental Protection Agency (EPA) Appropriations for FY2013: Debate During the 112th Congress*, coordinated by Robert Esworthy and CRS Report R43207, *Environmental Protection Agency (EPA): Appropriations for FY2013 in P.L. 113-6*, by Robert Esworthy and David M. Bearden.

Environment, and Related Agencies bill that includes funding for EPA was passed in the House and Senate, and no bills were introduced for FY2014,[19] these regulatory actions remained prominent during the debate.

Other Enforcement Issues

Many other aspects of pollution control enforcement have been the subject of debate, and highlighted in congressional hearings and legislation. Some additional areas of continued interest include

- whether there is a need for increased compliance monitoring and reporting by regulated entities;
- impacts of environmental enforcement and associated penalties/fines on federal facilities' budgets (most notably the Department of Defense, or DOD, and Department of Energy, or DOE);
- how best to measure the success and effectiveness of enforcement (e.g., using indicators such as quantified health and environmental benefits versus the number of actions or dollar value of penalties);
- whether penalties are strong enough to serve as a deterrent and maintain a level economic playing field, or too harsh and thus causing undue economic hardship;
- how to balance punishment and deterrence through litigation with compliance assistance, incentive approaches, self-auditing or correction, and voluntary compliance;
- the effect of pollutant trading programs on enforcement; and
- the level of funding required to effectively achieve desired benefits of enforcement.

These issues result from disparate values and perspectives among stakeholders, but also from the factors that are the focus of this chapter: the statutory framework, those who work within this framework, and the tools and approaches that have been adopted for achieving compliance with pollution control laws.

The discussion below, beginning with identification of the principal statutes and key players, followed by an overview of integrated systems of administrative and judicial enforcement, compliance assistance, and incentive tools, is intended to provide a macro-perspective of environmental enforcement infrastructure and operations.

---

[19] Appropriations committees and several authorizing committees held hearings to consider the President's FY2014 budget request for EPA during the 113th Congress, but no bill to fund Interior, Environment, and Related Agencies (which includes EPA) for FY2014 was introduced in the House or the Senate prior to the start of FY2014. Following a temporary gap in funding, EPA and other federal departments and agencies operated under two continuing resolutions (P.L. 113-46 and P.L. 113-73) prior to the enactment of P.L. 113-76, the Consolidated Appropriations Act, 2014.

STATUTORY FRAMEWORK FOR ENFORCEMENT OF POLLUTION CONTROL LAWS AND KEY PLAYERS

As Congress has enacted a number of environmental laws over time, as well as major amendments to these statutes, responsibilities of both the regulators and the regulated community have grown. Organizational structures of regulatory agencies have evolved in response to their expanding enforcement obligations. Regulators also must adapt to an evolving, integrated system of administrative and judicial enforcement, compliance assistance, and incentive tools (see discussion under "Enforcement Response and Compliance Tools," later in this chapter).

Statutory Framework

The 11 laws listed in *Table 1* generally form the legal basis for the establishment and enforcement of federal pollution control requirements intended to protect human health and the environment.

The discussion in this chapter focuses on these federal environmental laws for which the U.S. Environmental Protection Agency (EPA) is the primary federal implementing agency. Since EPA was created in 1970, Congress has legislated a considerable body of law and associated programs to protect human health and the environment from harm caused by pollution. Those federal statutes, intended to address a wide range of environmental issues, authorize a number of actions to enforce statutory and regulatory requirements.

Enforcement of this diverse set of statutes is complicated by the range of requirements, which differ based on the specific environmental problem, the environmental media (e.g., air, water, land) affected, the scientific basis and understanding of public risks, the source(s) of the pollutants, and the availability of control technologies. Regulatory requirements range from health and ecologically based numeric standards, or technology-based performance requirements, to facility-level emission and discharge permit limits. Several of the pollution control laws require regulated entities to obtain permits, which typically specify or prohibit certain activities, or delineate allowable levels of pollutant discharges. These permits are often the principal basis for monitoring, demonstrating, and enforcing compliance. In recent years, an increasing number of administrative initiatives have favored incentive-based regulatory approaches, such as trading of permitted emissions, which can affect the applicability of traditional enforcement approaches.

Table 1. Major Federal Pollution Control Laws

Statute	**Major U.S. Code**
Comprehensive Environmental Response,	42 U.S.C. §§9601-9675
Compensation, and Liability Act (Superfund)	
Clean Air Act	42 U.S.C. §§7401-7671
Clean Water Act	33 U.S.C. §§1251-1387
Safe Drinking Water Act	42 U.S.C. §§300f-300j
Solid Waste Disposal Act/Resource Conservation and Recovery Act	42 U.S.C. §§6901-6991k
Oil Pollution Control Act (1990)	33 U.S.C. §§2701 et seq.
Environmental Planning and Community-Right-To-Know Act	42 U.S.C. §§11001-11050
Federal Insecticide, Fungicide, and Rodenticide Act	7 U.S.C. §§136-136y
Toxic Substances Control Act	15 U.S.C. §2601 et seq.
Pollution Prosecution Act of 1990	42 U.S.C. §4321

Note: This list is not comprehensive in terms of all laws administered by EPA, but rather covers the basic authorities underlying the majority of EPA pollution control programs. For a discussion of these statutes and their provisions, see CRS Report RL30798, *Environmental Laws: Summaries of Major Statutes Administered by the Environmental Protection Agency.*

Regulating authorities establish enforcement response and compliance assistance programs to address the enforcement provisions of particular federal pollution control statutes. These environmental statutes typically authorize administrative, civil judicial, and criminal enforcement actions for violations of statutory provisions. For example, Section 309 of the CWA, Section 113 of the CAA, and Section 1414 of the Safe Drinking Water Act (SDWA) cover enforcement provisions.[20] As provisions for specific actions vary from statute to statute, each EPA regulatory program office establishes detailed criteria for determining what sanctions are preferable (and authorized) in response to a given violation. The statutes often provide a level of discretion to regulators for addressing specific circumstances surrounding certain environmental problems or violations of national requirements.

Enforcement of the many provisions of the major environmental laws across a vast and diverse regulated community involves a complex coordinated process between federal (primarily EPA and DOJ), state, tribal, and local governments. Congress provided authority to states for implementing and enforcing many aspects of the federal statutory requirements. Citizens also play a role in ensuring that entities comply with environmental requirements, by reporting violations or filing citizen lawsuits, which are authorized under almost all pollution control laws. The following discussion highlights the roles of these key players.

[20] See 33 U.S.C. §1319, 42 U.S.C. §7413, and 42 U.S.C. §300g-3.

Key Players in Environmental Enforcement and Compliance

U.S. Environmental Protection Agency

Primarily through its program offices (e.g., air, water, solid waste), EPA promulgates national regulations and standards.[21] Other federal agencies (e.g., the Department of the Interior, Army Corp of Engineers) and states, tribes, various stakeholder groups, and citizens may contribute input to EPA at various stages of regulatory development (including required public comment). (States may also establish their own laws based on the national requirements; see the discussion later in the "States and "Delegated Authority"" section of this chapter.) EPA (and states) inform the regulated community of their responsibilities and administer permitting, monitoring, and reporting requirements. EPA also provides technical and compliance assistance, and employs a variety of administrative and judicial enforcement tools as authorized by the major environmental laws it administers, as well as incentive approaches, to promote and ensure compliance.

Since EPA's establishment, the agency's enforcement organization has been modified a number of times, and continues to evolve.[22] EPA's Office of Enforcement and Compliance Assurance (OECA) at headquarters and in the 10 EPA regional offices sets the general framework for federal enforcement activities in coordination with the agency's program offices, states and tribes, and other federal agencies, particularly DOJ. OECA serves as the central authority for developing and implementing a national compliance and enforcement policy, and coordinating and distributing policies and guidance.

EPA's National Enforcement Initiatives (NEI)[23] and OECA's National Program Managers (NPM)[24] Guidance are primary strategic planning tools that set out national enforcement program priorities and coordinate and monitor state, regional, and EPA headquarters implementation of environmental enforcement/compliance activities. EPA's 10 regional offices, in cooperation with the states, generally are responsible for a significant portion of the day-to-day federal enforcement activities. The NEI is developed every three years with the cooperation of EPA regions and states/tribes, identifying overall program directions as well as specific enforcement activities/priorities. In June 2013, EPA-OECA announced its decision to continue the initiatives that were the focus of the FY2011-FY2013 NEI released February 22, 2010,

---

[21] See CRS Report RL32240, *The Federal Rulemaking Process: An Overview*, coordinated by Maeve P. Carey.

[22] For more information regarding EPA's current organizational structure for enforcement, see the agency's website at http://www.epa.gov/compliance/about/index.html. Several references describe the historical evolution of EPA, including Joel A. Mintz, *Enforcement at the EPA: High Stakes and Hard Choices*, 1st ed. (University of Texas Press, Austin, 1995); and Clifford Rechtschaffen and David L. Markell, *Reinventing Environmental Enforcement & the State/Federal Relationship* (Environmental Law Institute, 2003).

[23] On February 22, 2010, EPA renamed its "National Enforcement Priorities" the "National Enforcement Initiatives," http://www.epa.gov/compliance/data/planning/initiatives/ initiatives.html.

[24] See EPA "Planning and Budget Results / FY2015 Draft Addendum to the FY2014 NPM Guidances," http://www2.epa.gov/planandbudget/national-program-manager-guidances.

for the FY2014- FY2016 NEI cycle.[25] OECA NPM Guidance and other agency NPM Guidance are issued annually based on a three-year cycle coinciding with the NEI, identifying allocation of resources and expected outcomes, and serves as the basis for the enforcement agreements ("commitments") with the regional offices. The OECA NPM Guidance applies to OECA, all EPA regional enforcement programs, and states and tribes implementing EPA-approved inspection and enforcement programs. The agency is currently operating under the FY2014 OECA NPM Guidance finalized June 14, 2013, but on March 14, 2014, released an FY2015 draft NPM Guidance Addendum.[26]

The EPA National Enforcement Investigations Center (NEIC) provides technical expertise to the agency and states. The center administers an investigative team that assigns investigators to the regional offices as needed.[27] OECA also facilitates EPA's National Enforcement Training Institute (NETI), established under Title II of the 1990 Pollution Prosecution Act (P.L. 101-593). NETI provides a wide spectrum of environmental enforcement training online to international, federal, state, local, and enforcement personnel, including lawyers, inspectors, civil and criminal investigators, and technical experts.[28]

OECA's headquarters personnel conduct investigations and pursue or participate in national enforcement cases, particularly those potentially raising issues of national significance. More often enforcement activities fall to the regional offices. EPA (and the states') enforcement actions often require coordination with other federal agencies, most frequently DOJ.

U.S. Department of Justice[29]

In coordination with EPA, the Department of Justice (DOJ)—at its headquarters and through the U.S. Attorneys' offices around the country—plays an integral role in judicial federal enforcement actions of environmental regulations and statutes. EPA refers cases (including some initiated by states) to DOJ for an initial determination of whether to file a case in federal court. DOJ represents EPA in both civil and criminal actions against alleged violators, maintaining close interaction as needed with EPA, states, and tribes during various stages of litigation. DOJ also defends environmental laws, programs, and regulations, and represents EPA when the agency intervenes in, or is sued under,

---

25 See footnote 23 and EPA-OECA's announcement at http://www2.epa.gov/sites/production/ files/2013-09/documents/ 2014-2016-nei-announcement.pdf.

26 EPA, FY 2015 Office of Enforcement and Compliance Assurance (OECA) DRAFT National Program Manager (NPM) Guidance Addendum, March 14, 2014, http://www2.epa. gov/planandbudget/national-program-managerguidances. See also EPA's OECA, FY2014, FY2013, and FY2012 NPM Guidance, and EPA's *Planning, Budget Results: Historical Planning, Budget, and Results Reports* website at http://www2.epa.gov/planandbudget.

27 See EPA's National Enforcement Investigations Center (NEIC). See http://www2. epa.gov/enforcement/national-enforcement-investigations-center-neic.

28 EPA's National Enforcement Training Institute (NETI) at http://www.epa.gov/comp liance/training/neti/index.html.

29 See http://www.justice.gov/.

environmental citizen suits. For FY2013, EPA-OECA reported the referral of 138 civil cases to DOJ, and that 297 criminal cases[30] were opened.[31] Many of these cases are handled by DOJ's Environment and Natural Resources Division (ENRD).[32] EPA and DOJ work conjunctively with the other federal agencies as cases warrant.

Other Federal Agencies

EPA and DOJ coordinate with a number of other federal agencies, particularly when taking criminal action. Key federal agencies include the Federal Bureau of Investigation (FBI), Department of Transportation (DOT), Department of Homeland Security (DHS, particularly the Coast Guard and U.S. Immigration and Customs Enforcement, or ICE), Fish and Wildlife Service, Army Corps of Engineers, Defense Criminal Investigative Service, National Oceanic and Atmospheric Administration (NOAA), U.S. Internal Revenue Service (IRS), and U.S. Securities and Exchange Commission (SEC). These agencies may provide support directly in response to violations of laws implemented by EPA, or, as is often the case, in circumstances where multiple laws have been violated.

States and "Delegated Authority"[33]

Most federal pollution control statutes, but not all, authorize EPA to delegate to states the authority to implement national requirements.[34] For a state to be authorized, or "delegated," to implement a federal environmental program, it must demonstrate the capability to administer aspects of the program's requirements, including the capacity to enforce those requirements. Delegated authority must be authorized under the individual statute, and states must apply for and receive approval from EPA in order to administer (and enforce) federal environmental programs. While many federal pollution control laws provide authority for states to assume primary enforcement responsibilities, there is significant variability across the various laws, including as to standards states must meet and EPA's authority in determining whether states are authorized or have primacy. In some cases, state primacy is almost automatic.

---

[30] EPA-OECA discontinued reporting "criminal referrals" beginning with reporting in FY2005. EPA reported the referral of 168 criminal cases to DOJ in FY2004, the last year criminal referrals were reported publicly by EPA; see *EPA Annual Results: Previous Years*, http://www.epa.gov/compliance/data/results/annual/index.html

[31] EPA-OECA, *Enforcement Annual Results for Fiscal Year 2013*, released February 7, 2014, http://www2.epa.gov/enforcement/enforcement-annual-results-analysis-and-trends-fiscal-year-fy-2013. The primary source for the data reported is the EPA Regions certified FY2013 end-of-year workbooks as of December 7, 2013.

[32] EPA's cases are typically handled by three of the DOJ Environmental and Natural Resources Division's 10 sections: the Environmental Crimes Section, the Environmental Enforcement Section, and the Environmental Defense Section (http://www.usdoj.gov/enrd/ About_ ENRD.html). See also DOJ, *ENRD Accomplishments Report Fiscal Year FY2013*, http://www.justice.gov/enrd/ENRD_Assets/ENRD_Accomplishments_Report_2013_2.pdf. This report and prior fiscal year ENRD summaries of litigation accomplishment reports are available at http://www.justice.gov/enrd/Current_topics.html.

[33] The term "delegated authority" has become the most commonly used when referring to EPA's authority to approve states' programs. Federal statutes more often use "primary enforcement responsibility," "primacy," "approved," or "authorized" states' responsibility.

[34] See CRS Report RL30798, *Environmental Laws: Summaries of Major Statutes Administered by the Environmental Protection Agency*, for references to sections of individual acts that provide state authority.

Some federal pollution control laws limit the authority to a specific provision, while others do not authorize delegation at all. For example, Section 1413 of the Safe Drinking Water Act (SDWA) authorizes states to assume primary oversight and enforcement responsibility (primacy) for public water systems,[35] and Section 402 of the Clean Water Act (CWA) authorizes state-delegated responsibilities under that act to issue and enforce discharge permits to industries and municipalities. Under CERCLA (Superfund), states are authorized to participate in the cleanup of waste, from taking part in initial site assessment to selecting and carrying out remedial action, and negotiating with responsible parties. Under FIFRA, states may have primacy for enforcing compliance requirements contained on labels of registered pesticides, but are not granted enforcement authority related to registering pesticides or pesticide establishments. Programs under other laws, such as the Toxic Substances Control Act (TSCA), do not provide authority for state delegation. EPA can also authorize state government officials to conduct inspections for environmental compliance on behalf of the agency, subject to the conditions set by EPA, even if a specific statute does not provide delegation authority. However, there must be authority under the specific statute for authorizing such inspections.[36]

Even if delegation is authorized under a federal statute, states may opt not to seek delegation of a particular environmental program, or they may choose only to implement a select requirement under a federal law. For example, as of November 2012, 46 states had obtained the authority to operate the national permitting program under Section 402 of the CWA, but EPA had only delegated authority to two states to operate the wetlands permitting program under a separate CWA provision, Section 404.[37]

A majority of states have been delegated authority to implement and enforce one or more provisions of the federal pollution control laws.[38] Authorized states generally implement the national laws and regulations by enacting their own legislation and issuing permits, which must be at least as stringent as the national standards of compliance established by federal law. States consider and approve environmental permits, monitor and assess environmental noncompliance, provide compliance assistance and information to the regulated community and the public, conduct inspections, and take enforcement actions. Local government authorities also play a role in permitting and monitoring. For example, EPA has delegated authority to implement Section 112 of the Clean Air Act (CAA) to at least three county governments. However, local governments generally act within the context of assuring states' requirements. For example, local authorities may

---

[35] See footnote 34, p. 48.

[36] See EPA guidance for issuing federal inspector credentials to state/tribal governments to conduct civil inspections: http://www.epa.gov/compliance/resources/policies/monitoring/.

[37] See CRS Report RL30030, *Clean Water Act: A Summary of the Law*, by Claudia Copeland.

[38] The Environmental Council of the States (ECOS) has tracked delegated authority by state and statute; see http://www.ecos.org/section/states.

incorporate land use and other issues as well as code requirements (fire, construction, building safety, plumbing, etc.) in their consideration of permits. A more detailed discussion of the many facets of local authorities is beyond the scope of this chapter.

A significant proportion of inspections and enforcement actions are conducted by the states. Comparable, comprehensive data from the same or similar sources are not readily available for purposes of directly comparing enforcement activities in states relative to EPA. While EPA routinely reports trends in its major enforcement actions in the annual OECA accomplishments reports and on its website, the agency does not generally include state-specific activities. There are a number of limitations with regard to states' information currently retained by EPA in its databases (e.g., not all states report relevant information into the EPA databases, reported data are not provided consistently from state to state, and reporting requirements are variable from statute to statute).[39]

EPA has made an effort to enhance and improve enforcement reporting by states. The agency has been implementing its State Review Framework (SRF) tool developed and introduced in 2004, to improve its oversight of state enforcement programs.[40] Under this SRF tool, EPA representatives visit and evaluate each state's compliance and enforcement program based on specified criteria. In February 2013, on its Enforcement and Compliance History Online website (ECHO; see *Appendix A),* EPA released interactive dashboards and comparative maps that include state-level enforcement data for the most recent five years.[41] Information includes the number of completed inspections, types of violations found, enforcement actions taken, and penalties assessed by state, and users can customize the presentation to view state activity as well as view comparisons with EPA activity.

Through discussions and reports, EPA provides feedback to each state and based on its review, outlines recommendations for improvement. Full implementation of SRF was initiated by EPA in July 2005 and the agency reported that reviews of all states and territories were completed in 2007. EPA began conducting Round 2 of reviews in 2008 and completed reviews for most states by the end of calendar year 2012. OECA, with its partners, has also conducted an evaluation of the implementation of the first cycle of SRF recommendations and initialed revisions to SRF guidance for conducting subsequent reviews. OECA continues to work with its partners in evaluating implementation of SRF recommendations.[42] Despite these efforts, there are still perceived differences between states, EPA regions, and EPA headquarters.

---

[39] For example, see GAO report, *Drinking Water: Unreliable State Data Limit EPA's Ability to Target Enforcement Priorities and Communicate Water Systems' Performance*, Report No. GAO-11-381 June 17, 2011, http://www.gao.gov/products/GAO-11-381.

[40] EPA-OECA, http://www.epa.gov/compliance/state/srf/index.html; see also EPA-OECA report *Best Practices and Program Improvements Expected to Result from SRF*, September 12, 2007, http://www.epa.gov/compliance/resources/ reports/state/index.html.

[41] EPA ECHO Comparative Maps & Dashboards Home, http://echo.epa.gov/comp_maps_home; see also EPA's February 7, 2013, press release, "EPA Releases State Enforcement Performance Information and Comparative Maps / Agency to host webinar demonstrating new data features," http://yosemite.epa.gov/opa/admpress.nsf/Press%20Releases%20 issued%20by%20OECA?OpenView.

[42] See footnote 40.

In recent years, ECOS[43] has served as a forum to improve coordination and promote joint strategic planning between the states and EPA. In addition to other strategic planning tools, EPA and states established the National Environmental Performance Partnership System (performance partnerships, or NEPPS)[44] in 1995 in an effort to improve the effectiveness of EPA-state coordinated environmental management. Under this system, which includes elements of compliance and enforcement, EPA and states enter into individual partnerships (performance partnership agreements) to address jointly agreed-upon priorities based on assessments of localized environmental conditions. The partnerships can be broad in scope or comprehensive strategic plans, and often serve as work plans for funding through EPA grants.

Absent delegation, EPA continues to enforce the federal law in the state, although a state can enforce its own environmental laws where not preempted by federal law. Even with delegation, EPA retains the authority and responsibility as determined by each statute to take enforcement measures, generally taking action when there is a violation of an EPA order or consent decree, or when the federal government deems a state to have failed to respond to a major violation in a "timely and appropriate" manner. Additionally, when a noncompliance case involves an emergency or matters of potential national concern, such as significant risk to public health and safety, the federal government will typically intercede. There are cases where states request the federal government to step in and other cases where the federal government on its own initiative acts on violations that are the subject of state enforcement action or settlement, known as "overfiling." EPA contends that overfiling occurs infrequently and that certain environmental statutory provisions preclude EPA from overfiling. These provisions are not explicit in all the pollution control statutes, and are limited to specific subsections and violations.[45] Although overfiling of states' enforcement actions has occurred under various pollution control statutes, historically, overfiling of Resource Conservation and Recovery Act (RCRA) violations has been the subject of considerable debate and litigation. States have strongly objected to overfiling, and the utility and extent of overfiling with respect to environmental enforcement has been the subject of considerable litigation, debate, and literature.[46]

---

[43] The Environmental Council of the States (ECOS) is a national nonprofit (501(c)(6)), nonpartisan association of state and territorial environmental commissioners.

[44] See http://www.epa.gov/ocirpage/nepps/for information regarding NEPPS.

[45] Provisions of the Clean Water Act (CWA) under Section 309 are often cited as an example of legislation limiting EPA's authority to overfile. EPA's authority to enforce under this section is only limited when a state has commenced an "appropriate enforcement action" in response to and within 30 days of EPA's issuance of a notice of violation to the state (33 U.S.C. §1319(a)(1)); when the state has "commenced and is diligently prosecuting" an action under comparable state law; or when a penalty assessed under a state-issued final order has been paid, the violation will not be subject of a civil penalty action under §1319(d) §1321(b) or §1365 (33 U.S.C. §1319(g)(6)).

[46] Ellen R. Zahren, *Overfiling Under Federalism: Federal Nipping at State Heels to Protect the Environment*, 49 Emory L. J. 373, 375 n. 18 (2000); Joel A. Mintz, *Enforcement "Overfiling" in the Federal Courts: Some Thoughts on the Post-Harmon Cases*, 21 Virginia Envtl. L. J. 425, 427 (2003). Jeffrey G. Miller, Theme & Variations in Statutory Preclusions Against Successive Environmental Enforcement Actions by EPA & Citizens, Part II Statutory Preclusions on EPA Enforcement, 29 *Harvard Envtl. L. Rev*. 1, 3 (2005).

Overfiling

The term "overfiling" applies to situations when federal enforcement actions are filed during or after a state enforcement action against the same entity for violation of a federal statute. Some states and regulated entities use the term more broadly in reference to assertion of federal authority. Overfiling or the threat of overfiling sometimes strains EPA-state relations and cooperation, sometimes implying criticism of a delegated state's effectiveness.

Tribal Governments[47]

EPA and states increasingly have recognized the role of tribal governments in environmental enforcement, where tribes, rather than states, have primary jurisdiction.[48] Indian tribes, as sovereign governments, can establish and enforce environmental programs under their own laws, but must obtain approval from EPA to administer federal environmental programs on their land. As with states, some of the federal statutes authorize tribes,[49] with EPA approval, to assume responsibility for implementing certain federal pollution control programs. To obtain EPA approval, tribes must demonstrate adequate authority and jurisdiction over the activities and lands to be regulated. Where there is no approved tribal program, EPA exercises its federal authority and may undertake direct program implementation. In some instances, particularly when there are criminal violations, EPA may retain a role in compliance and enforcement even when there is an approved tribal program.

In addition to the federal statutes, a tribal government's authority for environmental protection can arise from federal executive orders, treaties, and agreements with the United States and/or state and local governments,[50] some of which explicitly reserve rights pertaining to the environment. When addressing environmental issues within tribal lands, EPA abides by the January 24, 1983, American Indian policy statement,[51] which reaffirmed the government-to-government relationship of Indian tribes with the United States.[52]

Relatively few tribes have obtained authority for implementing federal pollution control laws, and EPA identified tribal environmental compliance as a national

---

[47] See EPA's American Indian Environmental Office Tribal Portal, http://www.epa.gov/tribal/, and the Compliance & Enforcement through Tribal Government website at http://www. epa.gov/compliance/tribal/.

[48] EPA-approved/authorized state programs generally do not apply in Indian country.

[49] Some pollution control laws have been amended to clarify the role of tribal governments in the implementation of federal environmental programs. For example, from 1986 to 1990, Congress amended the Clean Water Act (33 U.S.C. §1377(e)(2)), Safe Drinking Water Act (42 U.S.C. §300j-11(b)(1)(B)), and Clean Air Act (42 U.S.C. §7601(d)(2)(B) to authorize EPA to treat Indian tribes in the same manner as states for purposes of program authorization.

[50] For example, see Executive Order No. 13175 on Consultation and Coordination With Indian Tribal Governments, 65 *Federal Register* 67249 (November 9, 2000); Executive Memorandum on Government-to-Government Relations with Native American Tribal Governments, April 29, 1994.

[51] Issued by President Ronald Reagan, the policy expanded the 1970 national Indian policy of self-determination for tribes, See EPA's Indian Policies website at http://www.epa.gov/ tribalportal/basicinfo/presidential-docs.html.

[52] In conjunction with the 1983 overall federal policy statement, EPA consolidated existing agency statements into a single policy statement to ensure consistency. See *EPA Policy for the Administration of Environmental Programs on Indian Reservations*, http://www. epa.gov/superfund/community/relocation/policy.htm.

enforcement and compliance priority in its FY2005-FY2007 and its FY2008-FY2011[53]enforcement strategic plans in an effort to enhance tribal governments' capabilities to implement federal environmental statutes. The primary focus was public drinking water systems, federal pollution control statutes applicable to schools, and unregulated dumping of solid waste. EPA's National Enforcement Initiatives (NEIs) for FY2011-FY2013, which were retained for the FY2014-FY2016 cycle,[54] did not designate Indian Country as one of the six national enforcement initiatives; however, the sector will continue to be addressed through the structure established under the previous designated initiative or through the regular program compliance assistance, inspections, investigations, and enforcement conducted in regional offices and states.[55] EPA continues to provide enforcement and regulatory guidance via the Tribal Compliance Assistance Center.[56]

Citizens

Private individuals play an important role in enforcing certain aspects of federal pollution control laws. Citizen participation, specifically authorized by Congress in many of the federal pollution control statutes, occurs in several ways. Individuals can identify and report violations of the laws, provide comments on settlements that are reached between the federal government and violators of the environmental laws in enforcement cases, and initiate enforcement proceedings directly in response to alleged violations. In addition, individuals may bring actions against EPA for failing to execute nondiscretionary duties required under federal environmental laws.[57]

To further enhance public participation and reporting of potential environmental violations, EPAOECA introduced the "National Report a Violation" website in January 2006.[58] The website provides access to OECA's online citizens' tips and complaints form and related contact information. EPA reported that the number of citizen tips and complaints increased from 1,485 in FY2005 to 3,274 in FY2006. According to EPA, more than 18,000 total tips were reported to date, including more than 7,800 received in FY2008[59] (not reported for FY2009 through FY2013). Additionally, in FY2009 EPA introduced the "EPA Fugitives" website to solicit public assistance in locating alleged

---

[53] EPA-OECA, National Enforcement Initiative (NEI) for Fiscal Years 2008-2010*: Indian Country*, http://www.epa.gov/compliance/data/planning/priorities/tribal.html.

[54] See the announcement at http://www2.epa.gov/enforcement/announcement-epas-selection-national-enforcement-initiatives-fy-2014-2016.

[55] EPA-OECA, National Enforcement Initiatives (NEI), http://www.epa.gov/compliance/data/planning/initiatives/index.html. See general discussion under "Key Players in Environmental Enforcement and Compliance," pp.7-9, earlier in this report.

[56] See http://www.epa.gov/tribalcompliance/.

[57] Although not strictly speaking "enforcement," citizens may also petition for review of agency actions under a program statute or the Administrative Procedure Act.

[58] EPA-OECA, *Report an Environmental Violation*, http://www.epa.gov/tips. See also *Report Spills and Environmental Violations*, http://www.epa.gov/epahome/violations.htm#who.

[59] EPA-OECA, *Compliance and Enforcement Annual Results FY2008: Report a Violation*, http://www.epa.gov/compliance/resources/reports/endofyear/eoy2008/2008-sp-report violations.html.

environmental criminal fugitives. (See brief overview of the website in the "Criminal Judicial Enforcement" section of this chapter.)

Regulated Community

The size and diversity of the regulated community are vast, spanning numerous industrial and nonindustrial entities, small and large, and their operations. The following discussion provides an overview of the regulated community, and highlights the role and activities of the key regulated entities in the enforcement of the primary pollution control statutes.

The universe of the regulated community as a whole is very large (see discussion below). The majority of those in the regulated community are required to comply with multiple statutes because of the nature of their activities and operations. The regulated community includes a diverse range of entities and operations, including utilities, refineries, manufacturing and processing facilities, agriculture producers and processors, mobile sources (e.g., private and commercial vehicles), and others. Local, state, tribal, and federal governments are also part of the regulated community, as they are engaged in a range of activities and operations—utilities, construction, waste and wastewater management, drinking water management, transportation, and pest management—that generate pollution similar to nongovernment sectors.

Regulated entities vary in their activities and operations, and in size—ranging from small individual business operations such as dry-cleaners to facilities and operations that are part of large corporations and conglomerates. Regulatory agencies generally categorize regulated entities into minor and major emitters/dischargers based on factors such as total earnings, number of employees, production volume, and amount of emissions, for purposes of implementing and enforcing the various statutes. In certain circumstances, some of the pollution control statutes make specific distinctions with regard to major and minor emitters/dischargers. A designation of "major" generally applies to those entities that, because of their size or operations, have the potential to have a significant impact on the environment. Most of the statutes and accompanying regulations include authorities for reducing the stringency, and in some cases providing exemptions from regulatory requirements to minimize their impacts on small businesses and operations.

There is no readily available, current, comprehensive list and description of the complete universe of those who are regulated under all of the major pollution control statutes. EPA has been criticized for not adequately defining the regulated universe, a step that GAO determined to be a critical component necessary to evaluate the effectiveness of enforcement.[60] EPA-OECA compiled data regarding the size of the regulated community in September 2001, and estimated a total universe of more than 41

---

[60] EPA-OIG, *Limited Knowledge of the Universe of Regulated Entities Impedes EPA's Ability to Demonstrate Changes in Regulatory Compliance*, 2005-P-00024, September 19, 2005 (http://www.epa.gov/oig/); and GAO, *Human Capital: Implementing an Effective Workforce Strategy Would Help EPA to Achieve Its Strategic Goals*, GAO-01-812, pp. 24-25, July 2001, http://www.gao.gov/docsearch/repandtest.html.

million.[61] Although cited by EPA subsequently from time to time, most commonly in strategic planning documents, the agency has not updated the estimate.

There are, however, data and information that provide some indications of the size and diversity of this universe—for example, in EPA's primary enforcement and compliance databases (see additional discussion in *Appendix A*). EPA's publicly available Enforcement and Compliance History Online (ECHO) provides for integrated searches of data for more than 800,000 facilities for compliance with CWA, CAA, and RCRA.[62] The data are primarily based on permitted facilities. Another EPA centrally managed database is the Facility Registry System (FRS), which primarily identifies "facilities, sites or places" subject to federal pollution control requirements; it contains more than 2.5 million unique facility records.[63] The FRS database is primarily based on permit information for CWA, CAA, and RCRA, but includes information reported regarding CERCLA sites. It does not include information indicating the universe regulated under other statutes. In yet another source, the ECOS indicated that states reported that more than 3 million regulated facilities required state agency oversight for environmental compliance in 2003.[64] The differences in the various sources are an indication of the difficulty involved in accurately and consistently tracking the size of the regulated populations.

EPA's various program offices (e.g., air, water, and waste) maintain and publish information and profiles regarding characterizations of regulated entities and their operations. Generally included are estimates of the types and amounts of emissions and discharges, or wastes being handled. For example, EPA's Office of Air and Radiation (OAR) maintains a national database of air emissions estimates for individual point- or major-source categories.[65] The database contains information on stationary and mobile sources that emit common ("criteria") air pollutants[66] and their precursors, as well as hazardous air pollutants (HAPs).[67] The categories presented in these sources do not reflect 100% of the total number of facilities being regulated.

Another source for characterizing the sectors of the regulated community is EPA's "Sector Notebooks."[68] EPA has defined sectors as distinct parts of the economy that share similar operations, processes or practices, environmental problems, and compliance

---

[61] EPA-OECA, OECA Regulatory Universe Identification Table. Internal EPA memorandum November 15, 2001, EPA-OIG, Limited Knowledge of the Universe of Regulated Entities Impedes EPA's Ability to Demonstrate Changes in Regulatory Compliance, 2005-P-00024, September 19, 2005, http://www.epa.gov/oig/.

[62] Enforcement and Compliance History Online (ECHO), http://www.epa-echo.gov/ echo/about_site.html

[63] EPA, Federal Registry System (FRS) Overview, http://www.epa.gov/enviro/html/ fii/index.html.

[64] ECOS, *State Environmental Contributions to Enforcement and Compliance: 2000-2003*, June 2006, http://ecos.org/ section/publications.

[65] EPA, *National Emissions Inventories for the U.S.*, http://www.epa.gov/ttn/chief/ index.html.

[66] Under §106 of the Clean Air Act, EPA has set National Ambient Air Quality Standards for six principal pollutants classified by the EPA as "criteria pollutants": sulfur dioxide (SO2), nitrogen dioxide (NO2), carbon monoxide (CO), ozone, lead, and particulate matter.

[67] Under §112 of the Clean Air Act, EPA is to establish technology-based emission standards, called "MACT" standards, for sources of 188 pollutants listed in the legislation, and to specify categories of sources subject to the emission standards.

[68] For more information regarding EPA's Sector Compliance Assistance and Sector Notebooks, see http://www.epa.gov/compliance/assistance/sectors/index.html.

issues. EPA recognizes that there are likely a number of circumstances where regulated entities within specific geographic regions may have unique characteristics that are not fully reflected in the profiles contained in the sector notebooks. In addition, some of the notebooks were completed several years ago. Nevertheless, notebook profiles provide fairly comprehensive characterizations of key sectors included within the regulated community.

Table 2 lists industry and government sectors for which the agency has completed sector notebooks and developed compliance assistance tools.

Table 2. EPA Industry and Government Sectors

Available Sector Notebooks		
Aerospace	Health Care	Prisons and Correctional Institutions
Agriculture	Local Government Operations	Pulp/Paper/Lumber
Automotive	Marinas	Ready Mix/Crushed Stone/Sand and Gravel
Chemicals	Metals	Retail
Computers/Electronics	Minerals/Mining/Processing	Rubber/Plastics
Construction	Paints and Coatings	Shipbuilding and Repair
Dry Cleaning	Petroleum	Textiles
Education	Pharmaceuticals	Transportation
Federal Facilities	Ports	Tribal
Food Processing	Power Generators	
Furniture	Printing	

Source: Table generated by CRS with information from EPA's Sector Compliance Assistance and Sector Notebooks website, http://www.epa.gov/compliance/ assistance/sectors/index.html.

Enforcement at Federal Facilities[72]

Unless a statutory exemption exists, federal facilities are subject to the federal pollution control statutes,[73] and generally also must adhere to the environmental laws and regulations of the states and municipalities in which they are located, to the same extent as others in the regulated community. EPA concluded 61 enforcement actions against

---

[72] See EPA, "Compliance and Enforcement at Federal Facilities," http://www.epa.gov/ compliance/federalfacilities/index.html.

[73] Most federal environmental laws contain provisions that subject federal facilities to federal, state, and local requirements, and allow such facilities to be sued just as a nongovernmental entity. In addition, such provisions generally grant the President authority to exempt federal facilities from such requirements when in the "paramount interest" or (less commonly) the "national security interest" of the United States. See Clean Air Act (42 U.S.C. §7418), Clean Water Act (33 U.S.C. §1323), Resource Conservation and Recovery Act (42 U.S.C. §6961), and Safe Drinking Water Act (42 U.S.C. §300j-6). A more limited federal facility provision and presidential exemption is found in the Comprehensive Environmental Response, Compensation and Liability Act (42 U.S.C. §9620). The Toxic Substances Control Act nowhere expressly states that federal facilities are subject to the statute, but nonetheless does authorize presidential exemptions (15 U.S.C. §2621). For discussion of exemptions, particularly as they pertain to DOD, see CRS Report RS22149, *Exemptions from Environmental Law for the Department of Defense (DOD)*, by David M. Bearden.

federal agencies for alleged violations of federal pollution control laws during FY2013, the same as FY2012.[74]

This is compared to 57 enforcement actions during FY2011[75] and 52 during FY2010.[76] Federal agencies are also subject to relevant requirements of executive orders.[77] Nearly $866.5 million in penalties was assessed for federal facility violations during FY2013, compared to $615.0 million assessed in FY2012, $9.0 million in FY2011, and $749,000 in FY2010. In addition, violators agreed to invest more than $128.0 million in cleanup and improved operations to comply with environmental laws in FY2013, compared to roughly $160.0 million in FY2012, more than $5.0 billion in FY2011 and an estimated $163.0 million in FY2010.[78] EPA no longer provides aggregated summaries of federal facility enforcement actions in its annual results report; however, the agency provides an interactive mapping tool of concluded enforcement cases that includes the capability to search information for individual federal facilities based on location.[79]

Regulating federal facilities under pollution control laws presents certain unique challenges. Although all are potentially subject to pollution control laws and regulations, a majority of federal agencies and their facilities are not involved in activities that would generally warrant compliance requirements. According to EPA, facilities operated by DOD and DOE make up a significant portion of the universe of "major" federal facilities.[80] Major federal facilities generally refer to those facilities that, because of their size or operations, have the potential to have a significant impact on the environment. Compliance/enforcement information for DOD and DOE is reported individually, while other federal agencies are generally categorized together as Civilian Federal Agencies.[81]

The major federal pollution control laws provide EPA with authorities to enforce requirements and impose penalties at federal facilities that are not in compliance. The Federal Facility Compliance Act of 1992 specifically amended RCRA to clarify that DOD and all other federal facilities are subject to penalties, fines, permit fees, reviews of plans or studies, and inspection and monitoring of facilities in connection with federal,

---

[74] Based on information provided upon request directly to CRS by EPA's Office of Congressional and Intergovernmental Relations (OCIR), March 18, 2014.

[75] EPA-OECA, *Compliance and Enforcement Annual Results 2011 Fiscal Year: Federal Facilities*, http://www.epa.gov/compliance/resources/reports/endofyear/eoy2011/programs/ federalfacilities.html.

[76] EPA-OECA, *Compliance and Enforcement Annual Results 2010 Fiscal Year: Federal Government Compliance*, http://www.epa.gov/compliance/resources/reports/endofyear/eoy 2010/fedfacilities.html.

[77] For examples of executive orders with directives addressing environmental management at federal facilities, see EPA's *Federal Facilities Sector Notebook: A Profile of Federal Facilities*, EPA 300-B-96-03, January 1996, pp. 2-11 and 2-12, available at http://www.epa.gov/compliance/assistance/sectors/index.html.

[78] See footnote 72 and footnote 73.

[79] See EPA, *Enforcement Annual Results Concluded EPA Enforcement Cases Map for Fiscal Year (FY) 2013*, http://www2.epa.gov/enforcement/enforcement-annual-results-concluded-cases-map-fiscal-year-fy-2013.

[80] EPA Federal Facilities Enforcement Office, *The 2008 State of Federal Facilities An Overview of Environmental Compliance at Federal Facilities,* EPA 305-R-09-001, August 2009; and *The State of Federal Facilities An Overview of Environmental Compliance at Federal Facilities FY 2005-2007,* EPA 305R08002, September 2008, http://cfpub.epa.gov/ compliance/resources/reports/accomplishment/federal.cfm?templatepage=4.

[81] See footnote 77.

state, interstate, or local solid or hazardous waste regulatory programs.[82] The SDWA includes similar language regarding federal facilities, but most of the other federal environmental laws do not include such specific provisions. CERCLA (Superfund) Section 120 requires federal agencies with NPL sites to investigate and clean up the contamination, and significantly contaminated federal facility sites have been listed on EPA's National Priorities List (NPL).

Whether other pollution control laws should be amended to clarify their applicability to federal facilities has been an issue of debate in Congress.

ENFORCEMENT RESPONSE AND COMPLIANCE TOOLS

EPA and states apply a set of environmental enforcement tools to identify and correct noncompliance, restore environmental damage, and impose penalties intended to deter future violations. Compliance with pollution control laws is addressed through a continuum of response mechanisms, ranging from compliance assistance to administrative and civil enforcement, to the stronger criminal enforcement. The spectra of tools, which escalate in terms of their level of severity and intensity, are authorized in each of the environmental statutes. The following sections of this chapter provide a brief overview of the various enforcement response mechanisms.

Over the years, EPA and states have sought to effectively balance the provision of guidance and assistance to prevent violations or achieve compliance by regulated entities with federal pollution control requirements, with the imposition of strong enforcement actions in response to violations. Some critics have depicted environmental enforcement as overly litigious, or requiring unwarranted remedies. Others counter that actions are not pursued with enough rigor and frequency, or that penalties are not severe enough to deter noncompliance. EPA officials have countered that, in some instances, the agency is relying more on settlements and focusing on requiring increased expenditures on pollution control technologies, and that it is focusing judicial actions on larger and more complex cases that are expected to result in larger environmental benefits.

EPA and states maintain a considerable degree of flexibility in determining how to respond to potential violations, to the extent authorized by individual statutes. Initially, a potential violation is identified through monitoring, inspecting, citizen reporting, or through self-reporting by the regulated entity. As a first step in the enforcement process, unless an imminent danger or hazard has been determined, EPA and states may attempt to obtain corrective actions by simply issuing a warning or notifying a facility that minor violations may exist, and granting reasonable time for compliance. EPA or a state may then (or sometimes as a first step) initiate a civil administrative action under its own

---

[82] 42 U.S.C. §6961.

authority without involving the judicial process, or file formal civil or criminal[83] judicial actions in court.

Sanctions imposed, whether through negotiated settlements or decisions by the court, generally include required actions to achieve compliance and to correct environmental damage (injunctive relief), and may include monetary penalties (and incarceration in the case of criminal violations). During the last 10 years, settlements increasingly have also included requirements that violators undertake mutually agreed-upon environmentally beneficial projects supplemental to other sanctions.[84]

As noted, EPA, states, and the courts have considerable discretion in determining sanctions and remedies on a case-by-case basis so that the individual circumstances of each case are appropriately addressed. A majority of environmental violations are addressed and resolved administratively by states and EPA, and many of these cases are settled through negotiations between the government and the alleged violator. For example, during FY2013, EPA issued 873 administrative compliance orders and filed 1,440 final administrative penalty order complaints. In comparison, also during FY2013, 137 civil judicial cases were filed with the court, and 176 civil judicial enforcement cases were concluded.[85] Civil judicial cases constitute the second-largest category of environmental enforcement actions. Historically, judicial actions focused on violation of a single environmental statute. In recent years, EPA and states have increased the frequency of reliance on a multimedia (multi-statute) approach and multimedia investigations.

The number of administrative and judicial enforcement actions and penalties often fluctuate significantly from year to year. These fluctuations are generally a reflection of a combination of factors, including statutory deadlines; new or amended requirements in response to new scientific information or amended and new regulations; increased or decreased resources; environmental priority changes at the federal or state levels; and increased or improved monitoring/reporting. For example, EPA reported that the number of administrative penalty order complaints issued by the agency more than doubled, from 2,229 complaints in FY2005 to 4,647 in FY2006, then declined to 2,237 in FY2007, 2,056 in FY2008, 1,914 in FY2009, 1,901 in FY2010, and 1,735 in FY2011.[86] The number of administrative penalty order complaints increased to 1,760 in FY2012,[87] then declined to 1,407 in FY2013.

EPA reported that the combined $1.15 billion in civil penalties (administrative and judicial) assessed in FY2013 was an all-time high, due to a record settlement of $1.0

---

[83] When "... persons willfully or knowingly disregard the law."

[84] See the discussion later in this report on Supplemental Environmental Projects (SEPs).

[85] See EPA, *Enforcement Annual Results Numbers at a Glance for Fiscal Year 2013*, http://www2.epa.gov/enforcement/enforcement-annual-results-numbers-glance-fiscal-year-fy-2013.

[86] See EPA-OECA, Annual Results for various fiscal years at http://www.epa.gov/compliance/data/results/annual/index.html.

[87] See EPA, *Enforcement Annual Results for Fiscal Year 2012*, http://www2.epa.gov/ enforcement/enforcement-annualresults-fiscal-year-2012.

billion reached with the Transocean defendants[88] to resolve alleged violations associated with the *Deepwater Horizon* Gulf of Mexico oil spill.[89] Additionally, the total dollar amount of penalties collected in a given year could reflect the completion of one or two large cases. For example, the settlement with the Transocean entities represented nearly 90% of the civil penalties assessed for FY2013. Illustrations of the frequency of enforcement actions by type over time are presented in *Appendix B*; this appendix also includes illustrations of administrative, judicial, and criminal penalties assessed over time by statute.

Monitoring, Inspections, and Evaluations

Critical steps in enforcing environmental laws include the compilation of monitoring data, and inspection and evaluation of the activities of the regulated community to determine who is complying with applicable regulatory requirements and permit conditions, and who is not. Compliance monitoring, evaluations, and investigations all serve to identify violations and provide insights into potential priority issue areas that may need to be addressed more broadly. Monitoring and reporting can be both media program-based (e.g., air, water, waste) and sector-based (e.g., industrial, mobile source, utilities), and are often included in permit requirements. Data reported and obtained, as well as observations and evidence collected by inspectors, enable EPA and states to identify specific environmental problems and determine whether a facility is in compliance. The information and evidence could eventually be used in an enforcement action. The mere collection of information or threat of inspection itself often creates an awareness of the regulators' interest, and can encourage compliance.

EPA identifies several forms of compliance monitoring that are used differently by the agency and states, depending upon the statute, the nature of the pollutants, and the types of facilities being regulated:

- *Self-Monitoring/Reporting*. Most environmental laws require (typically through permitting) regulated entities/facilities to monitor and record their own compliance status and report some or all of the tracking results to the responsible regulating authority. In addition to informing the regulators, self-monitoring also

[88] The Transocean corporate entities covered by the settlement are Transocean Deepwater Inc., Transocean Offshore Deepwater Drilling Inc., Transocean Holdings LLC, and Triton Asset Leasing GmbH.

[89] See EPA, *Enforcement Annual Results Analysis and Trends for Fiscal Year (FY) 2013*, http://www2.epa.gov/enforcement/enforcement-annual-results-analysis-and-trends-fiscal-year-fy-2013, and EPA, *Enforcement: Deepwater Horizon—BP Gulf of Mexico Oil Spill*, http://www2.epa.gov/enforcement/deepwater-horizon-bp-gulf-mexico-oil-spill. The *2010 Gulf Oil Spill* is also referred to as the *Deepwater Horizon Spill*, the *BP Spill*, and the *Macondo Well Blowout*. This oil release occurred during the summer of 2010, and was the largest oil spill in U.S. history. See also CRS Report R42942, *Deepwater Horizon Oil Spill: Recent Activities and Ongoing Developments*, by Jonathan L. Ramseur and Curry L. Hagerty.

allows a company to measure its performance and evaluate its strategies for achieving or maintaining compliance.

- *Review of Records.* Regulatory agencies review data and information reported or otherwise compiled and collected.
- *Full and Partial Inspections/Evaluations.*[90] Individual facility environmental inspections, conducted by EPA regional staff and the states, are the primary tool used by regulators for initial assessment of compliance. Through sampling, emissions testing, and other measures, inspections examine environmental conditions at a facility to determine compliance (or noncompliance) with specific environmental requirements, and to determine whether conditions present imminent and substantial endangerment to human health and the environment. Inspections/evaluations can be conducted all at once or in a series of partial inspections.
- *Area Monitoring.* Area monitoring looks at environmental conditions in the vicinity of a facility, or across a certain geographic area. Examples of methods used for area monitoring include ambient monitoring and remote sensing.

According to EPA's most recent reported trends data, a total of 18,000 EPA enforcement inspections and evaluations were conducted under the various statutes during FY2013.[91] Although most inspections are carried out by the states, annual data for the total number of inspections conducted by states are not readily available due to data-reporting variability and other limitations. Based on a subset of states surveyed, ECOS reported that roughly 136,000 compliance inspections were conducted by states in 2003 for the major federal environmental programs—air, drinking water, surface and groundwater, hazardous waste, and solid waste.[92] The total number of inspections reported by ECOS does not account for all inspections conducted by states under federal pollution control programs—for example, inspections under FIFRA are not included. In reports to EPA by states under the Pesticide Enforcement Grant program, states, tribes, and territories reported between 90,000 and 100,000 FIFRA inspections each fiscal year for FY2006 through FY2008. These FIFRA activities, typically administered by states' departments of agriculture, are not reflected in the EPA or the ECOS totals.

To put the ECOS number of inspections into perspective, in 2003, the ECOS survey identified 440,000 regulated facilities under these five major environmental programs. EPA's Facility Registry System (FRS), which identifies facilities and sites subject to federal environmental regulation, currently contains unique records for more than 2.5 million facilities (see the above discussion under the heading Regulated Community).

---

[90] The CWA requires evaluations instead of investigations, which include reviewing reports, records, and operating logs; assessing air pollution control devices and operations; and observing visible emissions or conducting stack tests.

[91] See footnote 82.

[92] The Environmental Council of the States (ECOS), "State Environmental Agency Contributions to Enforcement and Compliance: 2000-2003," June 13, 2006.

Appendix B presents data on the number of inspections conducted annually by EPA over time.

Civil Administrative Actions

As noted earlier, a majority of environmental pollution control violations are addressed and resolved administratively by states and EPA without involving a judicial process. EPA or a state environmental regulatory agency may informally communicate to a regulated entity that there is an environmental problem, or it may initiate a formal administrative action in the form of a notice of violation or an Administrative Order to obtain compliance. An Administrative Order imposes legally enforceable requirements for achieving compliance, generally within a specified time frame, and may or may not include sanctions and penalties.

An initial step in the enforcement process is often a Notice of Violation, or in some instances, a warning letter. Warning letters are issued mostly for first-time violations that do not present an imminent hazard. These notifications are intended to encourage regulated entities to correct existing problems themselves and come into compliance as quickly as possible. According to EPA, in many cases, these notices are not escalated to further formal enforcement action because a facility corrects problems and returns to compliance in response to the notice.

Through administrative enforcement actions, EPA and states may (1) require that the violator take specific actions to comply with federal environmental standards, (2) revoke the violator's permit to discharge, and/or (3) assess a penalty for noncompliance. As indicated previously, administrative actions frequently end in negotiated settlements. These mutually agreed-upon resolutions are typically in the form of a Consent Agreement or Final Administrative Order/Penalty. According to EPA's reported FY2013 annual results, during FY2013, EPA initiated 873 administrative compliance orders and 1,407 administrative penalty order complaints. EPA imposed penalties in 1,440 final administrative penalty orders during FY2013, representing a total value of $48.0 million.[93]

Federal administrative orders are handled through an administrative adjudicatory process, filed before an administrative law judge (ALJ), or, in the regions, by EPA's regional judicial officers (RJOs). The EPA Office of Administrative Law Judges (OALJ) is an independent office within the agency.[94] ALJs, appointed by the EPA Administrator,[95] perform adjudicatory functions and render decisions in proceedings between EPA and individuals, entities, federal and state agencies, and others, with regard to administrative actions taken to enforce environmental laws and regulations. RJOs, designated by each of

---

[93] See footnote 82.
[94] Administrative Procedure Act, 5 U.S.C. §557.
[95] 5 U.S.C. §3105.

the EPA Regional Administrators,[96] perform similar adjudicatory functions in the EPA regions. Decisions issued by ALJs and RJOs are subject to review and appeal to the Environmental Appeals Board (EAB), which also functions independently of EPA.[97] Environmental Appeals Judges are appointed by the EPA Administrator.[98] Federal pollution control laws and regulations specify who may raise an issue before the EAB, and under what circumstances. EAB decisions often involve reviews of the terms of federal environmental permits and the amount of assessed financial administrative penalties.

Civil Judicial Enforcement

After civil administrative enforcement actions, civil judicial cases constitute the next-largest category of environmental enforcement. These are lawsuits filed in court against persons or entities who allegedly have not complied with statutory or regulatory requirements, or, in some cases, with an Administrative Order. Authorities for pursuing civil judicial actions and penalties are specified in each of the individual environmental statutes. Civil judicial cases are brought in federal district court by DOJ on behalf of EPA, and, for the states, by State Attorneys General. Not all of the cases referred to DOJ are filed with the court. The length of a civil case from its initiation to completion is highly variable, often extending across several years and sometimes across different presidential administrations. Like administrative enforcement actions, many civil judicial actions end as negotiated settlements, typically in the form of Consent Decrees. During FY2013, EPA-OECA referred 138 civil judicial cases to DOJ; 137 civil judicial complaints were filed with the court; and 176 cases were concluded (cases filed prior to and during FY2013).[99]

Criminal Judicial Enforcement

States and EPA may initiate criminal enforcement actions against individuals or entities for negligent or knowing violations of federal pollution control law. Criminal actions are especially pursued when a defendant knew, or should have known, that injury or harm would result. Knowing criminal violations of pollution control requirements are considered deliberate, and not the result of accident or error.

---

[96] Title 40 C.F.R. Part 22 Subpart A §22.4(b).
[97] Title 40 C.F.R. Part 22 Subpart A §22.8.
[98] Title 40 C.F.R. §1.25(e). EAB judges are Senior Executive Service ("SES")-level career agency attorneys (U.S. EPA, *The Environmental Appeals Board Practice Manual*, June 2004, available at http://www.epa.gov/eab/pmanual.pdf).
[99] See footnote 82.

In addition to the imposition of monetary fines and requirements to correct a violation and restore damages, conviction of a criminal environmental violation can result in imprisonment. EPA reported that 297 new environmental crime cases were opened during FY2013,[100] roughly 7% fewer than the 320 criminal cases opened in FY2012.[101] Authorities for pursuit of criminal actions vary under each of the statutes. For example, under the SDWA (42 U.S.C. §300h-2(b)), the criminal violations must be deemed willful—that is, they were committed with intent to do something prohibited by that law; the CWA (33 U.S.C. §1319(c)) authorizes criminal sanctions against those who have knowingly or negligently violated that statute.

Recent examples of criminal actions include the illegal disposal of hazardous waste; importation of certain banned, restricted, or regulated chemicals; the export of hazardous waste without prior notification or permission of the receiving country; the removal and disposal of regulated asbestos-containing materials inconsistent with requirements of the law and regulations; tampering with a drinking water supply; and negligent maintenance resulting in discharge of hazardous materials.[102] EPA maintains a database, *Summary of Criminal Prosecutions Database*, which provides the capability to conduct searches for information on concluded criminal enforcement cases by fiscal year.[103]

The EPA-OECA Office of Criminal Enforcement, Forensics, and Training (OCEFT), the office to which the agency's criminal investigators are primarily assigned, oversees implementation of the agency's federal environmental crimes investigation program.[104] Within DOJ, the U.S. Attorneys Offices and ENRD's Environmental Crimes Section (ECS) prosecute criminal cases and work closely with EPA's OCEFT investigators.

State and local law enforcement agencies and their environmental protection-related agencies, and other federal agencies, are also often key participants in federal environmental criminal actions. To facilitate investigations and cases, environmental crime task forces have been established nationally.[105] These task forces are composed of representatives from federal (including representatives from DOJ-ECS and special agents from EPA), state, and local law enforcement, and environmental regulatory enforcement. The FBI, DOT, Coast Guard, Fish and Wildlife Service, Army Corps of Engineers, SEC, IRS, and other relevant federal agencies also may play significant roles.

An increased emphasis on criminal enforcement of the pollution control laws occurred in the mid-1970s with the issuance of extensive guidelines for proceeding in criminal cases, and in 1981 with the creation of an Office of Criminal Enforcement and

---

[100] See footnote 82.

[101] See footnote 84.

[102] For major criminal case highlights for FY2013, see http://www2.epa.gov/enforcement/2013-major-criminal-cases. See also *EPA's Criminal Case Activities* website at http://www2. epa.gov/enforcement/criminal-investigations.

[103] EPA-OECA, *Summary of Criminal Prosecutions Database*, http://www2.epa.gov/ enforcement/summary-criminalprosecutions.

[104] See *EPA OCEFT At a Glance*, http://www2.epa.gov/sites/production/files/2013-11/documents/oceft-at-a-glance_0.pdf.

[105] For information regarding EPA's Criminal Investigations and associated Environmental Crimes Task Force Partners nationwide, see http://www2.epa.gov/enforcement/criminal-investigations.

the hiring of criminal investigators in EPA's regional offices. During the late 1980s, criminal environmental enforcement was further enhanced when Congress conferred full law enforcement powers upon EPA criminal investigators as part of the Medical Waste Tracking Act of 1988 (18 U.S.C. §3063). Further, under Title II of the Pollution Prosecution Act of 1990 (P.L. 101-593), Congress authorized the appointment of a director of a new Office of Criminal Investigations within EPA, and mandated the hiring of 200 criminal investigators by FY1996.[106]

Staffing levels of criminal investigators at EPA have been of interest to some Members, particularly during deliberations on appropriations. For example, the 1990 Pollution Prosecution Act (P.L. 101-593) required EPA to hire and maintain 200 criminal investigators. A provision in the House-passed FY2008 Interior and Environmental Agencies Appropriations bill (H.R. 2643) would have required EPA to bring the total number of investigators up to the level of 200 as statutorily required.[107] The provision was not included in the FY2008 appropriations (P.L. 110- 161, Title II of Division F). Congressional concerns regarding staff and funding for EPA's criminal (and civil) enforcement were also expressed in conference report language accompanying EPA appropriations for FY2003 through FY2005,[108] and were the topic of a congressionally requested EPA-OIG investigation.[109]

EPA's criminal enforcement agents are authorized law enforcement officers who, in addition to investigating federal environmental statutes, investigate U.S. Criminal Code (Title 18) violations often associated with environmental crimes, such as conspiracy, false statements, and interfering with federal investigations. As noted, Congress has been concerned with the staffing of criminal investigators.

Table 3 below shows the number of EPA investigators assigned to the criminal enforcement program for FY1997 through FY2015 (projected), as reported by EPA.[110] The number of EPA special agents increased from about 50 in 1990 to more than 200 by 1998.[111] As of September 2007, the number of EPA investigators had dropped to 168; this decline was an issue of concern in Congress and elsewhere. As a result of an EPA-

---

[106] For a historical perspective of EPA's criminal enforcement, see *Management Review of the Office of Criminal Enforcement, Forensics, and Training*, December 2003, http://www2.epa.gov/enforcement/report-management-reviewoffice-criminal-enforcement-forensics-and-training.

[107] Title VI—Additional General Provisions §605 of H.R. 2643 (June 28, 2007, placed on the Calendar of the Senate). See also H.Amdt. 441 to H.R. 2643, offered and agreed to, June 26, 2007, inserted under Title VI—Additional General Provisions §601 (CR H7221-H7222). The FY2008 omnibus appropriations (P.L. 110-161) did not include this general provision (Joint Explanatory Statement Accompanying Division F of the Consolidated Appropriations Act for FY2008 (P.L. 110-161, H.R. 2764), as presented in the *Congressional Record*, December 17, 2007, p. H16142).

[108] Conference Report (H.Rept. 108-792, p. 1563) accompanying the Consolidated Appropriations Act, 2005, P.L. 108- 447; Conference Report (H.Rept. 108-401, p. 1126) accompanying the Consolidated Appropriations Act, 2004, P.L. 108-199; and Conference Report (H.Rept. 108-10, p. 1445) accompanying Consolidated Appropriations Resolution, 2003, P.L. 108-7.

[109] See footnote 9.

[110] Based on information provided directly to CRS by EPA's Office of Congressional and Intergovernmental Relations, October 31, 2007, March 26, 2008, April 4, 2010, August 15, 2011, June 20, 2012, and March 27, 2014, and EPA's *Compliance and Enforcement Annual Results for 2010: Criminal Enforcement*, http://www.epa.gov/compliance/resources/ reports/endofyear/eoy2010/criminal/index.html#agents.

[111] See footnote 103.

OECA three-year hiring strategy to increase the number of criminal investigators, the number of investigators increased to 183 in FY2008, 186 in FY2009, and 206 in FY2010. The total number of criminal investigators declined to 202 in FY2011 and EPA expected the current 192 inspectors on board to be 191 by the end of FY2012. In recent years, there has been an across-the-board reduction in FTEs at EPA, and all programs—including the criminal program in OECA—have been affected. As of September 30, 2013, EPA had 177 special agents working on environmental crime investigations.[112]

In FY2009, EPA opened the "EPA Fugitives" website, which provides photographs and information about alleged violations of individuals who have avoided prosecution for allegedly committing environmental crimes.[113]

Table 3. Number of EPA Criminal Investigators: FY1997-FY2013

Fiscal Year	Number of Investigators
1997	200
1998	200
1999	192
2000	179
2001	181
2002	217
2003	217
2004	202
2005	189
2006	183
2007	168
2008	183
2009	186
2010	206
2011	202
2012	191
2013	177

Source: Prepared by CRS with data provided by EPA's Office of Congressional and Intergovernmental Relations, October 31, 2007, March 26, 2008, April 4, 2010, August 15, 2011, June 20, 2012, and March 27, 2014, and *EPA's Compliance and Enforcement Annual Results for 2010: Criminal Enforcement*, http://www.epa.gov/compliance/resources/ reports/endofyear/eoy2010/criminal/index.html#agents.

Note: The number of agents listed for 2002-2004 includes investigators who performed a mix of environmental crimes and homeland security work. The number of agents in FY2005 and FY2006 includes a cumulative total of 1.5 FTEs for specialized national security event response training of 25 agents so they can be deployed, if needed, to respond to a national security event.

The site solicits assistance from the public and from law enforcement agencies to help locate identified environmental "fugitives," and provides guidance on how to report

[112] See footnote 107.

[113] "EPA Fugitives," http://www2.epa.gov/enforcement/epa-fugitives.

information related to the fugitives' identity and/or current location. EPA reported that creditable reports received through the website assisted in the arrest or capture of three fugitives and the surrender of two others during FY2009; four were sentenced and one awaits trial.[114] Comparable information regarding "EPA Fugitives" was not reported by EPA in its FY2010, FY2011, FY2012, or FY2013 annual results;[115] however, the agency indicated that information obtained through the "Report a Violation" website[116] during FY2010 contributed to the opening of seven active criminal investigations, one of which resulted in an indictment.[117] EPA did not report any similar direct correlations with the "Report a Violation" website in its FY2011, FY2012, and FY2013 annual results.

Sanctions[118] and Penalties

Sanctions and penalties imposed for violation of federal environmental pollution control laws have been of interest to some in Congress and the topic of a 2010 EPA-OIG report.[119] Settlements often require that violators achieve compliance and remedy environmental damages (injunctive relief).[120] Monetary penalties may be included. Sanctions can also include permanent or temporary closure of facilities or specific operations, increased monitoring/reporting, revocation of existing permits or denial of future permits, and barring of receipt of federal contract funding or other federal assistance.[121] The settlement-required corrective and compliance actions, and the monetary penalties (and possibly incarceration for criminal violations), are intended to correspond directly with the specific violations (noncompliance) and the extent (or "gravity") of action committed.

Monetary penalties collected by the federal government as a result of an environmental enforcement agreement, order, or decision, are deposited with the U.S. Treasury.[122] However, under CERCLA (Superfund) and CWA, money recovered for the costs of replacing or restoring natural resources is used to restore the resources.[123] In the November

---

[114] "EPA Fugitive Highlights," http://www.epa.gov/compliance/resources/reports/endofyear/ eoy2009/ 2009criminalfugitives.html.

[115] For FY2013, see footnote 82; for FY2010, FY2011, and FY2012, see EPA-OECA *Data, Planning and Results: Annual Results* at http://www.epa.gov/compliance/data/results/ annual/index.html.

[116] See discussion under the "Citizens" heading earlier in this report.

[117] EPA, *Compliance and Enforcement Annual Results for FY2010: Criminal Enforcement*, http://www.epa.gov/compliance/resources/reports/endofyear/eoy2010/criminal/index.html# engage.

[118] In the context of this report, "sanctions" refer to adverse consequences imposed in response to noncompliance.

[119] EPA-OIG, *EPA Needs to Improve Its Recording and Reporting of Fines and Penalties*, Report No. 10-P-0077, March 9, 2010, http://www.epa.gov/oig/reports/enforcement.htm.

[120] For a brief description of injunctive relief, see http://www.epa.gov/region5/enforcement/ir.html.

[121] 2 C.F.R. Part 180, 2 C.F.R. Part 1532, and Executive Order 12549. Debarment is also authorized following criminal conviction under Clean Water Act (§508) and Clean Air Act (§306). See information regarding EPA's suspension and debarment program at http://www.epa.gov/ogd/sdd/debarment.htm.

[122] Under the Miscellaneous Receipts Act (31 U.S.C. §3302(b)), all court, or administratively imposed penalties must be paid by the government official receiving the monies to the U.S. Treasury.

[123] See, for example, 33 U.S.C. §1321(f)(5).

6, 2013, *Federal Register*,[124] EPA published its modifications to civil violation penalties in a final rule revising statutory penalties at 40 C.F.R. Parts 19 and 27. The final rule increased most of the civil penalties by adjusting for inflation under the Debt Collection Improvement Act of 1996,[125] which requires federal agencies to adjust civil monetary penalties for inflation every four years.[126] Adjusted civil penalty amounts for the applicable administered statutes are identified in Table 1 in 40 C.F.R. 19.4, as presented in the November 6, 2013, FR Notice.[127] These amounts apply to violations that occur after December 6, 2013. In addition to adjustments to the civil penalty amounts for the various applicable statutes, a technical revision in the November 2013 FR Notice breaks out the maximum civil penalty amounts imposed under Section 325(b) of the Emergency Planning and Community Right-to-Know Act (EPCRA; 42 U.S.C. 11045(b)). Previously grouped together, the maximum amounts under the three penalty authorities contained in Subsection (b) of EPCRA are now presented separately in Table 1 of 40 C.F.R. 19.4.[128]

States may have the explicit administrative authority to impose penalties under individual federal statutes. For example, the Safe Drinking Water Act requires (unless prohibited by a state's constitution) administrative penalty authority for states in certain dollar amounts as a condition of obtaining and/or retaining primacy for the Public Water System Supervision (PWSS) Program (§1413 (a)(6)).[129] As of February 2014, 55 of 57 states (49) and territories had primacy authority for the PWSS program.[130] Although authorized under several of the other federal pollution control laws, EPA has not required—and not all states have obtained—administrative penalty authority. In some states, unlike the federal government, penalties obtained (or shared) as a result of an environmental enforcement action can be used to directly fund activities for environmental agencies and programs in the state, and not always to fund the state's general treasury.

In certain cases where the federal government has led the enforcement action, a state or states involved in the action may "share" resulting civil monetary penalties to the extent that the division is permitted by federal, state, and local law. A number of critical factors must be considered in accordance with EPA guidance[131] when determining division of penalties, including the state's active participation in prosecuting the case and

---

[124] U.S. EPA, "Civil Monetary Penalty Inflation Adjustment (Final) Rule," 78 *Federal Register* 66643-66648, November 6, 2013, http://www.gpo.gov/fdsys/pkg/FR-2013-11-06/pdf/2013-26648.pdf.

[125] Section 4 of the Federal Civil Penalties Inflation Adjustment Act of 1990, 28 U.S.C. 2461 note, as amended by the Debt Collection Improvement Act of 1996 (DCIA), 31 U.S.C. 3701, requires each federal agency to adjust statutory civil monetary penalties for inflation every four years under the laws administered by that agency.

[126] EPA's initial inflationary adjustments were published in the December 31, 1996, 61 *Federal Register* 69360; subsequent adjustments were published in 69 *Federal Register* 7121, February 13, 2014, and 73 *Federal Register* 75340, December 11, 2008 (as corrected 74 *Federal Register* 626, January 7, 2009). EPA intends to review and adjust statutory civil penalties for inflation as necessary in 2017.

[127] See Table 1 in U.S. EPA, "Civil Monetary Penalty Inflation Adjustment (Final) Rule," 78 *Federal Register* 66647- 66648, November 6, 2013, http://www.gpo.gov/fdsys/pkg/FR-2013-11-06/pdf/2013-26648.pdf.

[128] U.S. EPA, "Civil Monetary Penalty Inflation Adjustment (Final) Rule," 78 *Federal Register* 66645, November 6, 2013, http://www.gpo.gov/fdsys/pkg/FR-2013-11-06/pdf/2013-26648.pdf.

[129] SDWA, 42 U.S.C. §300g-2(a)(6), added by SDWA Amendments of 1996 (P.L. 104-182).

[130] SDWA, 42 U.S.C. §300g-2(a)(6), added by SDWA Amendments of 1996 (P.L. 104-182).

[131] EPA guidance for the "Division of Penalties with State and Local Governments," memorandum from Courtney M. Price, Assistant Administrator for Enforcement and Compliance Monitoring, October 30, 1985.

its authority to collect civil penalties. EPA's guidance emphasizes that an agreement to include a division of civil penalties with states must be completed prior to issuance of a final settlement (order or consent decree).

The several statutes establish various factors to be considered in determining penalties: (1) the magnitude of environmental harm and the seriousness or gravity of a violation; (2) the economic benefit or gain to the violator as a result of illegal activity (noncompliance), including the gaining of a competitive advantage by the delaying or avoidance of pollution control expenditures that have been incurred by those in compliance; (3) violation history of the violator; and (4) in some circumstances, the ability of the violator to pay. Other factors, such as the degree of cooperation by the violator, whether the violation is self-reported, or the extent to which immediate action has been taken by the violator to mitigate potential harm, may also be considered. Precedents in previous cases involving similar violations are also a consideration when determining penalties.

The federal pollution control statutes include civil administrative and judicial penalty assessment authority and limits, which are to be considered by ALJs or the courts in determining the appropriate penalty. *Figure B-3* in *Appendix B* presents examples of dollar amounts of civil administrative, civil judicial, and criminal penalties assessed by EPA for the 20-year period FY1993-FY2013. According to EPA, a significant portion of the total annual dollar amount of all penalties assessed often reflects penalties assessed in a few cases, and, in some years, a single case. For example, EPA reported that a major RCRA case accounted for 26% of the total value of civil penalties reported in FY2006, and that, in FY2005, penalties assessed in a single RCRA corrective action case accounted for 53% of the reported assessed civil penalties for the year.[132] For FY2013 EPA reported that the settlement with Transocean defendants[133] accounted for $1.0 billion, nearly 90%, of the total reported assessed civil penalties.[134] Nearly $1.4 billion of $1.5 billion in criminal fines and restitution assessed, and the majority of the $3.0 billion in court-ordered projects for FY2013, were attributed to the *Deepwater Horizon* Gulf of Mexico FY2013 criminal case, according to EPA's FY2013 annual results report.[135]

EPA and DOJ have established several policies and guidelines to be considered by counsel when negotiating agreements and setting penalties.[136] EPA-OECA has also developed five computer models for calculating economic advantage, costs of Supplemental Environmental Projects (SEPs; see discussion under "Supplemental Environmental Projects (SEPs)," below), and for measuring the ability to afford

---

[132] EPA, Office of Enforcement and Compliance Assurance, *FY2006 OECA Accomplishments Report*, Spring 2007, EPA-300-R-07-001, http://cfpub.epa.gov/compliance/resources/ reports/accomplishment/details.cfm.

[133] See footnote 85.

[134] See footnote 86.

[135] See footnote 86.

[136] For example, EPA's revised *Consolidated Rules of Practice* ("CROP") (64 FR 40138, July 23, 1999) contains procedural rules for the administrative assessment of civil penalties, issuance of compliance or corrective action orders, and the revocation, termination or suspension of permits, under most environmental statutes, at http://www.epa.gov/ enforcement/waste/ documents/policies/final-crop-fr.pdf. See EPA Enforcement web page, *Policy, Guidance & Publications,* http://www.epa.gov/enforcement/documents/index.html.

compliance requirements and penalties.[137] The latter models vary depending on whether a violator is an individual, municipality, individual facility, or business entity (small business, large corporation, or conglomerate partnership). Findings of limited ability or inability to pay are one factor under which an enforcement case may be settled for less than the economic benefit of noncompliance. The models are to be used in conjunction with the policies and guidelines for calculating civil penalties.[138] The available models are

- ABEL, for measuring a noncompliant entity's (e.g., a corporation's) ability to afford compliance costs, cleanup costs, and civil penalties;
- BEN, for calculating economic advantage/savings from delaying or avoidance of compliance and pollution control expenditures;
- INDIPAY, for measuring an individual violator's ability to afford compliance costs, cleanup costs, and civil penalties;
- MUNIPAY, for measuring a noncompliant municipality's ability to afford compliance and cleanup, and civil penalties; and
- PROJECT, for calculating cost to a violator of undertaking a SEP (see the discussion regarding SEPs later this chapter).

Penalties Assessed to Federal Facilities[139]

Most federal pollution control statutes contain a provision expressly subjecting federal facilities to federal (and state and local) environmental regulation, and waiving sovereign immunity (thereby allowing federal agencies to be sued by nonfederal entities). Further, many federal environmental statutes authorize (or arguably authorize) EPA, states, and local governments to assess civil monetary penalties against federal agencies. (The Supreme Court rejected state authority to do so under the CWA.)[140] DOJ has issued opinions concluding that the CAA and RCRA underground storage tank provisions give EPA authority to assess civil money penalties against federal facilities. However, DOJ limits these conclusions to *administrative* assessment of penalties. Citing its constitutional theory of the "unitary executive," DOJ has historically refused to allow EPA to enforce *judicially* against other federal agencies, though case law has consistently been to the contrary. In contrast with EPA enforcement, there is no longer serious doubt that the Constitution allows states and other nonfederal entities to use the citizen suit

---

[137] U.S. EPA, Penalty and Financial Models, http://www2.epa.gov/enforcement/penalty-and-financial-models.

[138] Additionally, see EPA, *Guidance on Calculating the Economic Benefit of Noncompliance by Federal Agencies*, issued on February 13, 2006, Memorandum from Granta Y. Nakayama, Assistant Administrator, Office of Enforcement Compliance and Assurance, http:// www.epa.gov/Compliance/resources/policies/federalfacilities/enforcement/cleanup/guid-econ-ben-noncomp-2-13-06.pdf.

[139] Prepared initially by Robert Meltz, Legislative Attorney, American Law Division.

[140] *U.S. Dept. Of Energy v. Ohio*, 503 U.S. 607 (1992).

provisions in federal environmental statutes to judicially enforce those laws against federal facilities.

During FY2013, EPA concluded 61 enforcement actions against federal facilities, and assessed nearly $866.5 million in penalties.[141] Federal agencies committed to invest more than $128.0 million in FY2013 in injunctive relief to improve their facilities and operations to remedy (clean up) past violations, to comply with federal laws, and to prevent future violations/pollution. Agencies also committed to an additional $5.4 million in the form of supplemental environmental projects (see discussion in the following section) during FY2013. As indicated earlier in this chapter, EPA no longer provides aggregated summaries of federal facility enforcement actions in its annual results report. However, the agency provides an interactive mapping tool of concluded enforcement cases that includes the capability to search information for individual federal facilities based on location.[142]

Supplemental Environmental Projects (SEPs)

In addition to requiring violators to achieve and maintain compliance, and imposing appropriate sanctions and penalties, enforcement settlements may also include Supplement Environmental Projects (SEPs).[143] SEPs are projects that provide environmental and human health benefits that a violator may voluntarily agree to undertake in exchange for mitigation of penalties. A project must be related to the violation, and cannot be an activity the violator is legally required to take to achieve compliance. Penalties are to be mitigated by a SEP only during settlement negotiation, prior to imposition of the final penalty.

EPA has established a SEPs policy and developed guidance for their legal requirements and applicability, and has specified eight categories of acceptable projects.[144] These include pollution prevention, public health, and emergency and preparedness planning. EPA reported that 110 civil settlement cases during FY2013 included SEPs at an estimated value of $22.0 million, compared to 124 cases valued at $43.6 million during FY2012.[145]

---

[141] Based on information provided upon request directly to CRS by EPA's Office of Congressional and Intergovernmental Relations (OCIR), March 18, 2014.

[142] See EPA-OECA, *Enforcement Annual Results for Fiscal Year 2013: Concluded EPA Enforcement Cases Map*, http://www2.epa.gov/enforcement/enforcement-annual-results-concluded-cases-map-fiscal-year-fy-2013.

[143] For more information regarding SEPs, see http://www2.epa.gov/enforcement/supplemental-environmental-projectsseps.

[144] Policy and guidance documents related to EPA's Supplemental Environmental Projects Policy are available at http://www.epa.gov/enforcement/sep.html#policy.

[145] EPA-OECA, *Fiscal Year 2013 EPA Enforcement & Compliance Annual Results*, February 7, 2014, http://www2.epa.gov/enforcement/enforcement-annual-results-numbers-glance-fiscal-year-fy-2013, and *Fiscal Year 2012 EPA Enforcement & Compliance Annual Results*, December 17, 2012, http://www.epa.gov/enforcement/data/ eoy2012/fy2012annualresults-analysistrends.pdf#page=7.

The incorporation of SEPs into enforcement actions became more common during the last decade, particularly by federal regulators, because of the potential for direct environmental benefit from such projects, versus the use of a monetary fine or penalty alone. Some states with administrative penalty authority have also employed the use of SEPs in their settlements.[146] Although these projects are required to be supplemental to other requirements, some contend that, in practice, inclusion of SEPs may result in lower monetary fines. The extent to which specific SEPs may have resulted in reduced monetary fines and penalties is not easily calculable.

Environmental Justice[147] and Enforcement/Compliance

Environmental justice (EJ) has been an area of debate among industry and public interest groups, and ongoing concern highlighted in congressional hearings and legislation. For example, an administrative provision included in the Omnibus Appropriations Act, 2009 (P.L. 111-8, Title II), specified that none of the funds made available by this act may be used in contravention of, or to delay the implementation of, Executive Order No. 12898 relating to federal actions to address environmental justice in minority populations and low-income populations.[148] On August 4, 2011, the White House announced the signing of a Memorandum of Understanding (MOU) on Environmental Justice and E.O. 12898 by the heads of 17 federal agencies.[149] The MOU included a Charter to add more structure and efficiency to the Federal Interagency Working Group on Environmental Justice; processes and procedures to more efficiently assist communities, and guidance for agencies to better coordinate their EJ activities; and various commitments each agency will be responsible for meeting. Discussion of the full scope of issues and concerns regarding environmental justice is beyond the scope of this chapter. However, the following discussion briefly highlights environmental justice in the context of enforcement and compliance.

The terms "environmental justice (or injustice)" and "environmental equity (or inequity)" may be interpreted broadly to describe the perceived level of fairness in the distribution of environmental quality across groups of people with different characteristics. In this sense, the environmental impact of any human activity might be evaluated to determine the distribution of environmental amenities and risks among

---

[146] The Environmental Council of States (ECOS), *State Environmental Agency Contributions to Enforcement and Compliance 2000-2003*, June 2006.

[147] See EPA, Environmental Justice, http://www.epa.gov/environmentaljustice/index.html.

[148] Executive Order 12898, 49 *Federal Register* 7629, February 16, 1994, http://www.archives.gov/federal-register/executive-orders/1994.html.

[149] Executive Office of the President, Council on Environmental Quality, August 4, 2011, press release, Memorandum of Understanding (MOU) on Environmental Justice and E.O. 12898, http://www.whitehouse.gov/administration/eop/ ceq/Press_Releases/August_04_2011. A copy of the MOU is available at http://epa.gov/environmentaljustice/ resources/publications/ interagency/ej-mou-2011-08.pdf See also EPA's "Federal Interagency Working Group on Environmental Justice," http://www.epa.gov/compliance/ej/ interagency/index.html.

people categorized according to any population characteristic, including gender, age, race, place of residence, occupation, income class, or language. In the political context, however, emphasis generally is more on the distribution of health risks resulting from exposure to toxic substances in residential or occupational environments of different racial, ethnic, or socioeconomic groups.

The 1994 Executive Order 12898, Federal Actions to Address Environmental Justice in Minority Populations and Low-Income Populations, directs each federal agency to "make achieving environmental justice part of its mission."[150] EPA is the federal agency with lead responsibility for implementing the executive order. EPA's Office of Environmental Justice (OEJ), located in OECA, is responsible for coordinating efforts to include environmental justice into policies and programs across the agency's headquarters and regional offices.[151] EPA's OEJ provides information and technical assistance to other federal agencies for integrating environmental justice into their missions, engages stakeholders to identify issues and opportunities, and administers EPA environmental justice grants.[152]

EPA incorporated Environmental Justice throughout its FY2014-FY2018 Strategic Plan.[153] EPA initiated implementation of "Plan EJ 2014," a strategy for expanding its efforts to integrate environmental justice into its various programs and strengthen "the Agency's effort to improve the environmental conditions and public health in overburdened communities."[154] EPA has issued annual reports documenting the progress toward meeting the commitments outlined in Plan EJ 2014 beginning with FY2013.[155]

Additionally, EPA released an "informational publication" entitled *Creating Equitable, Healthy, and Sustainable Communities: Strategies for Advancing Smart Growth, Environmental Justice, and Equitable Development*, in February 2013.[156] EPA's OEJ developed the Environmental Justice Strategic Enforcement Assessment Tool

---

[150] See footnote 145.

[151] EPA's Office of Civil Rights (OCR) is responsible for the agency's administration of Title VI of the Civil Rights Act of 1964, including processing and investigating administrative complaints under implementing regulations (40 C.F.R. Part 7) prohibiting EPA-funded permitting agencies from "... permitting actions that are intentionally discriminatory or have a discriminatory effect based on race, color, or national origin." See http://www.epa.gov/ civilrights/t6home.htm.

[152] EPA reports that since 1994, more than $23.0 million has been awarded to nearly 1,300 community-based organizations, and local and tribal organizations in the form of Environmental Justice Small Grants. For more information regarding EPA's Environmental Justice Program activities, see http://www.epa.gov/compliance/environmentaljustice/ index.html, for information on these grants and other environmental justice grants and cooperative agreements.

[153] Available at http://www2.epa.gov/planandbudget/fy-2014-2018-strategic-plan. EPA had previously incorporated Environmental Justice in its FY2011-FY2015 Strategic Plan as a Cross-Cutting Fundamental Strategy "Working for Environmental Justice and Children's Health." EPA FY2011-2015 Strategic Plan, http://www.epa.gov/planandbudget/ strategicplan.html.

[154] EPA Plan EJ 2014 was released September 2011; see http://www.epa.gov/compliance/environ mentaljustice/plan-ej/index.html.

[155] EPA, *Plan EJ 2014: Progress Reports*, http://www.epa.gov/compliance/ environmental justice/plan-ej/ index.html#reporting; and the "FY2014 Plan EJ 2014 Progress Report (PDF), http://www.epa.gov/compliance/ environmentaljustice/resources/policy/plan-ej-2014/plan-ej-progress-report-2013.pdf.

[156] Available at http://www.epa.gov/smartgrowth/equitable_development_report.htm.

(EJSEAT).[157] EJSEAT remains a draft tool in development for internal EPA use only. OECA expects to use EJSEAT to "consistently identify possible environmental justice areas of concern," where potentially disproportionately high and adverse environmental and public health burdens exist, and assist EPA in making "fair" enforcement and compliance resource deployment decisions.[158] OEJ published a "Toolkit for Assessing Potential Allegations of Environmental Injustice," primarily to assist agency staff in assessing allegations of environmental injustice.[159] Citizens can evaluate overlap between environmental conditions and demographic characteristics by using EPA's EJView (formerly Environmental Justice Geographic Assessment Tool).[160]

Compliance Assistance and Incentive Approaches

A frequent criticism regarding implementation and enforcement of federal environmental requirements has been an emphasis, historically, on a "command and control" approach. In response to these criticisms, since the 1990s EPA and states have relied increasingly on compliance assistance to help the regulated community understand its obligations to prevent violations and reduce the need for enforcement actions, as well as to assist violators in achieving compliance. Many states have advocated compliance assistance and developed assistance programs designed to address specific environmental issues at the local level.[161]

EPA's Office of Compliance (OC) within OECA has introduced a number of compliance assistance programs, many of them developed in conjunction with support from the regions, states, and tribes.[162] Each EPA region has a designated Compliance Assistance Coordinator who serves as an "expert" within the region on compliance assistance priorities, strategies, and performance measurement. The coordinators work with subject-matter experts in the regions and at headquarters in the development of compliance assistance guides and workshops, and contribute to other assistance activities such as conducting compliance assistance visits.[163]

In addition to providing compliance assistance across the individual pollution control statutes, sector-based assistance is also provided. Developed and introduced in partnership between EPA, states, academia, environmental groups, industry, and other

[157] EPA-OECA, "Environmental Justice Strategic Enforcement Assessment Tool (EJSEAT)," http://www.epa.gov/environmentaljustice/resources/policy/ej-seat.html.

[158] See footnote 154.

[159] EPA-OECA, Office of Environmental Justice, EPA 300-R-04-002, http://www.epa. gov/environmentaljustice/resources/policy/ej-toolkit.pdf.

[160] EPA-OECA, *EJView*, http://www.epa.gov/compliance/environmentaljustice/mapping.html.

[161] ECOS, State Agency Contributions to Enforcement and Compliance: 2000-2003, ECOS 06-004, June 2006, http://www.ecos.org/files/3895_file_ENF_Report_2006_Final_Doc.pdf.

[162] For more information regarding EPA's compliance assistance programs, see http://www.epa.gov/compliance/assistance/index.html.

[163] EPA-OECA, Compliance Assistance: Information for Compliance Assistance Providers, http://www.epa.gov/compliance/assistance/provider.html.

agencies, the National Compliance Assistance Centers provide sector-specific assistance.[164] There are currently 16 sector-specific web-based compliance assistance centers. As shown in *Table 4* below, the sector-specific centers include, among others, agriculture, auto repair, federal facilities, local governments, and transportation.

Table 4. Sector Web-Based Compliance Assistance Center

Sectors	
Agriculture	Federal Facilities
Automotive Recycling	Food Processing
Automotive Repair	Health Care
Beneficial Use	Local Government
Border Compliance	Metal Finishing
Chemical	Paints and Coatings
Colleges/Universities	Printing
Combustion Portal	Transportation
Construction	Transportation

Source: Table created by CRS with information from the National Compliance Center website, available at http://www.assistancecenters.net/.

The use of compliance incentive approaches has been evolving. Incentives generally are policies and programs that may reduce or waive penalties and sanctions under specific conditions for those who voluntarily take steps to evaluate, disclose, correct, and prevent noncompliance. Examples include self-disclosure programs and related tools such as environmental audit protocols, Environmental Management Systems, and other innovation projects and programs designed to achieve environmental benefits.

One of the earliest formal EPA incentive approaches is the EPA Audit Policy—"Incentives for Self-Policing: Discovery, Disclosure, Correction and Prevention of Violations"—in effect since 1995.[165] Under the policy, certain violations are voluntarily reported after being discovered through self-audit. In many cases EPA eliminates civil penalties, and may offer not to refer certain violations for criminal prosecution. In early 2007, EPA solicited comments on the question of to what extent, if any, the agency should consider providing incentives to encourage new owners of recently acquired facilities to discover and disclose environmental violations, and to correct or prevent their reoccurrence.[166]

To further promote compliance through the use of various incentive approaches, EPA encouraged incentive approaches as part of its core program guidance included in the

---

[164] For more information regarding the National Compliance Assistance Centers, see http://www.epa.gov/compliance/assistance/centers/index.html or http://www.assistance centers.net/.

[165] For more information regarding EPA's incentive programs and initiatives, see http://www. epa.gov/compliance/incentives/index.html.

[166] 72 *Federal Register* 27116, May 14, 2007.

OECA FY2012 and FY2013[167] National Program Manager Guidance. However, EPA also continues to evaluate its various voluntary programs.

In March 2009, for example, EPA notified participating stakeholders of its decision to discontinue the National Performance Track Program, which had been in place since July 2000.[168] Designed as a private-public partnership to supplement EPA's existing regulatory activities, Performance Track encouraged facilities who met certain criteria[169] to voluntarily work toward environmental goals that were beyond the legal requirements. EPA reported that at the time of its termination on May 14, 2009,[170] the program had a total membership of nearly 547 facilities (including 82 new members in 2008) in 49 states and Puerto Rico. In her March 2009 memorandum to Performance Track stakeholders,[171] former EPA Administrator Jackson announced her decision to halt the program "with the intent of refining those concepts that can lead us to a stronger system of environmental protection." EPA had conducted reviews of the Performance Track and the agency's environmental leadership programs in general.

EPA's reliance on incentive approaches has been met with some skepticism by those who favor more traditional enforcement. Critics are concerned that incentive and voluntary approaches subtract resources from an already limited pool of enforcement resources. EPA and other supporters of these approaches contend that they result in cost savings by reducing burdens on investigators, achieve desired environmental improvements, and allow for the leveraging of additional resources through partnerships. Aspects of EPA's incentive approaches have been the subject of reviews by EPA-OIG and GAO.[172]

---

[167] See footnote 26.

[168] Memorandum from Lisa P. Jackson, EPA Administrator, to Performance Track Members, State Program Contacts, and Network Partners, March 16, 2009; and Memorandum from Charles W. Kent, Director of Cross-Media Programs, EPA's Office of Policy, Economics, and Innovation, to Performance Track Members, State Program Contacts, and Network Partners, March 25, 2009, http://www.epa.gov/performancetrack/index.htm. See also 74 *Federal Register* 22742 (May 14, 2009), http://www.epa.gov/performancetrack/downloads/PTtermination_notice5-14- 09_74FR22741.pdf, and memorandum from Catherine R. McCabe, OECA Principal deputy Director, and Marcia E. Mulkey, Acting Associate Administrator, Office of Policy Planning and Compliance Assurance: Enforcement and Compliance Guidance on the termination of the National Environmental Performance Track Program, June 25, 2009, http://www.epa.gov/performancetrack/downloads/OECA-09-000-9690memo.pdf.

[169] According to EPA's Performance Track website, to be eligible for membership in the program, applicants must have "... implemented an independently assessed environmental management system, have a record of sustained compliance with environmental laws and regulations, commit to achieving measurable environmental results that go beyond compliance, and provide information to the local community on their environmental activities." Detailed criteria can be found at http://www.epa.gov/performancetrack/ (Note: As of October 2009, EPA no longer updates this information, but the link is still available as a reference or resource).

[170] See also *Performance Track Final Progress Report*, May 2009 on EPA's National Performance Track website, http://www.epa.gov/performancetrack/index.htm. (Note: as of October 20, 2009, the website is no longer being updated.)

[171] See footnote 165.

[172] EPA-OIG: Performance Track Could Improve Program Design and Management to Ensure Value, Report No. 2007-P-00013, March 29, 2007. GAO: Environmental Protection: Challenges Facing EPA's Efforts to Reinvent Environmental Regulation, RCED-97-155, July 2, 1997.

FUNDING FOR ENFORCEMENT/COMPLIANCE ACTIVITIES

The adequacy of resources needed by EPA, DOJ, and the states to effectively enforce the major federal environmental pollution control laws is often highlighted during congressional debate of fiscal year appropriations. Historically, Congress has specified funding levels for certain aspects of EPA enforcement activities, or required the agency to undertake certain actions under annual appropriations; an example is the previously mentioned provision included in the House-passed FY2008 Interior and Environmental Agencies Appropriations bill (H.R. 2643) that would have required EPA to hire criminal investigators to bring the total number of investigators up to the statutory requirement of 200, pursuant to the Pollution Prosecution Act of 1990.[173](The provision was not included in the FY2008 consolidated appropriations.)[174]

The President's FY2015 budget request included $583.0 million for EPA's enforcement program activities,[175] compared to $560.9 million enacted for FY2014, $553.1 million enacted (post-sequestration) for FY2013, and $583.4 million enacted for FY2012. As of the date of this chapter, no regular appropriations acts for FY2015 had been enacted. On September 19, 2014, President Obama signed into law the Continuing Appropriations Resolution, 2015 (P.L. 113-164, H.J.Res. 124). Section 101 of the act continues appropriations for federal departments and agencies generally at FY2014 enacted levels minus a 0.0554% rescission. The continuing resolution (CR) is authorized until December 11, 2014, or until the enactment of FY2015 appropriations. Funding for EPA under the CR is subject to the authority and conditions provided in the Interior, Environment, and Related Agencies Appropriations Act, 2014 (Division G, P.L. 113-76). Section 104 of the CR further states that continuing funding for all federal departments and agencies cannot be used to initiate or resume any project or activity that did not receive appropriations for FY2014.

Table 5, which follows, illustrates the distribution of funding and full-time equivalents (FTEs) among various enforcement activities across the agency's appropriations accounts for the FY2013 President's request and the three prior fiscal years enacted. Because of differences from fiscal year to fiscal year in the scope of the activities included within each of the accounts, apt direct comparisons below the appropriations account level are often difficult. These differences include the addition and discontinuation of program activities, as well as the reorganization and consolidation of certain activities.

---

[173] Title VI—Additional General Provisions, §605 of H.R. 2643 as placed on the Calendar of the Senate (June 28, 2007).

[174] Joint Explanatory Statement Accompanying Division F of the Consolidated Appropriations Act for FY2008 (P.L. 110-161, H.R. 2764), as presented in the *Congressional Record*, December 17, 2007 (p. H16142).

[175] Information as provided upon request directly to CRS by EPA's Office of Congressional and Intergovernmental Relations (OCIR), April 14, 2014. See also EPA's *FY2015 Justification of Appropriation Estimates for the Committee on Appropriations*, available at http://www2.epa.gov/planandbudget/fy2015.

DOJ's resource (funding/staff) requirements and outlays associated with its litigation activities under the major federal pollution control statutes are, in the main, a subset of the funding (proposed and previously appropriated) for the Environment and Natural Resources Division (ENRD) in its annual budget justifications. As discussed previously, ENRD is responsible for the majority of DOJ's support of the federal pollution control laws, as well as many other responsibilities, including representing the United States in matters regarding natural resources and public lands, acquisition of real property by eminent domain for the federal government, and cases under wildlife protection laws.

The President's FY2015 budget request for DOJ included $112.5 million for ENRD and an estimated 526 FTEs. An additional $21.8 million for up to 115 FTEs was included in the President's FY2015 request for EPA to be transferred by EPA to DOJ/ENRD through a reimbursable agreement for Superfund work. The FY2014 enacted levels for DOJ included $107.6 million and 526 FTEs for ENRD, plus an additional $23.1 million for up to 115 FTEs transferred from EPA for Superfund work. FY2012 enacted levels for ENRD were $108.0 million and 582 FTEs, plus $24.6 million for up to 115 FTEs transferred from EPA.

Of the FY2014 enacted amount, including the transfers from EPA, roughly $67.0 million and 312 FTEs were for environmental litigation activities; $9.9 million and 46 FTEs were for criminal litigation conducted by DOJ's Environmental Crimes Section; and $57.1 million and 266 FTEs were for civil environmental enforcement and defensive litigation conducted by the Environmental Enforcement and the Environmental Defense Sections.[176]

Detailed reporting of federal funding to states and states' funding contributions for pollution control enforcement/compliance activities is not readily available. ECOS has tracked a broader category of state funding and expenditures that it defines as annual "environmental and natural resource spending," which, in more recent years, has been primarily based on survey data reported by states.[174] The data, which include state and federal funding, are limited for purposes of enforcement of federal pollution control laws in that they combine environmental and natural resource spending. Also, states vary in how they track and report this type of spending. The data do provide a source of state funding from a national perspective. In its March 2008 state expenditures report, ECOS only analyzed regulatory environmental agency work, which includes work pursuant to the primary federal pollution control laws addressed in this chapter and related activities. For FY2008 ECOS projected a total expenditure for all states combined of $12.65 billion for these regulatory environmental protection activities.

---

[176] DOJ, Justice Management Division, Budget and Performance, Congressional Budget Submissions and Performance Summaries for FY2015 and FY2014, http://www.justice.gov/about/bpp.htm; EPA's Congressional Justifications for FY2014 and FY2013, http://www2.epa.gov/planandbudget; and information provided by DOJ ENRD staff in written communication to CRS, April 29, 2014. See also *Budget Trend Data 1975 Through the President's 2003 Request to the Congress*, pp. 55-61, Spring 2002, http://www.usdoj.gov/archive/jmd/1975_2002/btd02tocpg.htm.

[174] ECOS, *ECOS Budget Survey: Budgets are bruised, but Still Strong*, R. Steven Brown and Michael J. Kiefer, Summer 2003 *ECOStates*, http://www.ecos.org/section/states/spending.

Table 5. EPA-OECA's FY2012-FY2014 Enacted and FY2015 Requested Appropriations and FTEs by EPA Appropriations Account and Program Activity (dollars in thousands)

EPA Appropriation Account / Program Activity	FY2012 Enacted		FY2013 Enacted (post-sequestration)		FY2014 Enacted		FY2015 Requested	
	Dollars	FTEs	Dollars	FTEs	Dollars	FTEs	Dollars	FTEs
Total	$533,358.0	3,294.1	$553,122.0	3,009.3	$560,942.0	2,959.1	$583,007.0	2,882.9
Environmental Programs Management (EPM)	$361,997.0	2,220.1	$343,359.0	2,131.7	$352,459.0	2,028	$379,817.0	1,998.5
Brownfields	$478.0	2.6	$415.0	2.1	$424.0	2.2	$476.0	2.1
Civil Enforcement	$177,465.0	1,184.0	$167,484.0	1,141.6	$173,573.0	1,083.8	$180,641.0	1,067.8
Compliance Assistance and Centers	$0.0	0	$0.0	0	$0.0	0	$0.0	0
Compliance Incentives	$0.0	0	$0.0	0	$0.0	0	$0.0	0
Compliance Monitoring	$107,151.0	613.9	$101,548.0	568.3	$103,205.0	554.7	$118,405.0	532.0
Criminal Enforcement	$48,006.0	257.7	$46,292.0	256.1	$47,829.0	239.5	$50,885.0	237.7
Enforcement Training	$0.0	0	$0.0	0	$0.0	0	$0.0	0
Environmental Justice	$5,841.0	28.8	$6,526.0	29.8	$6,737.0	29.1	$7,936.0	36.9
Geographic Program: Chesapeake Bay	$2,583.0	8.1	$2,447.0	9.5	$2,285.0	7.6	$1,493.0	7.4
Homeland Security (critical infrastructure protection)	$0.0	0	$0.0	0	$0.0	0	$0.0	0
NEPA Implementation	$17,298.0	116.1	$16,545.0	115.7	$16,360.0	104.0	$17,841.0	106.6
Facilities Infrastructure and Operations	$2,175.0	8.9	$2,102.0	8.7	$2,046.0	8.0	$2,140.0	8.0
Science and Technology (S&T)	$15,265.0	89.6	$14,484.0	84.6	$14,125.0	75.4	$14,149.0	75.4
Forensics Support	$15,265.0	89.6	$14,484.0	84.6	$14,125.0	75.4	$14,149.0	75.4
State and Tribal Assistance Grants (STAG)	$23,725.0	—	$22,448.0	—	$22,969.0	—	$22,969.0	—
Categorical Grant: Pesticides Enforcement	$18,644.0	—	$17,672.0	—	$18,050.0	—	$18,050.0	—
Categorical Grant: Toxics Substances	$5,081.0	—	$4,816.0	—	$4,919.0	—	$4,919.0	—
Categorical Grant: Sector Program	$0.0	—	$0.0	—	$0.0	—	$0.0	—
Evidence-Based Enforcement	$0.0		$0.0					
Leaking Underground Storage Tanks (LUST)	$789.0	4.7	$746.0	4.3	$746.0	3.4	$639.0	3.4

EPA Appropriation Account / Program Activity	FY2012 Enacted		FY2013 Enacted (post-sequestration)		FY2014 Enacted		FY2015 Requested	
	Dollars	FTEs	Dollars	FTEs	Dollars	FTEs	Dollars	FTEs
Total	$583,358.0	3,294.1	$553,122.0	3,009.3	$560,942.0	2,959.1	$583,007.0	2,882.9
Compliance Assistance and Centers	$0.0	0	$0.0	0	$0.0	0	$0.0	0
Civil Enforcement	$789.0	4.7	$746.0	4.3	$746.0	3.4	$639.0	3.4
Oil Spills Response	$2,424.0	17.3	$2,423.0	15.6	$2,552.0	14.3	$2,661.0	14.3
Civil Enforcement	$2,286.0	16.4	$2,291.0	14.6	$2,413.0	13.4	$2,514.0	13.4
Compliance Assistance and Centers	$0.0	0	$0.0	0	$0.0	0	$0.0	0
Compliance Monitoring	$138.0	0.9	$132.0	1.0	$139.0	0.9	$147.0	0.9
Hazardous Substances Superfund	$179,158.0	962.4	$169,622.0	863.1	$168,091.0	837.1	$162,772.0	791.3
Compliance Assistance and Centers	$0.0	0	$0.0	0	$0.0	0	$0.0	0
Compliance Incentives	$0.0	0	$0.0	0	$0.0	0	$0.0	0
Compliance Monitoring	$1,221.0	1.9	$1,269.0	2.0	$998.0	1.2	$1,083.0	1.2
Criminal Enforcement	$7,895.0	37.2	$7,906.0	34.0	$7,488.0	31.2	$7,438.0	31.2
Enforcement Training	$0.0	0	$0.0	0	$0.0	0	$0.0	0
Environmental Justice	$583.0	3.9	$598.0	4.1	$604.0	3.7	$597.0	3.7
Forensics Support	$2,419.0	14.3	$2,381.0	5.5	$2,344.0	4.9	$1,112.0	4.9
Homeland Security (critical infrastructure protection)	$0.0	0	$0.0	0	$0.0	0	$0.0	0
Superfund: Enforcement	$156,744.0	839.2	$149,704.0	770.1	$149,167.0	750.9	$145,137.0	708.6
Superfund: Federal Facilities Enforcement	$10,296.0	65.9	$7,764.0	47.4	$7,490.0	45.2	$7,405.0	41.7

Source: Compiled by CRS with data received from the Environmental Protection Agency's Office of Congressional and Intergovernmental Relations (OCIR) in written communications, April 14, 2014. The FY2013 enacted amounts reflect automatic spending reductions triggered by the Budget Control Act (BCA; P.L. 112-25), as amended by the 0.2% across-the-board rescission required by the Consolidated and Further Continuing Appropriations Act, 2013 (P.L. 113-6). Differences from fiscal year to fiscal year in the scope of the activities included within each of the accounts make apt direct comparisons difficult below the appropriations account level. This includes adding or discontinuing program activities, as well as reorganizing and consolidating certain activities.

ECOS reported that $3.06 billion, or less than 25% of the projected amount, was from federal funding to states for these purposes.[175] Based on a February 2010 survey of 36 states and Puerto Rico, ECOS reported that of the agencies represented by the survey respondents, 2,112 positions were eliminated or were being held vacant due to budget limitations in FY2010.[176]

In an August 2010 report,[177] ECOS found that

> Overall, state environmental agencies' budgets have decreased from FY2009 to FY2011. That said, there is a budget increase from FY2009 to FY2010 among states that included SRF/ARRA [State Revolving Loan Funds/American Recovery and Reinvestment Act of 2009] administrative funds. There has been a slight overall drop in money coming from state general funds, while funds from the federal government have increased marginally, and funds from "other" sources, such as those raised through permitting fees, have remained relatively constant.

Federal appropriations, in particular allocations to states, for adequate staffing and effective enforcement of federal environmental statutes to protect human health and the environment, will likely continue to be an issue of concern.

CONCLUSION

Fully evaluating and measuring the overall effectiveness of current (and past) pollution control enforcement and compliance activities can be quite complicated. Discussion throughout this chapter highlights the difficulties inherent in characterizing the many facets of environmental enforcement at a macro level, and identifies many of the factors that may contribute to its perceived successes and shortcomings. However, several indicators do provide insight into a better understanding of the complexities associated with elements of enforcement, such as the vastness and diversity of the regulated community, the multiplicity of the activities and priorities across many regulating entities, and variability across statutes.

---

175 ECOS, *March 2008 Green Report: State Environmental Expenditures, 2005-2008*, by R. Steven Brown, Executive Director, Environmental Council of the States, March 2008, http://www.ecos.org/files/3057_file_March_2008_Green_Report.pdf; see also ECOS's Publications on State Spending website at http://www.ecos.org/section/states/spending.

176 ECOS, *March 2010 Green Report: Impacts of Reductions in FY 2010 on State Environmental Agency Budgets*, March 2010 (http://www.ecos.org/files/4011_file_March_2010_ ECOS_Green_Report.pdf), by Victoria Phillips, ARRA Reporting Manager, Massachusetts Department of Environmental Protection; see also ECOS's Publications on State Spending website at http://www.ecos.org/section/states/spending. David Emme, Chief, Bureau of Administrative Services, Nevada Division of Environmental Protection; and Beth Graves, Senior Project Manager, Environmental Council of the States.

177 ECOS, *August 2010 Green Report: Status of State Environmental Agency Budgets, 2009-2011*, by R. Steven Brown, Executive Director, and Adam Fishman, Intern, Environmental Council of the States, August 2010, http://www.ecos.org/files/4157_file_August_ 2010_Green_Report.pdf; see also ECOS's Publications on State Spending website at http://www.ecos.org/section/ states/spending.

Since the establishment of EPA in 1970, Congress has been concerned with a number of crosscutting issues associated with the enforcement of pollution control statutes and regulations, as reflected in provisions of enacted and amended environmental legislation over time. Congressional interest remains heightened, particularly with regard to the substance of intergovernmental relations, EPA-state relations, and fiscal requirements. Continued involvement for the remainder of the 113th Congress with these issues could take several directions. One likely result could be additional oversight hearings. Alternatively, relevant appropriations legislation may contain provisions or language regarding funding for specific enforcement activities.

Congressional interest might focus on further statutory approaches to establish changes in the EPA-states' partnership, such as legislation similar to past proposals concerning refinement of the National Environmental Performance Partnership System (performance partnerships, or NEPPS) and the associated grants award process. The 113th Congress is also considering other statute-specific legislation to address other long-standing concerns that affect certain aspects of EPA enforcement/compliance activities under the various pollution control laws.

The regulated community, public interest groups, federal and state officials, and Congress are often divided on whether to pursue legislation that would further expand or constrain enforcement/compliance. They are similarly divided with respect to proposals that would expand states' authority for implementing and enforcing certain aspects of the major federal pollution control laws.

Views and congressional involvement with respect to these issues are likely to continue to evolve in the years ahead.

APPENDIX A. ENFORCEMENT/COMPLIANCE DATABASES AND EXAMPLES OF REPORTED RESULTS

Enforcement/Compliance Databases and Reporting

Compliance monitoring data are used to manage the compliance and enforcement program, and to inform the public of enforcement actions taken and penalties imposed. EPA and the states collect and maintain compliance/enforcement data in many forms. According to EPA's OECA FY2006 Accomplishments Report, ECOS reported that states collect about 94% of environmental quality data contained in EPA's databases, primarily from state-issued permits and monitoring programs.[178] Information is often entered into multiple databases or transferred from state databases. Historically, the databases were often incompatible, making cross media/statute queries difficult. In recent years, EPA has

---

[178] EPA, *FY2006 OECA Accomplishment Report*, EPA-300-R-07-001, Spring 2007, http://cfpub.epa.gov/compliance/ resources/reports/accomplishment/details.cfm.

been working to expand and modernize the integration of several of the individual databases to allow more cross referencing of compliance data by regulators and to provide querying capabilities to the public.[179]

EPA compiles data from the various databases and provides various statistics in the form of annual accomplishment and multi-year trends reports.[180] Reporting has traditionally focused on statute-by-statute results, including actions initiated and concluded, and penalties and other sanctions assessed. The reliability and consistency of EPA and state databases, and how effectively the reported information can be used as an indicator of environmental progress and the impacts of environmental enforcement has been an issue of some debate, and questioned in reviews conducted by EPA-OIG[181] and GAO.[182] Critics contend, and EPA has long recognized, that while somewhat indicative of the failure to comply with environmental requirements, counting enforcement actions alone ("bean counting") does not provide a complete measure of the effectiveness of the national environmental enforcement/compliance program.

EPA has expanded its reporting by including estimates of environmental benefits (pollution reduction and impacts avoided) in its "Annual Results" for the most recent fiscal years. Additionally, EPA developed a new web-based tool and interactive map that allow the public to obtain detailed information by location about the environmental pollution control enforcement actions taken at approximately 4,600 facilities.[183] Released initially in December 2009, the interactive maps show facilities in the United States where EPA concluded an enforcement action between October 1, 2008, and September 30, 2013 (FY2013). The maps are provided on EPA's Annual Results website.[184] The maps do not include environmental pollution control enforcement actions taken by state or local governments.

EPA has taken steps to expand and modernize its primary publicly available enforcement reporting database Enforcement and Compliance History Online, or ECHO. In February 2013,[185] EPA released interactive dashboards and comparative maps on the

---

[179] For a more complete list and descriptions of EPA's enforcement/compliance databases, see http://www.epa.gov/compliance/data/systems/index.html.

[180] See EPA Enforcement "Data and Results" at http://www2.epa.gov/enforcement/data-and-results.

[181] EPA-OIG *ECHO Data Quality Audit—Phase 2 results: EPA Could Achieve Data Quality Rate with Additional Improvements*, Report No. 10-P-0230, September 22, 2010*; EPA Needs to Improve Its Recording and Reporting of Fines and Penalties*, Report No. 10-P-0077, March 9, 2010; *Overcoming Obstacles to Measuring Compliance: Practices in Selected Federal Agencies*, Report No. 2007-P-00027, June 20, 2007; *EPA Performance Measures Do Not Effectively Track Compliance Outcomes*, Report No. 2006-P-00006; *Congressional Request on Updating Fiscal 2003 EPA Enforcement Resources and Accomplishments*, Report 2004-S-00002, all available at http://www.epa.gov/oig/.

[182] GAO report: *Drinking Water: Unreliable State Data Limit EPA's Ability to Target Enforcement Priorities and Communicate Water Systems' Performance*, Report No. GAO-11-381 June 17, 2011, http://www.gao.gov/products/ GAO-11-381; GAO*: Environmental Enforcement: EPA Needs to Improve the Accuracy and Transparency of Measures Used to Report on Program Effectiveness*, GAO-08-1111R, September 18, 2008, http://gao.gov.

[183] For more information, see EPA's website "Questions About the Annual Results Maps," http://www.epa.gov/compliance/resources/reports/endofyear/eoy2010/2010-map-questionsabout.html.

[184] See EPA, "Concluded Enforcement Cases Map for 2013," at http://www2.epa. gov/enforcement/enforcement-annual-results-concluded-cases-map-fiscal-year-fy-2013.

[185] See also EPA, February 2, 2013, press release at http://yosemite.epa.gov/opa/admpress.nsf/2467feca60368729852573590040443d/0f482ac14c83e34285257b0b005c70a1%21OpenDocument.

ECHO website that include state-level enforcement data for the most recent five years. As part of its ECHO modernization[186] and redesign effort, following the September 2013 release of an ECHO Beta test site, EPA launched the initial phase of the modernized ECHO database in December 2013 (ECHO 2.0). The original ECHO website launched in 2002 was retired, and at the present time certain features are not available, but EPA has been continuously updating the website, and features from the original website have been modernized and phased in throughout 2014.[187] On September 11, 2014, EPA announced its most recent upgrade "ECHO 2.3" which includes the addition of Resource Conservation and Recovery Act (RCRA) and EPA Civil Enforcement Case compliance data searching capabilities.[188] The modernization of the ECHO database has also subsumed EPA's Online Tracking Information System (OTIS), which was available only to government employees (including tribal governments). OTIS was retired at the time of the December 2013 release of the modernized ECHO website.

Overview of Enforcement/Compliance Databases

A number of EPA's single- and multi-media national databases include enforcement and compliance data elements. While these databases are generally available to EPA staff, and in some cases state and local governments, most are not readily available to the public. The Enforcement and Compliance History Online, or ECHO, developed and maintained by OECA is the most prominent publicly accessible database.[189] Introduced in 2002, ECHO queries provide a snapshot of the most recent three years of a facility's environmental compliance record, but are limited primarily to certain requirements under the CAA, CWA, and RCRA. EPA continues to expand the integration and capabilities of this and other databases. Finally, several state environmental agencies maintain additional information about compliance and enforcement (beyond what is reported to EPA systems).[190]

For a more complete list and descriptions of EPA's enforcement/ compliance databases, see EPA's "Compliance and Enforcement Data Systems" web page at http://www.epa.gov/compliance/data/systems/ index.html.

The following brief summaries of several of EPA's integrated national databases are a consolidation of descriptions provided on the agency's website:

Enforcement and Compliance History Online (ECHO)

See http://echo.epa.gov/.

---

[186] See EPA, ECHO Modernization Information http://echo.epa.gov/resources/general-info/modernization-information. Also see the EPA December, 17, 2013, webinar at http://echo.epa.gov/system/files/Public_webinar-061913.pdf.

[187] See EPA ECHO What's New at http://echo.epa.gov/resources/general-info/whats-new.

[188] See EPA ECHO Release Notes at https://echo.epa.gov/resources/general-info/release-notes.

[189] See EPA, ECHO Frequently Asked Questions, http://echo.epa.gov/echo_faq.

[190] Several states have provided direct links to related websites, which EPA posts on the ECHO website at http://www.epa-echo.gov/echo/more_state_data.html.

Originally released in 2002, ECHO is an interactive website that allows users to query permit, inspection, violation, enforcement action, informal enforcement action, and penalty information for individual or multiple facilities. Initial queries return a list of relevant facilities, each linked to a "Detailed Facility Report," indicating

- whether a facility has been inspected/evaluated,
- occurrence and nature of violations (noncompliance),
- nature of enforcement actions (including penalties) that have been taken, and
- contextual information about the demographics surrounding the facility.

The ECHO database does not contain violations and enforcement information about all environmental laws for which EPA has primary jurisdiction, instead focusing only on actions for Clean Air Act (CAA) stationary source facilities, Clean Water Act (CWA) major direct discharge facilities, Resource Conservation and Recovery Act (RCRA) hazardous waste handlers, and systems violations of the Safe Drinking Water Act (SDWA). As noted by EPA, examples of data not generally available from ECHO include violations at CWA direct discharge minor facilities, CAA mobile source and asbestos violations, Toxic Substances Control Act (TSCA) and Federal Insecticide, Fungicide and Rodenticide Act (FIFRA) violations, Superfund violations, and CWA wetlands and pretreatment violations.

A "modernized" version of the ECHO website, ECHO 2.0, was launched on December 3, 2013, and the original ECHO website was retired (as was the nonpublicly available OTIS database). Not all features and data will be immediately available. The central features of the initial release were cross-statute (multimedia) facility searching and state comparative maps and dashboards. EPA began phasing in program-specific (air, water, waste searching) and other enhancements in January 2014, and will continue to do so throughout 2014 until all features of the original ECHO website are replaced. To date EPA released a variety of enhancements: ECHO Release 2.1 on March 24, 2014, ECHO Release 2.1.1 on April 18, 2014, ECHO Release 2.1 update on April 24, 2014, ECHO Release 2.2 on June 24, 2014, and ECHO Release 2.3 on September 10, 2014. EPA reports these updates, including a summary of upgrades and enhancements, on its ECHO "What's New" website at http://echo.epa.gov/whats_new. During the modernization process, some searches will be unavailable from time to time. As of the publication of this report, EPA proposed the following estimated schedule for completing additional specific capabilities:

- Clean Air Act facility and pollution search — February 2015
- EPA and State Annual Enforcement Actions Map — To be determined

Prior to the modernization effort, in February 2013, EPA had released interactive dashboards and comparative maps on the ECHO website that include state-level enforcement data for the most recent five years.[191] The feature remains available on the modernized version of the ECHO website.[192] Information includes the number of completed inspections, types of violations found, enforcement actions taken, and penalties assessed by state, and users can customize the presentation to view state activity as well as view comparisons with EPA activity. EPA initially added an interactive maps feature for tracking federal enforcement activities in FY2011.

Envirofacts

See http://www.epa.gov/enviro/.

Envirofacts provides public access to information about environmental activities, such as releases, permit compliance, hazardous waste handling processes, and the status of Superfund sites, which may affect air, water, and land anywhere in the United States. Data are retrieved from various EPA source databases. Users can develop online queries, create reports, and map results.

Facility Registry System (FRS)

See http://ofmpub.epa.gov/sor_internet/registry/facilreg/home/basicin formation/.

FRS (a companion to the integrated facility searches in Envirofacts) can be used to create facility identification records, including geographical location, and to locate sites or places subject to environmental regulations or oversight (e.g., monitoring sites). Records are based on information from EPA program national systems, state master facility records, and data collected from EPA's Central Data Exchange.

Integrated Compliance Information System (ICIS)

See http://www.epa.gov/compliance/data/systems/icis/index.html. ICIS integrates data that are currently located in several separate data systems. ICIS contains information on federal administrative and federal judicial cases under the following environmental statutes: the CAA, CWA, RCRA, EPCRA, TSCA, FIFRA, CERCLA (Superfund), SDWA, and MPRSA. ICIS also contains information on compliance assistance activities conducted in EPA regions and headquarters. The web-based system enables states and EPA to access integrated enforcement and compliance data. The public can only access some of the federal enforcement and compliance information in ICIS by using the EPA Enforcement Cases Search and EPA Enforcement SEP Search through ECHO.

---

[191] EPA's February 7, 2013, press release, "EPA Releases State Enforcement Performance Information and Comparative Maps / Agency to host webinar demonstrating new data features," http://yosemite.epa.gov/opa/admpress.nsf/Press%20Releases%20issued%20by% 20OECA?OpenView.

[192] EPA ECHO Comparative Maps, http://echo.epa.gov/node/16 and ECHO - Performance and Trends: State Air Dashboard; see also EPA's February 7, 2013, press release, "EPA Releases State Enforcement Performance Information and Comparative Maps / Agency to host webinar demonstrating new data features," http://yosemite.epa.gov/opa/admpress.nsf/ Press%20Releases%20issued%20by%20OECA?OpenView.

Integrated Data for Enforcement Analysis (IDEA)

See http://www.epa.gov/compliance/data/systems/multimedia/idea/in dex.html.

IDEA maintains copies of EPA's air, water, hazardous waste, and enforcement source data systems that are updated monthly. An internal EPA database, IDEA uses "logical" data integration to provide a historical profile of inspections, enforcement actions, penalties assessed and toxic chemicals released, for EPA-regulated facilities.

Online Tracking Information System (OTIS)

The OTIS web link is no longer updated, http://www.epa.gov/compliance/data/systems/ multimedia/aboutotis.html.

Note: The OTIS website, available only to federal, state, and tribal government staff, was retired with the December 2013 release of the modernized ECHO website. Former OTIS content and "government-only features" will be accessible internally to government staff via EPA LAN ID and password, or through an EPA Portal ID log-in on the "ECHO Gov" portion of echo.epa.gov.

OTIS was a collection of search engines that enabled EPA, state/local/tribal governments and certain other federal agencies to access a broad range of data relating to enforcement and compliance. No public access was available. This web interface application sent queries and drew data from the IDEA (discussed above) system, which integrates facility data from different EPA databases. IDEA copies many EPA and non-EPA databases, and organizes the information to facilitate cross-database analysis. See http://www.epa.gov/compliance/data/systems/multimedia/ aboutotis.html (website no longer supported).

A number of other databases, mostly for single media, also include compliance/enforcement data. Many of these databases were the basis for certain data elements in the various IDEA and OTIS integrated databases, and typically were not directly available to the public. Other databases included the Air Facility System (AFS); Permit Compliance System (PCS); Resource Conservation and Recovery Act Information System (RCRAInfo); National Compliance Data Base System and Federal Insecticide, Fungicide, and Rodenticide Act/Toxic Substances Control Act Tracking System (NCDB/FTTS); and Safe Drinking Water Information System/Federal (SDWIS/FED).

Although the information on the integrated systems did not represent "real time" data, it was updated once a month when OECA refreshed the source data systems. Additionally, EPA routinely made improvements to existing aspects of OTIS, IDEA, and ECHO, and often enhanced the system by adding new search capabilities and tools. For example, in December 2012 EPA announced that it had completed a major, multi-year database modernization for searching Clean Water Act actions. EPA decommissioned the previous Permit Compliance System (PCS) and replaced it with the new Integrated Compliance Information System for the National Pollutant Discharge Elimination System (ICIS-NPDES). As another example, in September 2012 EPA began adding greenhouse gas (GHG) emission data to the OTIS database, adding GHG Emission column in the Air and Multimedia search results, and new GHG rows to the Detailed Facility Report "Facility and Permits and Identifiers" section.

APPENDIX B. EXAMPLES OF REPORTED ENFORCEMENT ACTIONS AND PENALTIES OVER TIME

The following figures and tables provide examples of the type of enforcement data collected, compiled, and reported over time. They are intended to show proportional relationships of the various types of enforcement actions (e.g., administrative vs. judicial) in a given year and by statute, not annual or long-term enforcement trends. To compare the reported activities from year to year requires more detailed information regarding the specific circumstances in those years.

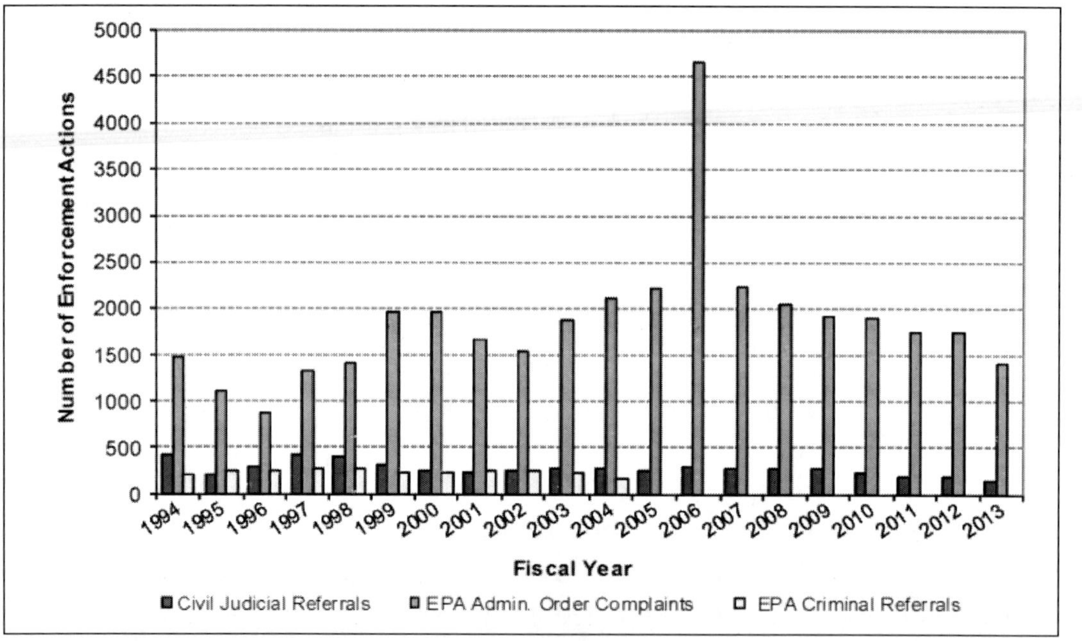

Source: Compiled by CRS using data from EPA's Enforcement Annual Results for various fiscal years http://www.epa.gov/compliance/data/results/annual/index.html. Long-term trends data previously available but no longer maintained by EPA on its website, EPA's National Enforcement Trends: EPA Long Term Trends; and National Enforcement Trends (FY2004)—EPA Long Term Trends: Criminal Referrals and Penalties. EPA terminated the count of criminal referrals as an internal Criminal Enforcement program measure in FY2005 and discontinued reporting these data in its annual results reports.

Figure B-1. EPA Civil Judicial Referrals, Administrative Order Complaints, and Criminal Referrals, FY1994-FY2013.

Table B-1. EPA Civil Administrative, Civil Judicial, and Criminal Enforcement Actions, FY2009-FY2013

Enforcement Action	FY2009	FY2010	FY2011	FY2012	FY2013
Administrative Compliance Orders	1,588	1,302	1,324	1,088	873
Administrative Penalty Order Complaints	1,914	1,901	1,760	1,760	1,407
Final Administrative Penalty Orders	1,916	1,830	1,735	1,780	1,440
Civil Judicial Referral	277	233	199	179	138
Civil Judicial Cases Concluded	201	200	182	144	176
Criminal Judicial Referral	NR	NR	NR	NR	NR
Criminal Judicial Cases Initiated (Opened)	387	346	371	320	297

Source: Compiled by CRS using data from EPA's Enforcement Annual Results for various fiscal years, http://www.epa.gov/compliance/data/results/annual/ index.html.

Notes: NR = not reported. EPA terminated the count of criminal referrals as an internal Criminal Enforcement program measure in FY2005 and discontinued reporting these data in its trends reports. EPA no longer maintains the previously available website National Enforcement Trends (FY2004)—EPA Long Term Trends: Criminal Referrals and Penalties.

Table B-2. Number of EPA Enforcement Inspections and Evaluations by Statute, FY2009-FY2013

Statute	FY2009	FY2010	FY2011	FY2012	FY2013
CAA	3,384	3,690	3,558	4,055	3,695
CWA	3,488	3,446	3,364	3,439	3,071
EPCRA	969	871	901	775	648
FIFRA	346	372	364	584	385
MPRSA	0	0	0	0	0
RCRA	3,171	4,231	3,181	3,384	2,800
SDWA	6,927	7,034	5,955	6,161	6,163
TSCA	1,439	1,368	1,640	1,436	1,183
Total	19,724	21,012	18,963	19,834	17,945

Source: Compiled by CRS. FY2012 and FY2013 data provided directly to CRS by EPA's Office of Congressional and Intergovernmental Relations (OCIR), most recently on March 18, 2014, for FY2013. EPA discontinued specifying the number of inspections and evaluations by statue in its annual results reports beginning in FY2012.

Abbreviations: CAA: Clean Air Act. CWA: Clean Water Act. EPCRA: Emergency Planning and Community Right-to-Know Act. FIFRA: Federal Insecticide, Fungicide and Rodenticide Act. MPRSA: Marine Protection, Research, and Sanctuaries Act. RCRA: Resource Conservation and Recovery Act. SDWA: Safe Drinking Water Act. TSCA: Toxic Substances Control Act

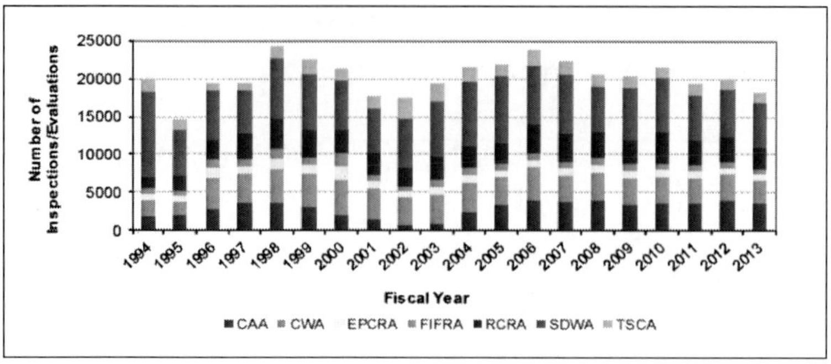

Source: Compiled by CRS. FY2012 and FY2013 data provided directly to CRS by EPA's Office of Congressional and Intergovernmental Relations (OCIR), most recently on March 18, 2014, for FY2013. Data for fiscal years prior to FY2012 are from National Enforcement Trends: EPA Long Term Trends: FY1994-FY2011 Federal Inspections and Evaluations, which is no longer maintained by EPA on its website.

Figure B-2. Number of EPA Federal Inspections and Evaluations by Statute, FY1994-FY2013.

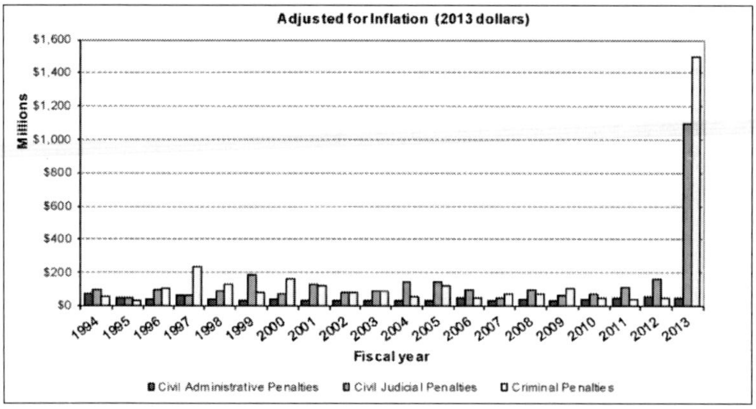

Source: Compiled by CRS using data from EPA's Enforcement *Annual Results* for various years. Long-term historical trends data from *National Enforcement Trends—EPA Long Term Trends: FY1974-FY2011 Enforcement Penalties*; previously available but no longer maintained by EPA on its website. Amounts converted to 2013 dollars using the GDP Chained Price Index from the Office of Management and Budget, *Budget of the U.S. Government for FY2015*, Historical Tables, Table 10, http://www.gpo.gov/fdsys/browse/collectionGPO.action?collection Code=BUDGET. See note b in *Table B-3* below regarding FY2013 amounts.

Figure B-3. Environmental Enforcement Penalties Assessed by EPA: Administrative, Civil Judicial, and Criminal, FY1994-FY2013.

There can be significant variability from year to year in how data were reported and which entities reported. EPA has refined terms and definitions in the data elements from year to year. Other factors that result in variability include the introduction of new regulatory requirements in a given year, and fruition of statutory deadlines.

The figures presented reflect longer-term data (15 to 20 years) through FY2013, whereas the tables generally provide data for the most recent five years, depending on the availability of data for the most recent fiscal year. Results are presented by action (e.g., administrative, civil judicial, criminal judicial), and where readily reported, for actions by statute (e.g., Clean Water Act, Clean Air Act). The data have been compiled from EPA's annual fiscal year results reports,[193] National Enforcement Trends charts and tables,[194] and data provided directly to CRS by EPA.[195]

Table B-3. Environmental Enforcement Penalties Assessed and Court-Ordered Projects by EPA: Administrative, Civil Judicial, and Criminal, FY2009-FY2013 (nominal dollars in thousands—not adjusted for inflation)

Fiscal Year	Administrative	Civil Judicial	Criminal[a]	Criminal: Court-Ordered Projects	Total
FY2009	$31,609	$58,497	$96,000	$10,000	$186,105
FY2010	$33,359	$70,249	$41,000	$18,000	$144,600
FY2011	$44,881	$104,391	$35,000	$2,000	$184,272
FY2012	$52,023	$155,539	$44,000	$14,000	$251,562
FY2013	$48,000	$1,100,000[b]	$1,500,000[b]	$3,000,000	$5,648,000

Source: Compiled by CRS using data from EPA's Enforcement Annual Results for various years.

[a] Criminal penalties represent "fines and restitution."

[b] EPA reported that $1.0 billion of the total $1.1 billion in civil judicial penalties, nearly $1.4 billion of $1.5 billion in criminal fines and restitution assessed, and the majority of the $3.0 billion in court-ordered projects for FY2013 were attributed to the Deepwater Horizon Gulf of Mexico civil and criminal cases, EPA Enforcement Analysis and Trends for Fiscal Year (FY) 2013, http://www2.epa. gov/enforcement/enforcementannual-results-analysis-and-trends-fiscal-year-fy-2013. See also EPA Enforcement: Deepwater Horizon—BP Gulf of Mexico Oil Spill, http://www2.epa.gov/enforcement/deepwater-horizon-bp-gulf-mexico-oil-spill, Enforcement: Criminal Press Releases—2013, http://www2.epa.gov/ enforcement/criminal-press-releases-2013.

Table B-4. Supplemental Environmental Projects (SEPs) Dollar Values as Reported by EPA: FY2009-FY2013 (dollars in thousands—not adjusted for inflation)

Statute	FY2009	FY2010	FY2011	FY2012	FY2013
CAA	$12,509.7	$8,160.0	$6,665.7	$6,244.5	$14,548.6
CERCLA	$449.7	$174.8	$352.1	$258.2	$478.3
CWA	$5,264.9	$12,419.0	$10,488.0	$27,023.9	$3,471.3
EPCRA	$2,338.9	$736.7	$914.9	$1,054.7	$1,776.6
FIFRA	$158.6	$46.0	$6.9	$1,973.4	$49.4
MPRSA	$0.0	$37.5	$0.0	$0.0	$0.0
RCRA	$7,123.7	$1,111.9	$4,946.3	$6,204.2	$983.0
SDWA	$3.8	$325.0	$1,212.8	$103.1	$0.0
TSCA	$13,281.8	$763.4	$800.1	$744.0	$852.9
Total	$41,121.1	$23,774.3	$25,386.8	$43,606.0	$22,160.0

[193] EPA-OECA, *Data and Results: Annual Results*, at http://epa.gov/compliance/data/ results/annual/index.html.

[194] This website is no longer available. EPA-OECA, *Data and Results: National Enforcement Trends*, at http://epa.gov/ compliance/data/results/nets.html.

[195] Information as provided upon request directly to CRS by EPA's Office of Congressional and Intergovernmental Relations (OCIR) in various years, most recently on March 18, 2014.

Source: Compiled by CRS using data from EPA's *Enforcement Annual Results* various fiscal years, http://epa.gov/ compliance/data/results/annual/index.html, and previously available EPA's *National Enforcement Long Term Trends: FY1999-FY2011 Supplemental Environmental Projects* previously available but no longer maintained by EPA on its website, and data provided directly to CRS by EPA's Office of Congressional and Intergovernmental Relations (OCIR) in various years, most recently on March 18, 2014; EPA discontinued reported SEP amounts by statue in the EPA's *Annual Results for Fiscal Year 2012: Analysis and Trends*, and subsequent fiscal years. Totals may not add due to rounding.

Abbreviations: CAA: Clean Air Act. CERCLA: Comprehensive Environmental Response, Compensation, and Liability Act (Superfund). CWA: Clean Water Act. EPCRA: Emergency Planning and Community Right-to-Know Act. FIFRA: Federal Insecticide, Fungicide and Rodenticide Act. MPRSA: Marine Protection, Research, and Sanctuaries Act. RCRA: Resource Conservation and Recovery Act. SDWA: Safe Drinking Water Act. TSCA: Toxic Substances Control Act

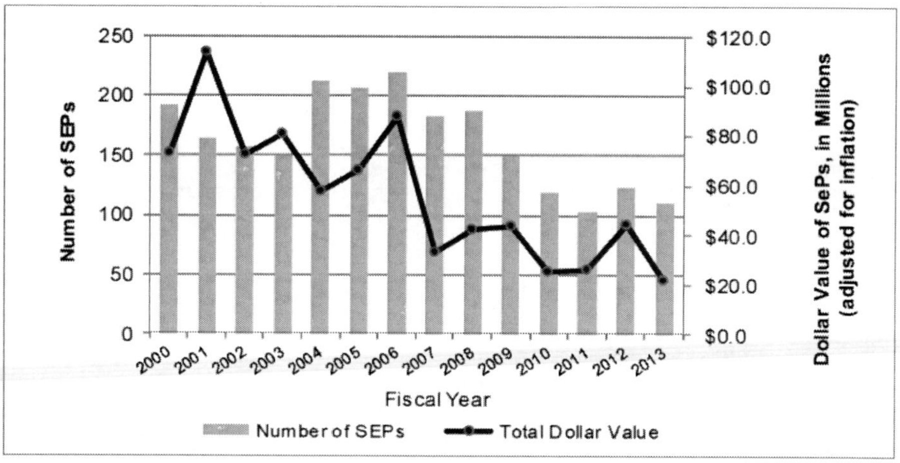

Source: Compiled by CRS using data from EPA's Enforcement Annual Results various fiscal years, http://epa.gov/ compliance/data/results/annual/index.html, and National Enforcement Long Term Trends: FY1999-FY2011 Supplemental Environmental Projects, previously available but no longer maintained by EPA on its website. Amounts converted to 2013 dollars using the GDP Chained Price Index from the Office of Management and Budget, Budget of the U.S. Government for FY2015, Historical Tables, Table 10, .http://www.gpo.gov/fdsys/ browse/collectionGPO.action?collectionCode=BUDGET.

Figure B-4. EPA Supplemental Environmental Projects: Number of Projects and Dollar Value (adjusted for inflation), FY2000-FY2014.

In: Environmental Protection Agency (EPA)
Editor: Coby Schaefer
ISBN: 978-1-53614-933-3

Chapter 3

EPA POLICIES CONCERNING INTEGRATED PLANNING AND AFFORDABILITY OF WATER INFRASTRUCTURE*

Jonathan L. Ramseur

SUMMARY

For several years, the U.S. Environmental Protection Agency (EPA) has been working with states and cities to develop and implement new approaches that will achieve water quality goals cost-effectively and in a manner that "addresses the most pressing public health and environmental protection issues first." Two recent EPA initiatives are an integrated planning policy and a framework policy for assessing a community's financial capability to meet objectives and requirements of the Clean Water Act (CWA).

Pressed by municipalities about the financial challenges that they face in addressing needs for wastewater and stormwater control projects, in 2012 EPA issued an integrated permitting and planning policy. The intention of the policy is to provide communities with flexibility to prioritize and sequence needed water infrastructure investments so that limited public dollars can be invested in ways that each municipality finds most valuable.

Water utilities and municipalities have welcomed the opportunity for flexibility under the integrated planning policy. But they have sought clarification of a number of issues, including EPA and state roles in developing integrated plans. A major point of contention between EPA and local government stakeholders has been the agency's reliance on administrative orders or judicially approved consent decrees to codify integrated pollution reduction plans, rather than through modification of CWA permits. City and town officials say that they would prefer that EPA allow compliance flexibility through permits, rather than subjecting cities and towns to legally binding consent decrees with penalties and fines for noncompliance. The agency takes the position that

* This is an edited, reformatted and augmented version of Congressional Research Service Publication No. R44223, dated March 14, 2017.

both enforcement and permits are necessary, depending on individual circumstances. EPA typically uses a consent decree after periods of permit noncompliance.

While integrated planning may be helpful in identifying communities' relative priorities, a longstanding concern for local governments is EPA's process for evaluating how much communities can afford for CWA-mandated and other water infrastructure improvements. EPA has worked with communities to refine how the agency determines when a project is affordable for individual communities, because affordability considerations can influence schedules established for a community to meet CWA requirements. In 2014, EPA released a Financial Capability Assessment Framework that identifies a range of information related to a community's financial strength that may help provide a complete picture of cities' financial capabilities in relation to water infrastructure investments.

State and local governments generally support EPA's efforts to encourage ways for communities to prioritize their CWA infrastructure investments and to consider a wide range of factors that affect affordability. Nevertheless, cities and states have continuing concerns with aspects of both policies. For example, some criticize EPA for relying in part on Median Household Income (MHI) as a measure of community affordability. Further, cities are critical that integrated plans that have been approved so far have been incorporated only into new or amended consent decrees, not CWA permits.

EPA's integrated planning process and water infrastructure affordability are issues of interest to legislators. In recent years, Members have offered bills that would address these issues in different ways. Some bills have sought to codify EPA's integrated planning approach—as written in the 2012 framework—into the CWA. Other proposals would go beyond codifying the integrated planning approach by including other provisions that alter the CWA's existing framework to varying degrees (e.g., H.R. 465 in the 115th Congress). In general, some environmental groups have opposed such changes proposed in bills from prior Congresses, but municipalities have generally supported the additional flexibility.

INTRODUCTION

Many state and local governments face difficult economic challenges with limited resources and financial capability to meet their water quality goals and compliance obligations. According to the most recent estimate by the Environmental Protection Agency (EPA) and the states,[1] the nation's wastewater treatment facilities will need $271 billion over the next 20 years to meet the water quality objectives of the Clean Water Act (CWA).[2]

For several years, EPA has been working with states and cities to develop and implement new approaches that will achieve water quality goals cost-effectively and in a manner that "addresses the most pressing public health and environmental protection issues first."[3] As funding for water infrastructure projects remains a perennial topic of

[1] EPA, *Clean Watersheds Needs Survey 2012*, Report to Congress, 2016, https://www. epa.gov/cwns.

[2] The Federal Water Pollution Control Act, commonly referred to as the Clean Water Act, is codified in 33 U.S.C. §1251 et seq.

[3] Nancy Stoner, Acting Assistant Administrator, EPA Office of Water, and Cynthia Giles, Assistant Administrator, EPA Office of Enforcement and Compliance Assurance, *Integrated Municipal Stormwater and Wastewater Planning Approach Framework*, memorandum, June 5, 2012, http://water.epa.gov/polwaste/npdes/stormwater/upload/ integrated_planning_framework.pdf.

congressional interest, EPA's recent initiatives have garnered attention from policymakers and led to legislative proposals in the 115th Congress and prior Congresses.

The first two sections of this chapter examine two recent initiatives by EPA: (1) an integrated planning policy and (2) a framework policy for assessing a community's financial capability to meet objectives and requirements of CWA. The third section of this chapter discusses congressional interest and activity. The fourth section provides some concluding observations.

EPA's Integrated Planning Policy

Local communities face challenges to fund wastewater infrastructure improvements that are needed to comply with CWA requirements. EPA's integrated planning approach is a voluntary tool that allows municipalities to propose to meet multiple CWA requirements by identifying efficiencies and sequencing investments so that the highest priority projects come first.

The prominence of these issues is due in part to EPA's use of governmental enforcement to achieve compliance by the municipal sector with CWA requirements. Since 1998, and more intensely since 2008, EPA has focused national enforcement actions on keeping raw sewage and contaminated stormwater discharges out of the nation's waters as part of the agency's National Enforcement Initiatives.[4] A number of communities are under federal consent decrees[5] or administrative orders,[6] and many others are under state consent decrees or orders requiring them to address uncontrolled discharges from combined sewer systems (combined sewer overflows, or CSOs)[7] and separate sewer systems (separate sewer overflows, or SSOs).[8] Consent decrees typically specify deadlines for complying with the CWA and involve significant investments from ratepayers to bring about compliance and to achieve associated improvements in water quality infrastructure.[9] Construction associated with consent decrees frequently involves long

[4] See https://www.epa.gov/enforcement/national-enforcement-initiative-keeping-raw-sewage-and-contaminated-stormwater-out-our.

[5] A consent decree is a negotiated settlement—contained in a court order—between the enforcing agency and the permittee. The consent decree can include injunctive relief, which requires actions to bring the defendant back into compliance, and monetary penalties.

[6] An administrative order is a non-judicial enforcement action—taken by EPA or a state under its own authority—directing an entity to take a specific action in order to come into compliance.

[7] EPA maintains a list of entities currently subject to a consent decree for CSO issues. See https://www.epa.gov/enforcement/status-civil-judicial-consent-decrees-addressing-combined-sewer-systems-csos.

[8] EPA tracks of the progress its enforcement initiative to address combined sewer overflows and separate sewer overflows here: https://www.epa.gov/enforcement/national-enforcement-initiative-keeping-raw-sewage-and-contaminated-stormwater-out-our.

[9] EPA's Inspector General estimated that it will cost communities more than $32 billion to complete the required actions and penalties that are contained in 47 federal consent decrees that address CSOs. EPA, Office of Inspector General, *EPA Needs to Track Whether Its Major Municipal Settlements for Combined Sewer Overflows Benefit Water Quality*, Project No. 15-P-0280, September 15, 2015, http://www.epa.gov/oig/reports/2015/20150916-15-P-0280.pdf.

time frames—more than a decade in many cases, according to EPA.[10] Other communities are not specifically subject to court-approved enforcement schedules, but are required by CWA permits to take actions, including infrastructure upgrades, to maintain their compliance status.

For some time, municipalities have pressed EPA for greater flexibility to meet the financial and compliance challenges that they face for wastewater, stormwater, and other CWA infrastructure improvements. Correcting CSO problems typically represents one of the most expensive CWA compliance issues.[11] Cities and towns have urged regulators to recognize how difficult and expensive it can be for them to implement wastewater upgrades, stormwater improvements, and related infrastructure projects, while also providing a range of other necessary municipal services on a day-to-day basis.

In response to such concerns, EPA officials in 2011 issued a memorandum describing the agency's support for an integrated approach to wastewater and stormwater management. Such an approach, the memorandum said, would allow communities to "maximize their infrastructure improvement dollars through appropriate sequencing of work." Implementing the integrated approach would require coordination between permit and enforcement actions and complementary state actions.[12]

EPA also began a round of talks with municipal officials about the costs of CWA requirements facing local governments. The talks resulted in a framework policy issued in 2012 for communities to pursue integrated planning to manage wastewater and stormwater and thus implement the approach outlined in the 2011 EPA memorandum.[13] The policy's intention is to reduce overall a community's compliance costs by considering all wastewater and stormwater management obligations in an integrated fashion. It allows communities to prioritize water management goals—such as water conservation or lower wastewater treatment costs—so that limited public dollars can be invested in ways that each city finds most valuable. The policy allows cities and towns to voluntarily seek modification of consent decrees and the terms of CWA discharge permits, and to gain the flexibility to prioritize water quality and infrastructure projects based on affordability. Under this integrated approach, municipalities can evaluate how best to meet all of their CWA obligations within their financial capability, while maintaining existing regulatory standards that protect public health and water quality, and to sequence wastewater and

---

[10] EPA, Office of Inspector General, *EPA Needs to Track Whether Its Major Municipal Settlements for Combined Sewer Overflows Benefit Water Quality*, Project No. 15-P-0280, September 15, 2015, http://www.epa.gov/oig/reports/ 2015/20150916-15-P-0280.pdf.

[11] See EPA, *Clean Watersheds Needs Survey 2012*, Report to Congress, 2016, https://www.epa.gov/cwns.

[12] Nancy Stoner, Acting Assistant Administrator, EPA Office of Water, and Cynthia Giles, Assistant Administrator, EPA Office of Enforcement and Compliance Assurance, *Achieving Water Quality Through Integrated Municipal Stormwater and Wastewater Plans*, memorandum, October 27, 2011, http://water.epa.gov/infrastructure/greeninfrastructure/ upload/memointegratedmunicipalplans.pdf.

[13] Nancy Stoner, Acting Assistant Administrator, EPA Office of Water, and Cynthia Giles, Assistant Administrator, EPA Office of Enforcement and Compliance Assurance, *Integrated Municipal Stormwater and Wastewater Planning Approach Framework*, memorandum, June 5, 2012, http://water.epa.gov/polwaste/npdes/stormwater/upload/ integrated_planning_framework.pdf.

stormwater projects in a way that allows the highest-priority environmental projects to come first.

Each integrated plan will be site-specific, but the policy identifies six elements that an integrated plan must include:

- Description of the water quality, human health, and regulatory issues to be addressed in the plan.
- Description of existing wastewater and stormwater systems under consideration, including description of current performance.
- A process for enabling public participation in development and implementation of the plan.
- A process for identifying, evaluating, and selecting alternative means of compliance, such as use of green infrastructure.
- A process for evaluating performance and measuring success.
- A process for identifying, evaluating, and selecting proposed new projects or modifications to ongoing or planned projects and implementation schedules based on changing circumstances.

A major point of contention between EPA and local government stakeholders has been the agency's reliance on administrative orders or judicially approved consent decrees to codify pollution reduction plans, including plans approved under the integrated planning policy, rather than through modification of CWA permits. Mayors, represented by the U.S. Conference of Mayors, have stated that they would prefer that EPA authorize compliance flexibility through permits, rather than subjecting cities and towns to legally binding consent decrees with penalties and fines for noncompliance. EPA typically uses a consent decree after periods of permit noncompliance. The agency takes the position that both enforcement and permits are necessary, depending on individual circumstances [14]

The 2012 policy provides that all or part of an integrated plan can be incorporated into a CWA discharge permit. For example, innovative practices for managing stormwater as a resource— green infrastructure practices and technologies—can be considered and incorporated into permits where such practices provide sustainable solutions for municipal wet weather control.[15]

The policy further provides that all or part of an integrated plan may be incorporated into the remedy of a federal or state enforcement action to address noncompliance with the CWA. EPA initially focused the policy on *new* enforcement, meaning that cities already

---

[14] Nancy Stoner, Acting Assistant Administrator, EPA Office of Water, and Cynthia Giles, Assistant Administrator, EPA Office of Enforcement and Compliance Assurance, *Integrated Municipal Stormwater and Wastewater Planning Approach Framework*, memorandum, June 5, 2012, http://water.epa.gov/polwaste/npdes/stormwater/upload/ integrated_planning_framework.pdf.

[15] 15 Green infrastructure uses or mimics natural processes to infiltrate, evapotranspire, or reuse stormwater runoff on the site where it is generated. It includes green roofs, rain gardens, and permeable pavements. For information, see CRS Report R43131, *Green Infrastructure and Issues in Managing Urban Stormwater*, by Claudia Copeland.

subject to consent decrees were unable to take advantage of flexibility to modify existing plans to resolve enforcement actions.[16] However, in 2012, EPA, the Department of Justice, and officials of one city, the District of Columbia, agreed to consider extending compliance deadlines in that city's 2005 consent decree in order to allow the city to test green infrastructure technologies. To address sewer overflows, the original consent decree called for the city to construct by 2025 three large tunnels capable of holding 31 million gallons of diluted sewage at any one time, allowing the city's wastewater treatment plant to process the effluent after storms. The 2012 agreement contemplated extending deadlines for some of the new construction required by the consent decree. Further, it potentially would allow the city to avoid some construction project elements (constructing one tunnel as planned, but reducing a second tunnel's size and eliminating the third tunnel), if implementation of a green infrastructure program can reasonably be expected to lead to CSO reductions required by the consent decree, while also satisfying the city's responsibility to mitigate stormwater runoff.[17] The DC plan illustrates integrated planning because it allowed the city to compare the effectiveness of alternatives—green infrastructure and constructed tunnels— to reduce rainwater flows into storm drains enough to stop overflows into nearby waterways.[18]

Under terms of the 2012 agreement, EPA indicated that it would support reopening the consent decree only if it were convinced that green infrastructure would perform as well as construction of tunnels that are designed to capture combined stormwater and wastewater and hold it until it can be processed by the city's wastewater treatment plant. EPA agreed to support DC's research into whether the extensive use of green infrastructure could reduce rainwater runoff into sewers enough to eliminate the need for the planned tunnel projects.

In May 2015, EPA, the Department of Justice, and the DC government revised the 2005 consent decree with the modifications outlined in the 2012 agreement.[19] The revised consent decree also extends the deadline for compliance by five years, to 2030. Officials in other cities have hoped that the EPA-DC agreement could provide a model to be applied elsewhere. Some environmental groups criticized EPA for granting DC additional compliance time for green infrastructure projects that the city could already have pursued, under the existing consent decree.

Utilities and municipalities have welcomed the opportunity for flexibility under the integrated planning policy. But they sought clarification on a number of issues, such as how communities can proactively ensure that the plan they develop will be acceptable to

---

[16] Many consent decrees are modified to reflect new information, improved design, or other factors. These modifications often result in extended compliance timelines.

[17] District of Columbia Water and Sewer Authority, EPA, and the Government of the District of Columbia, *Green Infrastructure Partnership Agreement*, December 10, 2012, http://www.epa.gov/reg3wapd/pdf/pdf_chesbay/GreenPartnshipAgreement.pdf.

[18] See, for example, testimony of George Hawkins (DC Water and Sewer Authority), in U.S. Congress, House Committee on Transportation and Infrastructure Subcommittee on Water Resources and the Environment, "Integrated Planning and Permitting," 112th Cong., 2nd sess., July 25, 2012.

[19] See https://www.epa.gov/sites/production/files/2015-05/documents/firstamendment-dcwasa-cd.pdf.

regulators; who determines a community's most pressing water quality needs; and whether a municipality can include ongoing needs for infrastructure rehabilitation under an integrated planning approach. Some stakeholders believe that clarification is needed regarding state and EPA roles. The agency's position is that it is the responsibility of cities to work and coordinate with state permitting agencies to develop integrated plans. However, some states are uncertain what EPA's oversight role would be if the agency disagrees with a plan that a state and municipality have developed.[20]

Some states believe that examples of integrated stormwater and wastewater plans are needed, in part to serve as models for states and communities. To do so, in October 2014, EPA awarded grants to five communities to assist them in developing integrated plans, as well as to provide transferable tools for other interested communities. Funding was awarded to Burlington, VT; Durham, NH; Santa Maria, CA; Springfield, MO; and Onondaga County, NY. The five were selected from 28 communities that had expressed interest in technical assistance from EPA.

With the planning policy in place, a number of communities have developed plans pursuant to it (e.g., Kansas City, KS; Seattle and King County, WA; and Cincinnati, OH). EPA officials have had discussions with other cities about writing and implementing integrated plans to manage stormwater and wastewater. But nearly five years after the framework policy was announced, one concern is that, so far, integrated plans have been incorporated only into new or amended consent decrees, not in CWA permits.

DETERMINING COMMUNITY AFFORDABILITY

A long-standing concern for local governments is EPA's process for evaluating how much communities can afford for CWA-mandated and other water infrastructure improvements. Affordability considerations can influence schedules established by EPA and states for communities to meet CWA requirements. In assessing municipalities' capability to finance infrastructure upgrades, EPA relies significantly on guidance issued in 1997.[21]

This guidance is intended to provide general boundaries to aid EPA, states, and cities in negotiating reasonable and effective schedules for implementing infrastructure upgrades. It uses a two-phase approach to assess financial capability. First, EPA identifies the combined impact of wastewater and CSO control costs on individual households, calculating average costs per household as a percentage of the local median

[20] In all but four states (Idaho, Massachusetts, New Hampshire, and New Mexico) and the District of Columbia, EPA has delegated CWA permitting to state permitting authorities. However, under Section 402(d) the act (33 U.S.C. §1342(d)), EPA retains an oversight role and may review individual permits that a state proposes to issue, and it may object to a permit outside the guidelines and requirements of the act. If the state does not revise the permit to meet the agency's objections, EPA may issue a permit that meets the act's requirements.

[21] EPA, Office of Water, Office of Wastewater Management, *Combined Sewer Overflows—Guidance for Financial Capability Assessment and Schedule Development*, EPA 832-B-97-004, February 1997, http://www.epa.gov/npdes/ pubs/csofc.pdf.

household income (MHI). This phase analyzes the residential share of current and planned controls needed to meet CWA requirements using a value range of whether the costs impose a "low" (less than 1% of MHI), "mid-range" (1%-2% of MHI), or "high" (more than 2% of MHI) financial impact on residential users, yielding a Residential Indicator. Second, EPA develops Financial Indicators to evaluate the debt, socioeconomic, and financial conditions that affect the community's financial capability as "weak," "mid-range," or "strong." The combined indicators measure a community's ability to afford compliance with CWA regulations. Many local government officials contend that EPA regions use the 2% of MHI measure as an absolute benchmark, or minimum threshold for determining that affordability is an issue for a given community.

For several years, municipal officials have urged EPA to revise the guidance, arguing that it should take into consideration a larger set of factors and that MHI is a misleading indicator of a community's ability to pay. In some cases, they say, cost impacts for an entire community may be in EPA's "mid-range," although impacts in portions of the community (e.g., low-income neighborhoods) are more than 2% of MHI.[22] Alternative household affordability metrics could include average water rates as a percentage of income for potentially vulnerable populations, or expected future water rate increases, or using other indicators of economic need such as the unemployment rate or poverty rate, or percentage of households receiving public assistance. Further, they say that affordability should be tailored to each local government.

Municipal officials also favor having EPA look at the overall financial impacts of both CWA and drinking water projects (which are not addressed in the 1997 guidance), rather than considering them separately. In 2013, local government officials issued a report with recommendations on alternative analyses that cities can present to federal and state regulators with data and information that demonstrate a range of additional factors relevant to evaluating affordability.[23] Further, a 2013 white paper issued by municipal water officials recommended that EPA assess financial capability to meet CWA obligations through an integrated planning framework that allows EPA to go beyond taking a "snapshot" of current financial indicators to also consider past or emerging trends that affect current conditions.[24]

EPA repeatedly insisted that it is not necessary to revise the guidance, which the agency believes already provides flexibility for financially disadvantaged municipalities,

---

[22] A report of the U.S. Conference of Mayors criticized EPA's use of MHI to assess affordability at the local level, arguing that MHI, which looks at household income across all parts of a community, masks the distributional cost impacts on below median income households. Based on a survey of California cities, the report found that the economic burden of water and stormwater costs falls disproportionately on such households. U.S. Conference of Mayors, *Public Water Cost Per Household: Assessing Financial Impacts of EPA Affordability Criteria in California Cities*, November 2014, http://www.usmayors.org/ pressreleases/uploads/2014/1202-report-watercostsCA.pdf.

[23] U.S. Conference of Mayors, American Water Works Association, and Water Environment Federation, *Affordability Assessment Tool for Federal Water Mandates*, May 2013, http://www.mayors.org/urbanwater/media/2013/0529-reportWaterAffordability.pdf.

[24] National Association of Clean Water Agencies, The Evolving Landscape for Financial Capability Assessment: Clean Water Act Negotiations and the Opportunities for Integrated Planning, May 2013, http://www.nacwa.org/images/ stories/public/2013-05-31fcalands apesipdesign.pdf.

including allowing a phased approach for implementing wastewater management controls. As noted above, EPA already has the flexibility and discretion to reopen consent decrees. Officials also point out that as a result of the integrated planning framework, in determining affordability, municipalities now can factor in the costs to manage stormwater flows, along with combined sewer overflows and wastewater treatment. The 1997 guidance did not include consideration of stormwater. To some extent, it appears that EPA and municipal officials fundamentally disagree on this point, with mayors and others contending that the 2% MHI metric is strictly enforced and is EPA's primary determinant, and EPA responding that it is one of many financial factors that should be evaluated to assess overall burden on a community.

Nevertheless, EPA committed to further dialogue with cities and states and to clarifying guidelines on how it considers financial capability. Based on those discussions and in consultation with its Environmental Financial Advisory Board, in November 2014, the agency released a Financial Capability Assessment Framework. The document identifies additional information that may help some communities provide a more complete picture of their financial capability than under the 1997 guidance alone. The new framework includes two sets of examples meant to supplement—but not revise—the existing guidance. One set of examples concerns residential impacts, such as service area poverty rates and trends, or the percent of households that own versus rent. The second set of examples concerns the financial strength of the community, such as unemployment data; state or local legal limitations on property taxes, other revenue streams, or debt levels; or other financial obligations, such as drinking water infrastructure, that significantly affect a city's ability to raise revenue.[25]

As mentioned above, communities' investment needs for clean water infrastructure are substantial. Funding needs for drinking water projects also are substantial; projects needs were estimated in 2011 to total $385 billion nationwide through 2031.[26] The 2014 Financial Capability Assessment Framework does allow for consideration of Safe Drinking Water Act (SDWA) obligations as relevant information about a community's overall financial capability. For example, a city's bond rating can be affected by both wastewater and drinking water obligations. The purpose of the assessment is to aid in establishing schedules for developing and implementing integrated plans and permit and consent decree implementation to meet CWA requirements. EPA's policies do not similarly address or integrate planning for communities' additional water infrastructure challenges of meeting SDWA requirements.

Using a wider range or more nuanced factors to assess financial capability would in all likelihood lead to additional determinations that affordability is a challenge for many U.S.

---

[25] Ken Kopocis, EPA Deputy Assistant Administrator for Water, and Cynthia Giles, EPA Assistant Administrator for Enforcement, *Financial Capability Assessment Framework for Municipal Clean Water Act Requirements*, memorandum, November 24, 2014, http://water.epa.gov/polwaste/npdes/cso/upload/municipal_fca_framework.pdf.

[26] EPA, *Drinking Water Infrastructure Needs Survey and Assessment*, EPA-816-R-13-006, April 2013, http://water.epa.gov/infrastructure/drinkingwater/dwns/index.cfm.

communities. Still, what response regulators and policymakers should make to such information is unclear. Some stakeholders would likely be satisfied with additional time to achieve compliance with applicable requirements, while others undoubtedly would prefer more time and relaxation of the requirements themselves. Many stakeholders seek flexibility in meeting regulatory requirements, as well as increased federal spending on local water and wastewater infrastructure projects. Federal and state regulators have considerable, but not unlimited, discretion to provide extended compliance time, but much less flexibility to alter standards or provide additional funds.

CONGRESSIONAL INTEREST

EPA's integrated planning process and community affordability are issues of interest to legislators, because they relate to policymakers' overall concern with funding needs for water infrastructure projects and the federal role in assisting communities. Both issues have been addressed in legislation and through congressional oversight.

In recent years, Members have offered bills that would address these issues in different ways. Some bills sought to codify EPA's integrated planning approach—as written in the 2012 framework—into the CWA. This approach might be considered largely symbolic: The 2012 framework did not provide new authority; it highlighted existing authority and opportunities within the CWA.

Other proposals would go beyond codifying the integrated planning approach by including other provisions that alter the CWA's existing framework to varying degrees. In general, some environmental groups have opposed such changes proposed in bills from prior Congresses.[27]

In the 115th Congress, H.R. 465, the Water Quality Improvement Act of 2017, would codify an integrated plan and permit approach into the CWA; direct EPA to carry out a pilot program to work with at least 15 communities desiring to implement an integrated plan, and require EPA to update the 1997 CSO affordability guidance document, which has been criticized by some stakeholders. The proposal also includes provisions that may alter the existing CWA framework. For example, the bill states that a "schedule of [permit] compliance would provide for *reasonable progress*" (emphasis added). This phrase may conflict with the regulatory requirement that permit schedules "shall require compliance *as soon as possible*" (40 C.F.R. 122.47(a)(1)). It is uncertain how these provisions, if enacted, would be implemented.

In the 114th Congress, H.R. 1093/S. 2358, the Clean Water Compliance and Ratepayer Affordability Act of 2015, would have directed EPA to carry out a pilot

---

[27] See, for example, letter to Congressmen Gibbs and Bishop from multiple environmental groups, "Water Resources and Environment Subcommittee Hearing entitled 'Integrated Planning and Permitting Framework: An Opportunity for EPA to Provide Communities with Flexibility to Make Smart Investments in Water Quality,'" July 23, 2014.

program to work with communities desiring to implement an integrated stormwater/wastewater management program. The bill would have allowed extended CWA discharge permit terms for such communities, and it would have allowed for modification of consent decrees to reflect terms of an integrated plan.[28]

A related proposal in the 114th Congress, the Clean Water Affordability Act of 2015, would have codified an integrated approach through CWA permits. This legislation also would have required EPA to update the 1997 CSO affordability guidance document. Bills in the 114th Congress that included these provisions are H.R. 1705, S. 2768, and S. 2848, which the Senate passed on September 15, 2016.[29] As described previously, communities generally support the financial capability framework issued by EPA in 2014, but some also support legislation that would further expand the criteria for determining affordability. In report language accompanying FY2016 appropriations legislation for EPA, the Senate Appropriations Committee directed EPA to contract with the National Academy of Public Administration to conduct an independent study to create a definition and framework for community affordability and report to Congress within one year.[30]

Some in Congress have sought financial support to test EPA's integrated planning policy. In 2014, Republican and Democratic leaders of the House Transportation and Infrastructure Committee's Water Resources Subcommittee wrote to House appropriators to request funds for a limited number of pilot projects. The letter requested $5 million in FY2015 appropriations for three to five projects per EPA region (30 to 50 projects in all) to assist communities in developing and implementing integrated plans. If successfully implemented, the letter said, EPA's framework "could help communities more affordably manage their clean water obligations while ensuring continuous progress toward water quality goals."[31] Some utility and municipal groups also have been seeking funding for pilot projects, and, as noted above, in October 2014, EPA awarded grants to five communities to assist them in developing integrated plans. In the report accompanying FY2016 appropriations legislation for EPA, the House Appropriations Committee said that it supports EPA's efforts to expand technical assistance for communities seeking to develop and implement an integrated planning approach, but the committee did not recommend specific funding for such assistance.[32]

In July 2014, the Water Resources Subcommittee held a hearing to review EPA's 2012 integrated planning policy. The subcommittee heard from witnesses who contended

---

[28] Related bills have been introduced previously, including H.R. 2707, H.R. 3862, and S. 2995 in the 113th Congress, and H.R. 1189 and S. 2094 in the 112th Congress.

[29] H.R. 3862 in the 113th Congress and S. 2094 in the 112th Congress also would have directed EPA to revise the 1997 affordability guidance.

[30] U.S. Congress, Senate Committee on Appropriations, *Department of the Interior, Environment, and Related Agencies Appropriations, 2016*, Report to accompany S. 1645, 114th Cong., 1st sess., June 23, 2015, S.Rept. 114-70, p. 54. See http://www.napawash.org/images/ WorkInProgress/EPA-Work_in_Progress-2-1-17.pdf.

[31] Letter from Representative Bob Gibbs and Representative Tim Bishop to Representative Ken Calvert, Chairman, and Representative Jim Moran, Ranking Member, Subcommittee on Interior, Environment and Related Agencies, House Committee on Appropriations, April 1, 2014.

[32] U.S. Congress, House Committee on Appropriations, *Department of the Interior, Environment, and Related Agencies Appropriations, 2016*, Report together with Minority Views to accompany H.R. 2822, 114th Cong., 1st sess., June 18, 2015, H.Rept. 114-170, p. 53.

that EPA has not provided enough regulatory flexibility in implementing the policy. Some witnesses urged Congress to pass legislation that would require EPA to speed its approval of integrated plans, and some urged EPA to develop "model permits" to serve as templates for integrated plans.[33] The Senate Environment and Public Works Committee held a hearing on April 7, 2016, that addressed issues of water infrastructure affordability.

The House Water Resources Subcommittee also held two hearings in the 112th Congress on EPA's efforts to provide flexibility to communities in addressing wastewater and stormwater project needs. The first hearing, in December 2011, examined EPA's integrated planning policy, then still in draft form, and the second hearing invited witnesses to discuss the policy document, which EPA had issued in June. Particularly at the second hearing, in July 2012, witnesses addressed concerns with EPA's approach to determining a community's ability to afford water infrastructure projects.

CONCLUSION

Local government stakeholders generally support EPA's efforts to encourage ways for communities to prioritize their investments in CWA infrastructure and to consider a wide range of factors that affect affordability of such investments. As described in this chapter, cities and states have some continuing concerns with aspects of both the integrated planning policy and the financial capability assessment framework. For example, EPA's 2014 Financial Capability Assessment Framework is focused on plans and implementation to meet CWA requirements. While it might be beneficial to municipalities if EPA's policies encompassed priorities, plans, and schedules for all of a community's water infrastructure obligations—including for drinking water projects, as the U.S Conference of Mayors has advocated—currently that is not the case. Nevertheless, these recent policies may partially help local governments address the economic challenges of meeting their CWA water infrastructure obligations.

The 2012 integrated planning framework highlighted existing authority and opportunities within the CWA without providing new authority or flexibility. Thus, legislative efforts to codify the framework into the CWA may be considered largely symbolic. However, some of the recent proposals would not only codify the integrated planning approach but also include provisions that would alter the CWA's existing framework to varying degrees. In general, these provisions have raised concerns from

[33] U.S. Congress, House Committee on Transportation and Infrastructure, Subcommittee on Water Resources and Environment, *Integrated Planning and Permitting Framework: An Opportunity for EPA to Provide Communities with Flexibility to Make Smart Investments in Water Quality*, 114th Cong., 2nd sess., July 24, 2014.

environmental groups but received support from municipalities seeking additional flexibility.

In: Environmental Protection Agency (EPA)
Editor: Coby Schaefer
ISBN: 978-1-53614-933-3

Chapter 4

EPA'S VESSEL GENERAL PERMITS: BACKGROUND AND ISSUES*

Claudia Copeland

SUMMARY

In November 2011 the Environmental Protection Agency (EPA) proposed two Clean Water Act (CWA) permits to regulate certain types of vessel discharges into U.S. waters. The proposed permits would replace a single Vessel General Permit (VGP) issued in 2008 that was due to expire in December 2013. As proposed, the permits would apply to approximately 71,000 large domestic and foreign vessels and perhaps as many as 138,000 small vessels. This universe of regulated entities is diverse as well as large, consisting of tankers, freighters, barges, cruise ships and other passenger vessels, and commercial fishing vessels. Their discharges are similarly diverse, including among other pollutants aquatic nuisance species (ANS), nutrients, pathogens, oil and grease, metals, and toxic chemical compounds that can have a broad array of effects on aquatic species and human health, many of which can be harmful.

EPA proposed two permits, one for large vessels to replace the 2008 VGP, and one for smaller vessels covered by a congressionally enacted temporary moratorium. Both were proposed well in advance of the VGP's expiration to provide ample time for the regulated community to prepare for new requirements. On March 28, 2013, EPA issued a final version of the VGP for large vessels. It took effect December 19, 2013. The permit for smaller vessels, the sVGP, was issued on September 10, 2014, and was scheduled to take effect on December 19, 2014. However, in December 2014, Congress passed legislation (S. 2444/P.L. 113-281) extending until December 18, 2017, the date when small vessels will need a CWA permit.

The CWA requires that all regulated discharges must meet effluent limitations representing applicable levels of technology-based control. The 2013 VGP largely retains the current permit's approach of relying on best management practices to control most discharges, because EPA concluded that it is infeasible to develop numeric effluent limits

* This is an edited, reformatted and augmented version of Congressional Research Service Publication No. R42142, dated October 12, 2016.

for most controlled discharges. However, the new VGP includes for the first time numeric ballast water discharge limits, which are consistent with standards in a 2012 Coast Guard rule and an international convention.

The 2013 VGP raises two key issues. One concerns inclusion of specific numeric ballast water discharge limits in the permit. At issue had been whether EPA would propose more stringent numeric limits, as some environmental groups have favored and a few states have already adopted. A second issue concerns the role of states in regulating vessel discharges.

Environmental groups and Canadian shippers challenged the 2013 permit in federal court. In October 2015, the court supported the environmentalists' challenge, ruling that the permit violated the CWA because it did not require the use of best available technology to control discharges of invasive species from ships' ballast water. The court remanded the permit to EPA, but did not vacate it; the permit remains in effect until EPA issues a new permit.

Congressional interest in this topic has been evident for some time. In 2008 Congress enacted two bills to exempt certain vessels from a CWA permit requirement, thus restricting the population of vessels subject to the VGP. One was a permanent permit moratorium for recreational vessels of all sizes. The other act was a temporary permit moratorium for small commercial vessels and commercial fishing vessels, which was extended twice by Congress and would have expired December 18, 2014, had Congress not enacted an additional three-year extension in P.L. 113-281. In the 114th Congress, bills addressing the temporary permit moratorium for small vessels and regulation of ballast water discharges have been introduced (S. 373/H.R. 980 and S. 371). The Senate Commerce Committee approved S. 373 in February 2015; the committee also included provisions of this bill in S. 2829, which the Senate passed in June 2016, and S. 1611. Further, the House has passed H.R. 4909, which includes the text of H.R. 980 as one title of that bill.

INTRODUCTION

In November 2011 the Environmental Protection Agency (EPA) proposed two Clean Water Act permits to regulate certain types of discharges from vessels into U.S. waters. The proposed permits would replace a single permit issued by EPA in 2008 that was due to expire in December 2013. As proposed, the two permits would apply to approximately 71,000 large domestic and foreign vessels and perhaps as many as 138,000 small vessels. This universe of regulated entities is diverse as well as large, consisting of tankers, freighters, barges, cruise ships and other passenger vessels, and commercial fishing vessels. Their regulated discharges are similarly diverse, including among other pollutants non-native aquatic nuisance species (ANS), nutrients, pathogens, oil and grease, metals, and toxic chemical compounds that can have a broad array of effects on aquatic species and human health, many of which can be harmful.

Developing and administering a regulatory program covering sources so numerous and different from one another is more complicated than for other currently regulated sources. Because the sources themselves are mobile and move between jurisdictions, the traditional mechanism of regulating through state-issued permits is problematic. Many regulated vessels are small entities; thus, the economic impacts of regulatory requirements are an important consideration. Identifying technology-based treatment systems and management practices that can control vessel discharges effectively and economically presents many challenges.

The two permits proposed by EPA in 2011 included one (draft VGP) for large vessels to replace the 2008 VGP, and one for smaller vessels covered by a congressionally enacted temporary moratorium (draft sVGP). On March 28, 2013, EPA issued a final version of the VGP for large vessels. It became effective on December 19, 2013. The permit for smaller vessels was issued on September 10, 2014; it was scheduled to become effective on December 19, 2014, but Congress passed a three-year extension as part of P.L. 113-281.

This chapter is an overview of the revised VGP for large vessels and two key issues: inclusion of numeric performance standards to limit ballast water discharges from vessels, and controversies about the role of states in regulating vessel discharges. It also reviews the final sVGP for small vessels and Congress's recent interest in these issues.

BACKGROUND: CLEAN WATER ACT REQUIREMENTS

The Clean Water Act (CWA) prohibits the discharge of pollutants from a point source into the navigable waters of the United States without a permit. Vessels are defined in the statute as point sources. In 1973, EPA promulgated a regulation that excluded discharges incidental to the normal operation of vessels (including ballast water, but not including vessel sewage discharges, which are regulated under CWA Section 312) from CWA permitting requirements. This long-standing regulation was challenged in federal district court by environmental advocacy groups who wanted EPA to address ballast water as a source of ANS in U.S. waters. In 2005 the court found that Congress had directly expressed its intention that discharges from vessels be regulated under the CWA, and that the 1973 regulation contradicted that intention. In September 2006 the court issued a final order vacating (revoking) the regulatory exclusion as of September 30, 2008, and remanding the ruling to EPA for further proceedings. The Ninth Circuit U.S. Court of Appeals upheld the district court's ruling on July 23, 2008.[1] On June 17, 2008, EPA proposed two CWA general permits in response to the court's 2006 order, one applicable to commercial vessels and one applicable to small recreational vessels.[2]

CWA permits are either individual permits issued to specific facilities or general permits. Both types of permit are issued for a specific period of time (not to exceed five years), after which the permit must be renewed. A general permit covers multiple facilities within a specific category having common elements, such as similar types of

---

[1] Northwest Environmental Advocates v. U.S. Environmental Protection Agency, 537 F.3d 1006 (9th Cir. 2008).

[2] U.S. Environmental Protection Agency, "Draft National Pollutant Discharge Elimination System (NPDES) General Permits for Discharges Incidental to the Normal Operation of Vessels," 73 *Federal Register* 34296-34304, June 17, 2008.

operations that discharge the same types of wastes. Because of the large number of potential sources of vessels,[3] EPA believed that it made administrative sense to use general permits, rather than individual permits. In August 2008, the federal district court agreed to EPA's request to delay vacating the regulatory exemption for three months, to ensure that permits could be issued before the exemption was eliminated. EPA finalized a Vessel General Permit for vessels subject to a permit requirement on December 18, 2008. The permit became effective on December 19, 2008. However, on the same day, the federal district court granted an EPA motion to delay vacating the existing regulatory exclusion until February 6, 2009. Thus, the effective date remained December 19, 2008, but regulated sources were not required to comply with terms of the permit until February 6, 2009.

The 2008 Vessel General Permit and 2011 Draft Vessel General Permits

In July 2008, Congress enacted two bills to exempt discharges incidental to the normal operation of certain types of vessels from CWA permitting, thus restricting the population of vessels subject to EPA regulation. The first measure, P.L. 110-288, the Clean Boating Act of 2008, exempted discharges incidental to the normal operation of recreational vessels of all sizes from CWA permitting requirements.[4] The legislation directed EPA and the Coast Guard to create a regulatory regime under new CWA Section 312(o); EPA is currently developing regulations for recreational vessels, as required by the legislation.[5]

The second measure, P.L. 110-299, provided a two-year moratorium on CWA permitting for certain discharges from commercial fishing vessels of all sizes and non-recreational vessels less than 79 feet in length. This moratorium has been extended three times. First, it was extended to December 18, 2013, by P.L. 111-215. During the moratorium, EPA was directed to study the discharges from these vessels and submit a report to Congress.[6] The 112th Congress extended the permit moratorium for one more year, until December 18, 2014, in P.L. 112-213. And in the 113th Congress, legislation providing an additional three-year extension, until December 18, 2017, was enacted as part of a Coast Guard reauthorization bill (S. 2444/P.L. 113-281). Ballast water

---

[3] EPA estimated that the universe of vessels potentially affected by the court's order and proposed permits could include over 13 million recreational boats and 98,000 commercial fishing, passenger, cargo and other vessels operating in U.S. waters.

[4] The CWA defines recreational vessels to mean any vessel that is manufactured or used primarily for pleasure, or that is leased, rented, or chartered to a person for the pleasure of that person. CWA Section 502(25); 33 U.S.C. 1362(25).

[5] See http://water.epa.gov/lawsregs/lawsguidance/cwa/vessel/CBA/about.cfm.

[6] U.S. Environmental Protection Agency, *Study of Discharges Incidental to Normal Operation of Commercial Fishing Vessels and Other Non-Recreational Vessels Less Than 79 Feet*, August 2010, http://water.epa.gov/polwaste/npdes/ vessels/Report-to-Congress.cfm.

discharges from vessels less than 79 feet in length are not affected by the moratorium (although EPA believes that few of these smaller vessels use or discharge ballast water) and are required to be authorized by permits. (See "Small Vessel Permit Moratorium" for discussion of legislation).

However, P.L. 110-288 and P.L. 110-299 did not exempt or provide a permitting moratorium for all discharges from all types of vessels. Thus, the Vessel General Permit (VGP) finalized by EPA in December 2008 gave permit coverage to an estimated 72,000 vessels larger than 79 feet in length used in a transportation capacity that were not affected by the moratorium in P.L. 110-299, including tankers, freighters, barges, and cruise ships, and it also applied to ballast water discharges from vessels covered by the moratorium.[7] It applied to pollutant discharges, including ballast water, that are incidental to the normal operation from non-recreational vessels that are 79 feet or more in length, and to ballast water discharges from commercial vessels of less than 79 feet and commercial fishing vessels of any length. Geographically, it applied to discharges into waters of the United States in all states and territories, extending to 3 miles from the baseline (i.e., shoreline).[8]

In the 2008 permit, EPA identified 26 types of waste streams or discharge types from the normal operation of covered vessels (some are not applicable to all vessel types). The types of pollutant discharges subject to the permit included ANS (also known as invasive species), nutrients, pathogens, oil and grease, metals, and pollutants with toxic effects.

The CWA requires that all point source discharges must meet effluent limitations representing applicable levels of technology-based control. Under the 2008 VGP, EPA concluded that, based on available information, it was not practicable to derive numeric effluent limits to achieve technology-based controls for many of the discharge types regulated under the permit. Thus, most discharges covered by the 2008 VGP were controlled by specific best management practices (BMPs), many of which were already in use.[9] Some vessel categories, such as cruise ships, were subject to more detailed requirements for discharges such as graywater (water from showers, baths, sinks, and laundry facilities) and pool and spa water. Monitoring, recordkeeping, and reporting requirements applied, as well.

Procedurally, vessels larger than 79 feet or more than 300 gross tons (an estimated 50,000 domestic and foreign vessels) were required to submit a Notice of Intent (NOI) to be covered by the permit. Smaller regulated vessels were automatically covered. There

---

[7] U.S. Environmental Protection Agency, "Final National Pollutant Discharge Elimination System (NPDES) General Permit for Discharges Incidental to the Normal Operation of a Vessel," 73 *Federal Register* 79473-79481, December 29, 2008.

[8] Discharges incidental to the normal operation of military vessels are not subject to the EPA permit; see Appendix B.

[9] BMPs include practices that generally are preventive in nature, such as vessel maintenance techniques or training personnel in use of on-board equipment to minimize pollutant discharges.

were no permit fees. Projected industry compliance costs (including paperwork requirements) ranged from a low of $8.9 million to $23.0 million annually; they varied based on assumptions of vessel populations affected and the number of instances in which incremental costs would be incurred.

In anticipation of the expiration of the 2008 VGP on December 18, 2013, in November 2011, EPA proposed two Vessel General Permits, one for large vessels (draft VGP) to replace the 2008 VGP, and one for smaller vessels to authorize discharges from vessels covered by the congressionally enacted temporary moratorium (draft sVGP).

Both draft permits proposed to regulate discharges from 26 types of waste streams (like the 2008 VGP), plus an additional waste stream category—fish hold effluent. Pollutants in these waste streams can include ANS, nutrients, pathogens, oil and grease, metals, and toxic chemical compounds. Both draft permits largely retained the 2008 permit's approach of relying on specific behaviors or BMP techniques to control most regulated discharges, as EPA again concluded that it is infeasible to develop numeric effluent limits for most controlled discharges covered by the permit. The draft VGP for larger vessels contained several changes that are discussed below— notably, including for the first time numeric ballast water discharge limits, more stringent effluent limits for oil-to-sea interfaces and exhaust gas scrubber washwater, as well as specifications to manage fish hold effluent. Both draft permits included streamlined recordkeeping and reporting requirements, modifying aspects of the 2008 VGP, such as allowing electronic recordkeeping and requiring an annual report in lieu of a one-time report and annual noncompliance report.

On March 28, 2013, after reviewing over 5,500 public comments on the draft permit, EPA issued a final permit to replace the 2008 VGP.[10] Requirements of the final permit are similar to the draft VGP. Smaller vessels continued to be covered by the congressional moratorium provided by P.L. 112-213 until December 18, 2014. If the moratorium had expired without further congressional action, smaller vessels would have been subject to the sVGP beginning on December 19.[11] However, as described below, Congress enacted an extension of the moratorium until December 18, 2017.

The following sections of this chapter provide an overview of EPA's revised VGP for large vessels and two issues of particular interest—requirements concerning ballast water management, and federal and state roles. It reviews recent congressional interest in these topics. Appendix A describes the sVGP for smaller vessels, now superseded by congressional action.

---

[10] U.S. Environmental Protection Agency, "Final National Pollutant Discharge Elimination System (NPDES) General Permit for Discharges Incidental to the Normal Operation of a Vessel; Notice of final permit issuance," 78 *Federal Register* 21928-21945, April 12, 2013. Full text of the permit and accompanying permit fact sheet and economic analysis are available at http://cfpub.epa.gov/npdes/vessels/vgpermit.cfm#final.

[11] U.S. Environmental Protection Agency, "Final National Pollutant Discharge Elimination system (NPDES) Small Vessel General Permit for Discharges Incidental to the Normal Operation of Vessels Less Than 79 Feet," 79 *Federal Register* 53702-53707, September 10, 2014. Full text of the sVGP and accompanying permit fact sheet and economic and benefits analysis are available at http://water.epa.gov/polwaste/npdes/vessels/Small-Vessel-General-Permit.cfm.

THE 2013 VGP FOR LARGE VESSELS

The 2013 final VGP applies to seven categories of vessels operating in a capacity of transportation that have discharges incidental to their normal operations into waters subject to the permit: commercial fishing including fish processing, freight barge, freight ship, passenger vessel, tank barge, tank ship, and utility vessel. Freight barges (such as open and covered dry cargo barges, 68% of total), tank barges (e.g., liquid cargo barges, 12%), and utility vessels (such as research vessels and tug vessels, 11%) account for the majority of the 58,600 domestic vessels eligible for coverage under the VGP. Of the 12,430 foreign vessels eligible for coverage, freight ships (e.g., container ships) account for 66%, and tank ships (such as oil tankers) account for 28% of the total. Like the 2008 VGP, "waters subject to the permit" means "waters of the United States," including the territorial seas as defined in the CWA and extending to 3 miles from the baseline.

EPA concluded that requiring all covered vessels to submit an NOI indicating coverage under the replacement VGP would be administratively impracticable, so the permit does not require operators of vessels smaller than 300 gross tons and with capacity to carry less than 8 cubic meters (2,113 gallons) of ballast water to submit NOIs. Consequently, more than 10,000 vessels would be automatically covered by the permit without submitting an NOI. This is essentially the same approach used in the 2008 VGP. However, all covered vessels are subject to the permit's requirements and must complete a Permit Authorization and Record of Inspection form and maintain that form on board at all times. The purpose of the form, according to EPA, is to confirm that vessels owners and operators have read the terms of the VGP and understand their obligation to comply.

As noted above, CWA permits normally are issued for a specific period of time, not to exceed five years, with a provision for reapplying for further permit coverage prior to the expiration date. EPA had proposed a four-year permit term for the draft VGP, as a way to ensure that the permit keeps pace with developing technologies, especially for ballast water treatment, but the final permit provides a five-year term, consistent with most EPA-issued CWA permits.[12]

The 2013 permit's principal ballast water and non-ballast modifications of the 2008 VGP are discussed next, along with EPA's economic and benefits analysis.

Ballast Water Requirements

Ballast water discharge has been identified as a major pathway for the introduction of ANS. Ships use large amounts of ballast water for stability during transport. Ballast water is often taken on in the coastal waters in one region after ships discharge wastewater or

---

[12] U.S. Environmental Protection Agency, *2013 Final Issuance of National Pollutant Discharge Elimination System (NPDES) Vessel General Permit (VGP) for Discharges Incidental to the Normal Operation of Vessels*, Fact Sheet, November 30, 2011, p. 12. Hereinafter, VGP Fact Sheet.

unload cargo, and then discharged at the next port of call, wherever more cargo is loaded, which reduces the need for compensating ballast. Thus, the practice of taking on and discharging ballast water is essential to the proper functioning of ships, because the water that is taken in or discharged compensates for changes in the ship's weight as cargo is loaded or unloaded, and as fuel and supplies are consumed. However, ballast water discharge typically contains a variety of biological materials, including non-native ANS that can alter aquatic ecosystems. Concern about harmful impacts of ballast water discharge was the core of the legal challenge by environmental groups to EPA's 1973 regulations, which ultimately led to issuance of the 2008 VGP.

The ballast water requirements of the 2008 VGP are minimal, largely requiring what current Coast Guard rules require—primarily use of ballast water exchange, or BWE. The 2008 permit mandates mid-ocean BWE for ships traveling outside the 200-nautical-mile exclusive economic zone (EEZ) of the United States. This requirement already applies under a 2004 Coast Guard rule (codified at 33 C.F.R. Part 151). EPA's VGP also requires BWE at least 50 nautical miles from shore for vessels engaged in Pacific nearshore voyages, which are not covered by the Coast Guard's mandatory exchange procedures. Further, the 2008 VGP requires vessels that declare they have "no ballast on board" either to seal the ballast tanks to prevent any discharge or to carry out saltwater flushing. The 2008 permit requires vessel operators to maintain a log book and records of ballast water management and submit reports of noncompliance to EPA annually.

The 2008 VGP does not include numeric limits on living organisms or pathogenic discharges, which some environmental groups have advocated that EPA issue. EPA explained this position in a fact sheet accompanying the 2008 permit.[13]

> EPA is not requiring any numeric treatment standards for the discharge of living organisms as part of this permit issuance and is instead requiring management practices (e.g., ballast water exchange) that decrease the risk of ANS introduction. EPA is proposing this approach because treatment technologies that effectively reduce viable living organisms in a manner that is safe, reliable, and demonstrated to work onboard vessels are not yet commercially available ... [R]equiring a numeric effluent limit for the discharge of living organisms is not practicable, achievable, or available at this time.... EPA will consider establishing treatment requirements in the next generation of permits that will provide for compliance with treatment standards that will be expressed as units of living or viable organisms per unit of volume in ballast water discharge.

While the 2013 VGP contains a number of ballast water BMP and recordkeeping requirements similar to the 2008 permit, the 2013 permit departs from the 2008 permit by specifying ballast water numeric discharge limits. By replacing the non-numeric

---

[13] U.S. Environmental Protection Agency, *2008 Final Issuance of National Pollutant Discharge Elimination System (NPDES) Vessel General Permit (VGP) for Discharges Incidental to the Normal Operation of Vessels Fact Sheet,* December 2008, pp. 62-63.

limitation for ballast water in the 2008 VGP with numeric limits, EPA expects that the changes will achieve significant reductions in the number of living organisms discharged via ballast water into waters subject to the permit. The VGP sets the numeric effluent limits for ballast water in terms of maximum acceptable concentration of living organisms per cubic meter discharged, as shown in the text box below. As discussed further below (see "Ballast Water Standards"), EPA now concludes that treatment technologies are available to meet limits in the VGP, and the requirements are economically practicable and economically achievable.

Ballast Water Numeric Discharge Limits in the 2013 VGP

1) For organisms greater than or equal to 50 micrometers in minimum dimension: discharge must include fewer than 10 living organisms per cubic meter of ballast water.
2) For organisms less than 50 micrometers and greater than or equal to 10 micrometers: discharge must include fewer than 10 living organisms per milliliter (mL) of ballast water.
3) Indicator microorganisms must not exceed:
   - For Toxicogenic *Vibrio cholerae* (serotypes O1 and O139): a concentration of less than 1 colony forming unit (cfu) per 100 mL.
   - For *Escherichia coli*: a concentration of fewer than 250 cfu per 100 mL.
   - For intestinal enterococci: a concentration of fewer than 100 cfu per 100 mL.

The numeric limits in the VGP are identical to performance standards specified in the International Maritime Organization's (IMO's) 2004 International Convention for the Control and Management of Ships' Ballast Water and Sediment.[16] They also are the same as standards finalized by the Coast Guard in 2012 under 33 C.F.R. Part 151 and 46 C.F.R. Part 162.[17] Many ballast water treatment systems produce or use biocides as a

[16] The IMO, a body of the United Nations, sets international maritime vessel safety and marine pollution standards. Numeric discharge limits in the IMO ballast water convention, referred to as the D-2 standards, will enter into force 12 months after ratification by at least 30 nations, representing 35% of the world merchant shipping tonnage. As of April 2016, this convention had been ratified by 44 nations, representing 34.79% of the world merchant shipping tonnage, but has not entered into force. The United States has not ratified the convention.

[17] U.S. Department of Homeland Security, Coast Guard, "Standards for Living Organisms in Ships' Ballast Water Discharged in U.S. Waters; Final rule," 77 *Federal Register* 17254-17320, March 23, 2012. This rule, as proposed in 2009, would have provided standards in two phases. Under that proposal, if a practicability review shows it is feasible, more stringent Phase 2 standards would apply by January 1, 2016, to new vessels constructed after that date and existing vessels that have not installed a ballast water management system by that date. The proposed Phase 2 standards would set concentration limits 1,000 times more stringent than Phase 1 standards for the smallest viable organisms and also would set limits on the discharge concentration for bacteria and viruses. However, the final Coast Guard rule promulgated in March 2012 defers the Phase 2 standard, pending assessment of additional data on available technology and development of a subsequent rule with an economic and environmental analysis to support a Phase 2 standard. The Coast Guard's authority for these rules is the Nonindigenous Aquatic Nuisance Prevention and Control Act, as amended by the National

disinfection agent to reduce living organisms present in the ballast water tank, but discharges of such substances may cause or contribute to violation of applicable water quality standards. Thus, the VGP also includes biocide effluent limitations to protect aquatic life.

The permit sets limits of 200 micrograms per liter (µg/l) of chlorine dioxide, 500 µg/l of peracetic acid, 100 µg/l of ozone, and 1,000 µg/l of hydrogen peroxide. Vessels may comply with the concentration-based numeric treatment limits in one of four ways: (1) discharge treated ballast water meeting the applicable numeric limits (i.e., by using treatment technology); (2) transferring the ship's ballast water to a third party for on-shore treatment; (3) use treated municipal/potable water as ballast water; or (4) by not discharging ballast water. EPA estimates that approximately 2,880 domestic and 5,270 foreign vessels are potentially subject to the ballast water standards because they operate with on-board ballast water tanks, and the agency anticipates that about 40% of covered vessels will comply by installing a ballast water treatment system. EPA has concluded that several treatment technologies capable of meeting the permit's numeric limits are commercially and economically available now for shipboard installation.

Under the VGP, new vessels constructed after December 1, 2013, must comply with the permit's numeric limits upon delivery. EPA determined that it would be infeasible to require all existing vessels to be fitted with ballast water treatment systems with a one-to-two year schedule. Thus, the permit requires existing vessels, constructed before December 1, 2013, to comply under a staggered schedule.

- Existing vessels with ballast water capacity of less than 1,500 cubic meters must comply by the time of their first scheduled drydocking after January 1, 2016.
- Existing vessels with ballast water capacity of more than 1,500 and less than 5,000 cubic meters must comply by the time of their first scheduled drydocking after January 1, 2014.
- Existing vessels with ballast water capacity greater than 5,000 cubic meters must comply by the time of their first scheduled drydocking after January 1, 2016.

This time schedule is consistent with the timelines in the Coast Guard's March 2012 rules, described above. The IMO D-2 standard includes a phased schedule for similar ballast water capacity sizes of vessels, but with slightly different implementation dates.

Certain vessel classes would not be subject to the ballast water numeric limits in the VGP. These include vessels engaged in short-distance voyages (e.g., they travel no more than 10 nautical miles), unmanned and unpowered barges, small inland and seagoing

Invasive Species Act (16 U.S.C. 4701 et seq.). The Coast Guard rule does not apply to ballast water discharges from military vessels; see Appendix B.

vessels (less than 3,000 gross tons), and existing bulk carrier vessels built before January 1, 2009, that operate solely within the Great Lakes (commonly known as Lakers). In general, according to EPA, these vessels face a number of challenges for managing ballast water, and in the case of existing Lakers there currently are no available treatment systems. Thus EPA has concluded that it is more appropriate to require these vessels to use BMPs such as avoiding discharge of ballast water in environmentally sensitive areas, but not require compliance with numeric limits. EPA will follow the state of technologies currently being tested for Lakers and will consider revising permit requirements during the term of the permit if technologies become available.[18]

Non-Ballast Water Requirements

The 2013 VGP contains several more stringent effluent limits/BMPs than in the 2008 VGP for certain vessel discharges. First, it requires all vessels to use "environmentally acceptable lubricants" on mechanical and other equipment that operate at the sea interface, such as wire rope or cables, unless technically infeasible. Vessel operators often use lubricants to maintain the functionality of such equipment, which can release quantities of oil or grease to water. The permit also requires maintenance BMPs to prevent leaks that could lead to oil discharges.

Second, the permit prescribes BMPs to reduce discharges of fish hold effluent, which was not covered by the 2008 VGP. Commercial fishing vessels use various methods to store seafood after it is caught. Fish hold effluent is composed of seawater, melted ice, or ice slurry that is collected inside fish hold tanks. It contains pollutants such as biological wastes and nutrients which result from seafood catch. In addition, because holding tanks often are cleaned or disinfected between catches, the resulting effluent can contain organic material, oils, nutrients, and bacteria and viruses. BMPs specified in the permit are intended to minimize the discharge of fish hold water and ice while vessels are stationary at a pier.

Third, the VGP includes numeric limits to control discharge to water of harmful exhaust emissions from engines that power ocean going vessels. The permit sets numeric limits and monitoring requirements for pH, turbidity, polycyclic aromatic hydrocarbons (PAHs), and nitrates plus nitrites. The limits are consistent with guidelines established by the IMO to implement engine and fuel standards in Annex VI of the International Convention for the Prevention of Pollution from Ships (MARPOL 73/78).[19]

---

[18] VGP Fact Sheet, p. 101. Lakers built after January 1, 2009, must meet ballast water treatment limits in the permit.

[19] MARPOL is an international treaty that regulates discharges from vessels. It includes several annexes that regulate specific types of vessel pollution, such as oil, garbage, and plastic. In the United States, MARPOL is primarily implemented by the Act to Prevent Pollution from Ships (33 U.S.C. 1901). The Coast Guard is the lead agency for implementing this statute.

The permit also includes certain administrative modifications of the 2008 VGP, which requires owners/operators to self-inspect their vessels routinely as well as annually in more detail and keep written records. The 2008 permit allowed use of electronic recordkeeping systems, and the 2013 permit includes provisions clarifying how such systems are to be maintained in forms as readable and legally dependable as a paper equivalent. All vessels must report electronically, unless specifically exempted. The 2013 VGP modifies the earlier permit's reporting requirements by consolidating requirements for an annual noncompliance report as part of an annual report, instead of calling for separate reports.

Additional Requirements for Certain Vessel Classes

Under the 2008 VGP, cruise ships are subject to more detailed requirements for certain discharges, such as graywater and pool and spa water, and additional monitoring and reporting. These additional requirements recognize that cruise ships generate considerably more graywater discharges than a container or cargo ship, and onboard amenities such as photo developing and dry cleaning produce chemicals that are toxic to the aquatic environment and, thus, are not authorized by the permit.[20] The 2008 VGP includes BMPs as well as numeric effluent limits for fecal coliform and residual chlorine in cruise ship discharges of graywater that are based on U.S. Coast Guard rules for discharge of treated sewage or graywater in Alaska. It also includes operational limits on cruise ship graywater discharges in nutrient-impaired waters, such as Chesapeake Bay or Puget Sound. The 2013 VGP retains the same numeric limits for graywater discharges, but tightens operational limits: cruise ships are prohibited from discharging graywater within 3 nautical miles of shore (rather than 1 nautical mile from shore under the 2008 VGP) unless it has been treated to the standards specified in the permit. In general, the 2013 permit includes the same requirements for large (more than 500 passengers) and medium cruise ships (carrying 100 to 499 passengers), but with some flexibility for the latter category because of differences in graywater holding capacity and operation.

The 2013 VGP also includes additional requirements for large ferries (to minimize potential spills, drips, and leaks associated with carrying vehicles), barges (to prevent contamination of condensation with oily or toxic materials), oil and petroleum tankers (to protect against environmentally harmful discharges of oil during cargo loading and unloading), research vessels (to authorize only discharges for the purpose of conducting research on the aquatic environment or its natural resources), and emergency vessels (specifically to allow discharges incidental to the public safety responsibilities of firefighting and similar boats).

---

[20] For background information, see CRS Report RL32450, *Cruise Ship Pollution: Background, Laws and Regulations, and Key Issues*, by (name redacted).

Economic Impacts, Benefits, and Uncertainties

EPA estimates that the total annual incremental costs of implementing the VGP will range from $7.2 million to $23.0 million (in 2010$) for domestic vessels.[21] For vessels covered by this permit, these costs would be in addition to previous costs for complying with the 2008 VGP, described above. About 90% of the costs of the permit are associated with requirements mandating the use of environmentally acceptable lubricants, followed by those for ballast water. Further, EPA's estimates of compliance costs do not include the capital costs of installing, operating, and maintaining ballast water treatment systems, as these costs were previously estimated by the Coast Guard in its 2009 regulatory proposal to be approximately $168 million per year and thus represent baseline for estimating costs of the 2013 VGP.

The average per vessel compliance costs of the 2013 permit range between $51 and $7,004, depending on the number of applicable discharge categories and existing baseline practices. Tank ships are expected to have the highest average compliance costs, due to potential incremental costs for oil tankers exclusively engaged in coastwise trade that may install onboard ballast water treatment systems to comply with the 2013 permit. Overall, EPA concludes that the estimated compliance costs generally represent a small share of per vessel operating revenue. Thus, meeting the 2013 VGP permit requirements is economically practicable and achievable for permittees.[22]

The principal benefits of the VGP will be reduced risk of ANS introduction and enhanced environmental quality from reduced pollutants, according to EPA. EPA concludes that the permit's ballast water management practices—including discharge standards, monitoring, and reporting—should reduce the number of ANS invasions, thus preventing significant future damages to fisheries, water-based recreation and tourism, biodiversity and ecosystems, threatened and endangered species, human health, and infrastructure. However, the agency cannot quantify these benefits

> [T]he complexity of analyzing the probability of ANS introduction and spread, the wide range and varied nature of impacts ANS invasions can cause, and the great breadth of the scope of this Permit prohibit EPA from developing a quantified estimate of these benefits.[23]

Likewise, EPA concludes that the permit's controls on specific discharges, as well as its general housekeeping requirements, can be expected to generate both monetized benefits—such as preventing fishery closures and adverse human health impacts and increasing recreation opportunities—and nonmonetized benefits—such as preventing

[21] EPA did not estimate compliance costs for foreign vessels subject to the VGP.

[22] U.S. Environmental Protection Agency, *Economic and Benefits Analysis of the Proposed 2013 Vessel General Permit (VGP)*, October 30, 2011, p. x. EPA updated overall compliance cost estimates but did not issue a revised economic analysis when it issued the final 2013 VGP.

[23] Ibid., p. 137.

further stress on biodiversity and ecosystems. The magnitude of benefits is not calculable, according to EPA.[24]

EPA acknowledged significant uncertainty about several assumptions affecting estimated costs of the VGP, including uncertainty regarding discharge control practices currently implemented and the number of vessels expected to implement new practices. There also is uncertainty, EPA said, about costs of certain treatment systems, such as for bilge water, and practices such as use of environmentally acceptable lubricants, because of limited data and unknowns about applicability to different vessels. As a result, EPA concluded that its estimates should be interpreted as illustrative of a range of incremental costs, not as a precise account of costs that a vessel owner may incur for any specific vessel.[25]

ISSUES

Two prominent issues raised by the 2013 VGP are questions about inclusion of specific numeric ballast water discharge limits in the permit, and controversies about the role of states in regulating vessel discharges.

Ballast Water Standards

Many observers expected EPA to propose numeric limits in the next iteration of the VGP after the 2008 permit, in view of the IMO and Coast Guard performance standards, which were promulgated in 2012. At issue had been whether EPA would propose more stringent numeric limits, as some environmental advocacy groups favor and a few states have already adopted.

Anticipating expiration of the 2008 VGP, in 2010 EPA requested two reports to advise the agency on possible changes to the permit's ballast water management requirements. First, EPA and the Coast Guard jointly asked the National Research Council (NRC) of the National Academy of Sciences to evaluate the state of the science to support a quantitative approach to setting ballast water discharge standards, that is, specific numeric limits. The two agencies sought advice to better understand the relationship between concentrations of living organisms in ballast water discharges and the probability of ANS successfully establishing populations in U.S. waters, that is, whether setting maximum permissible limits on live organisms in ballast effluent can adequately protect against establishment of ANS in aquatic systems. The resulting NRC report concluded that the density of organisms released in a ballast discharge is "but one

[24] Ibid., p. 145.
[25] Ibid., p. 114.

of scores of variables that can and do influence invasion outcome."[26] The NRC concluded that, while a benchmark discharge standard that reduces the concentration of organisms below levels achieved by open-sea BWE is an important first step, additional research is needed in order to focus on the relationship between the quantity, quality, and frequency of release and the risk of successful invasion by ANS.

Second, EPA asked its Science Advisory Board (SAB) to provide advice on technologies and systems to minimize the impacts of ANS in vessel ballast water discharge. EPA requested the SAB to assess whether existing shipboard treatment technologies can reach specified concentrations of organisms in vessel ballast water, how these technologies might be improved in the future, and how to overcome limitations in existing data. The SAB's overarching recommendation in its report[27] is that, rather than relying solely on numeric standards, the agency should adopt a risk-based approach to minimize impacts of invasive species in vessel ballast water discharge, including methods to reduce invasion events, process and environmental monitoring, containment, and eradication. The SAB found that several existing technologies have been demonstrated that are capable of meeting the IMO D-2/Coast Guard and more stringent standards, but that technology is not available to reliably test for standards 100 or 1,000 times more stringent than the IMO standard, such as the Phase 2 standard included in the 2009 Coast Guard proposal[28] or those adopted by California and New York.[29] Reaching a more stringent standard would require new treatment systems that have not been tested in order to determine their practicality and cost, according to the SAB.

These reports clearly influenced EPA's development of the 2011 draft VGP, which proposed to harmonize the permit's requirements for controlling ANS in ballast water discharge with the numeric limits in the IMO D-2 standards/Coast Guard rule, plus continued use of BMPs. EPA referenced both reports in explaining its conclusion that a more rapid implementation schedule than in the IMO D-2 standard is not economically achievable at this time, nor are more stringent numeric discharge limits practicable at this time. In particular, EPA concluded that data cited by the California State Lands Commission to justify that state's more stringent discharge limits "are not adequate to

[26] National Research Council, Committee on Assessing Numeric Limits for Living Organisms in Ballast Water, *Assessing the Relationship Between Propagule Pressure and Invasion Risk in Ballast Water*, June 2011, p. 5, http://www.nap.edu/catalog.php?record_id=13184.

[27] U.S. Environmental Protection Agency Science Advisory Board, *Efficacy of Ballast Water Treatment Systems: A Report by the EPA Science Advisory Board*, EPA-SAB-11-009, July 12, 2011.

[28] The March 2012 final Coast Guard rule does not include Phase 2 standards. See infra footnote 15.

[29] California Public Resources Code Section 71200, enacted in 2006. In response to the law, the California State Lands Commission promulgated standards 1,000 times more stringent than the IMO D-2/Coast Guard proposed Phase 1 standards and requiring compliance beginning January 1, 2009. California's rules also prohibit discharge of detectable living organisms for the largest size classes of organisms (greater than 50 micrometers in size). However, the SAB found that it is not reasonable to assume that current ballast water management systems are able to reliably meet or closely approach a "no detectable living organism" standard. In 2008 the New York Department of Environmental Conservation (DEC) promulgated rules requiring existing vessels to treat ballast waters to a level 100 times more stringent than the IMO D-2 standard in January 2013. Vessels built after that date must include technology to treat ballast water to the same level as California's standards. However, in February 2012, the DEC announced that it would delay the effective date of these more stringent standards until December 19, 2013 (the day after expiration of the 2008 VGP).

determine whether any of the treatment systems can meet a significantly more stringent limit than that proposed for this permit term."[30]

Nevertheless, EPA recognized that some commenters would urge the agency to require numeric limits more stringent than the IMO D-2/Coast Guard standards. Many environmental groups and some states have argued, for example, that setting a higher standard will better protect water quality from ANS invasion while also serving as incentive to industry to develop technology that meets the standard. Thus, EPA requested public comment on the appropriateness of the proposed ballast water controls in the draft VGP and whether to adopt alternative treatment limits (such as the California standards), as well as whether additional management measures discussed in the NRC or SAB reports (such as managing ballast uptake or reducing ballast water discharge volumes) should be incorporated in the permit.

In the 2013 final VGP, EPA retained the position taken in the 2011 draft VGP that, based on current treatment and monitoring technologies, more stringent standards, such as California's, cannot be supported. EPA continues to agree with the NAS that establishing precise, quantifiable ballast water discharge standards more stringent than the IMO D-2/Coast Guard standards is not possible at this time. The final permit does not include additional management measures as a general requirement. However, the 2013 permit acknowledges unique vulnerabilities of the Great Lakes system to ANS invasion through ballast water discharges, and it includes additional protection for these waters. It requires all vessels that operate outside the EEZ and more than 200 nm from any shore to conduct saltwater flushing of ballast tanks before entering Great Lakes waters through the Saint Lawrence Seaway System. Also, all vessels that are equipped to carry ballast water and that enter the Great Lakes must conduct open ocean ballast water exchange.

Federal Preemption and State Role

Preemption of state regulatory programs with a uniform national standard has been a key issue in dispute concerning efforts to regulate discharges from vessels, including ballast water discharges.

The CWA permits EPA to authorize qualified states to administer the act's principal permitting program under Section 402, and EPA has done so for 46 states. Pursuant to CWA Section 402(c)(1), after such authorization, EPA suspends issuance of permits in lieu of the state. In other situations when EPA has issued a CWA general permit covering a similar category of dischargers,[31] the EPA general permit only applies in non-

[30] U.S. Environmental Protection Agency, *2011 Proposed Issuance of National Pollutant Discharge Elimination System (NPDES) Vessel General Permit (VGP) for Discharges Incidental to the Normal Operation of Vessels*, Draft Fact Sheet, November 30, 2011, p. 83.

[31] For example, EPA's Multisector General Permit (MGSP) covers stormwater discharges from approximately 4,100 industrial facilities in 29 sectors. EPA's Pesticide General Permit (PGP), issued in October 2011,

authorized states where EPA retains permitting authority.[32] In such cases, the EPA general permit typically is the model for a general permit issued directly by the authorized state; the state-issued permit must be at least as stringent as the EPA permit, but can be more stringent. Further, CWA Section 510 allows states to adopt standards, discharge limitations, or other requirements more stringent than federal rules, meaning that if a state were to assume the responsibility to issue vessel permits under the CWA, it could do so with alternative requirements no less stringent than the federal requirements. States often want the flexibility to require standards more stringent than federal, and this general authority in the statute gives states the ability to tailor and strengthen their implementation of federal water quality programs to address local conditions and circumstances.

However, because vessels are mobile and frequently travel between jurisdictions, allowing individual states to issue CWA permits to vessels would be administratively more complex than issuing a permit to a factory or other stationary source. Thus, both the 2008 VGP and final 2013 VGP uniquely apply to vessel discharges into U.S. waters in all states and territories, regardless of whether a state is authorized to administer other aspects of CWA permitting.[33] By preempting states from issuing CWA permits for discharges incidental to the normal operation of vessels, the possibility of vessels being subject to potentially conflicting conditions as they move between the waters of different states is theoretically precluded.

However, even without issuing CWA permits, a number of states are effectively requiring vessels to meet their own discharge requirements beyond the VGP through a procedure called 401 certification. Under CWA Section 401, an applicant for a federal license or permit to conduct any activity that may result in a discharge to waters of the United States must provide the federal agency with a Section 401 certification. The certification, made by the state in which the discharge originates, declares that the discharge will comply with applicable provisions of the CWA, including state-established water quality standards. Section 401 provides states with two distinct powers: one, the power indirectly to deny federal permits or licenses by withholding certification; and two, the power to impose conditions upon federal permits by placing limitations on certification. Where states impose conditions on a federal permit—such as the VGP—the permittee must meet the additional state limitations as conditions of the federal permit. Prior to issuance of the 2013 VGP, 25 states certified the permit with additional permit conditions covering one or more of the 27 effluent streams. Of the 25 states, 14 certified the permit with conditions applicable to ballast water discharges, either with specific

---

authorizes point source discharges from the application of pesticides to U.S. waters; this permit will regulate discharges from about 365,000 pesticide applicators.

[32] Idaho, Massachusetts, New Hampshire, and New Mexico, plus the District of Columbia and most of the U.S. territories.

[33] EPA explains that it may issue permits in the 46 authorized states for discharges covered by the VGP because the discharges included in the VGP were formerly excluded from state CWA permitting programs (until the 2006 court ruling previously described, see "Background: Clean Water Act Requirements") and are not considered part of any currently authorized state program. 2013 VGP Fact Sheet, pp. 15-16.

numeric discharge standards, or with more general language prohibiting nuisance conditions or other conditions in order to protect state waters.[34]

A group of commercial shipping operators challenged the state certifications under the 2008 federal permit, contending that the shipping industry is placed in the difficult regulatory position of being subject to a single federal permit with multiple state requirements.[35] In federal court, the vessel operators argued that EPA should have provided notice and opportunity for comment before promulgating the final permit, which included the state certifications. They also argued that EPA erred by failing to consider possible effects and costs of compliance with state conditions. The court rejected the challenge, stating in its ruling that under the CWA, EPA does not have the power to amend or reject state certifications, which must be attached to the permit. The court wrote that petitioners do have recourse, including a challenge in state court to certification conditions imposed by a particular state, a challenge in federal or state court if they believe that a particular state's law imposes an unconstitutional burden on interstate commerce, or seeking modification of the CWA.[36]

States also have used their authority to issue state permits independent of the VGP. Both the commercial shipping industry and environmental groups have challenged these state actions, on differing grounds, but courts have generally upheld the permits. For example, a Minnesota appellate court upheld the state's permit despite challenges from an environmental group alleging that the state did not perform an adequate water quality impact review before issuing the permit and that the state failed to impose numeric limitations for ANS. Additionally, Michigan's permitting program and New York's 401 certification of the 2008 federal permit were upheld after challenges by shipping industry groups.[37]

The role of states in implementing the VGP is likely to remain an issue. EPA plans to provide a clearinghouse of information and other tools to track development of each state's 401 conditions.

LEGAL CHALLENGES TO THE VGP

Some environmental advocacy groups criticized the 2013 final permit, asserting that it does not include adequate requirements to stop or reduce the spread of invasive species. Challenges to the final permit were filed by the Natural Resources Defense Council, Inc.

---

[34] See U.S. Environmental Protection Agency, Final 2013 VGP, Vessel General Permit for Discharges Incidental to the Normal Operation of Vessels (VGP), March 28, 2013, Section 6, http://www.epa.gov/npdes/pubs/vgp_permit2013.pdf.

[35] For comparison, 10 states certified the 2008 VGP with ballast water conditions.

[36] Lake Carriers' Association v. EPA, 652 F.3d 1, 10 (D.C. Cir 2011).

[37] Mariel Yarbrough, "Ballast Water Permits Survive Challenges," *SandBar,* a publication of The National Sea Grant Law Center, vol. 8, no. 3 (2009), pp. 9-12.

and the Northwest Environmental Advocates, and the National Wildlife Federation, among others, in several federal appeals courts. All of the challenges were consolidated in the U.S. Court of Appeals for the Second Circuit in New York.

Another party in the consolidated legal challenge to the 2013 permit is the Canadian Shipowners Association (CSA), which asked the court to review the VGP's January 1, 2014, deadline for implementing best available technology for ballast water management systems. At issue is the fact that the Coast Guard has been granting compliance extensions to its ballast water rules because of unavailability of certified technologies.[38] The association argued that the deadline in the VGP, which does not provide for similar extensions, was not realistic—a point that EPA has conceded. EPA officials stated in congressional hearings that enforcement action against vessels that are unable to install Coast Guard-approved technology to meet the numerical limits in the 2013 VGP would have a low priority for the agency. In April 2014, the federal court granted CSA's request to stay the January 1, 2014, VGP deadline for vessels operated by CSA members, and subsequently the court agreed to a joint request by CSA and EPA to sever the shippers' petition for review from the environmental groups' challenge.

In October 2015, the court ruled on the environmentalists' challenge and found that EPA acted arbitrarily and capriciously in issuing parts of the 2013 VGP.[39] The court stated that, in choosing the IMO standard to control ballast water discharges, EPA did not adequately explain why standards higher/more stringent than the IMO standard should be used, given available technology. The court noted that the Science Advisory Board's 2011 report, discussed previously, identified a number of technologies, including onshore treatment, that can achieve standards higher than IMO for one or more organism sizes, which would require only "reasonable/feasible modifications." The court said that EPA should have adjusted its standard in light of available technologies discussed in the SAB report, or explained why it would not do so.

The court also agreed with environmentalists that EPA's decision to exempt Lakers built before 2009 from numeric effluent limits of the VGP was arbitrary and capricious. The court said that EPA's belief that there is a lack of supply of updated shipboard systems for Lakers to meet numeric standards was not a legitimate reason to exempt pre-2009 Lakers from the 2013 VGP.

The court remanded the permit to EPA for proceedings consistent with the opinion, but allowed the 2013 permit to remain in place until EPA issues a new VGP. The Canadian Shipowners Association's challenge to the 2013 VGP is in abeyance until EPA issues its response to the October 2015 remand order.

---

[38] Coast Guard rules require owners and operators of vessels to install treatment technologies that have been certified by two Coast Guard-approved, third-party laboratories. Because no technological system has yet attained approval, the Coast Guard has granted two-year extensions to nearly 350 vessels.

[39] Natural Resources Defense Council et al. v. EPA, No. 13-1745, 2nd Cir., Oct. 5, 2015.

Congressional Interest

Congressional interest in this topic has been evident for some time—as reflected in the bills enacted in 2008 and described previously to exempt certain vessels from a CWA permit requirement.

Small Vessel Permit Moratorium

As discussed above, a permit moratorium for small non-recreational vessels and commercial fishing vessels enacted in 2008 was temporary, but was extended twice by Congress until December 18, 2014. The sVGP was to apply to these vessels, if the moratorium had expired without congressional action.

Several bills that included provisions to make that moratorium permanent were introduced in the 113th Congress. These bills included H.R. 3464, which the House passed in May 2014 as a provision of H.R. 4005; S. 2094, approved by the Senate Commerce and Transportation Committee in July 2014; and S. 2963, approved by the Senate Environment and Public Works Committee in December 2014. Supporters of proposals in these bills argued that small vessels covered by the current moratorium—commercial fishing vessels and other small non-recreational vessels—contribute little pollution and are not sources of invasive species, so a permanent permit exemption is appropriate. Others disagree with this view and argued that because of environmental concerns over vessel discharges, a permanent permit moratorium is inappropriate.

Other legislation, to provide a one-year extension of the permit moratorium, also was introduced (H.R. 5769 and S. 2943). The House passed H.R. 5769 on December 3, 2014. Some supporters of the approach in these bills said that a short-term extension would allow for comprehensive congressional attention to a number of vessel issues, including ballast water standards, in the 114th Congress.

As the date for the expiration of the moratorium approached, on December 10, 2014, the Senate and House passed legislation (S. 2444/P.L. 113-281) that includes a three-year extension of the small vessel moratorium, until December 18, 2017. (See Appendix A for details of the sVGP that would have become effective without this congressional action).

Ballast water discharges from vessels less than 79 feet in length are not affected by the moratorium and are required to be authorized by permits. According to EPA, owners and operators of small vessels that discharge ballast water can seek coverage either under the general permit for large vessels, requiring compliance with numeric standards (see "Ballast Water Requirements" above), or the general permit for small vessels, which took effect December 18, 2014. As described in Appendix A, the sVGP prescribes best management practices (BMPs) for ballast water discharges, not numeric standards.

In the 114th Congress, several bills that would make the existing temporary permit moratorium for small vessels permanent have been introduced. These bills are S. 371, a bill that only addresses the permit moratorium, and S. 373/H.R. 980, bills that also address ballast water standards, as described next. The Senate Commerce, Science, and Transportation Committee approved S. 373 in February 2015, and it later included the text of the bill as titles of S. 2829, the Maritime Administration Authorization and Enhancement Act for Fiscal Year 2017, which the Senate passed by voice vote on June 29, 2016, and S. 1611, Coast Guard Authorization Act of 2015.

Ballast Water Standards Legislation

Legislation addressing regulation and management of ballast water discharges that can contribute aquatic invasive species into U.S. waters has been introduced in Congress a number of times since 2000. In the 112th Congress, two House Transportation and Infrastructure Committee subcommittees held a hearing that focused on how best to address invasive species problems. Since then, several legislative proposals have sought to harmonize ballast water management requirements in a single authority to be implemented by the Coast Guard, including provisions of H.R. 2838, which the House passed in 2011.This legislation would establish a single federal ballast water management standard (i.e., the IMO D-2/Coast Guard numeric standard described previously). The legislation would supersede existing state standards or permits for any discharge incidental to the normal operation of a commercial vessel, although states could develop a ballast water inspection and enforcement program. It also would supersede EPA's ballast water management requirements under the CWA. Upon enactment of the legislation, state 401 certifications for ballast water discharge in the 2008 VGP would expire. The Coast Guard would have the primary role for enforcing the ballast water performance standard.

In December 2012, Congress enacted H.R. 2838 with a number of modifications (P.L. 112-213). First, as noted previously, it extended for one additional year the moratorium on CWA permit requirements for small vessels that was first enacted in P.L. 110-299. Second, it deleted all of the comprehensive ballast water management requirements in the House-passed bill.

In the 113th Congress, similar legislation to establish nationally uniform ballast water discharge standards (S. 2094) was approved by the Senate Commerce, Science, and Transportation Committee in July 2014, but no further action occurred.

In the 114th Congress, legislation addressing ballast water standards similar to S. 2094 has been introduced: S. 373 and H.R. 980.[40] As noted above, these bills also would

---

[40] For additional discussion of the legislation, see CRS Report R44181, *The Vessel Incidental Discharge Act: Background and Issues*, by (name redacted).

make permanent the existing temporary permit moratorium for small non-recreational vessels and commercial fishing vessels.

The Senate Commerce Subcommittee on Oceans, Atmosphere, Fisheries, and Coast Guard held a hearing on issues concerning regulation and management of discharges incidental to the normal operation of vessels on February 4, 2015. Maritime industry witnesses discussed the overlapping Coast Guard and EPA regulatory requirements, and they also discussed concerns with additional state-imposed requirements. Two weeks later, the full committee approved S. 373, the Vessel Incidental Discharge Act. As approved, the bill is similar to the legislation that the committee approved in the 113th Congress, with some amendments. The amended bill would allow a state to adopt or enforce a more stringent ballast water performance standard if the Coast Guard determines that compliance with the state standard is achievable and is consistent with obligations under relevant international treaties or agreements. A second amendment would require vessels to conduct saltwater flushing of ballast water tanks prior to entering the Great Lakes (as they are required to do under the 2012 Coast Guard rule). As described above, the Commerce Committee subsequently included the text of S. 373 as titles of S. 2829, the Maritime Administration Authorization and Enhancement Act for Fiscal Year 2017, which the Senate passed by voice vote on June 29, 2016, and S. 1611, the Coast Guard Authorization Act of 2015.

Further, on May 18, the House passed H.R. 4909, the National Defense Authorization Act for FY2017. Title XXXVI of this bill, as passed, is identical to the text of H.R. 980. In the Statement of Administration Policy on H.R. 4909, the Administration indicated that it objects to this title of the bill.[41] The Senate-passed companion bill to H.R. 4909 (S. 2943) does not contain similar provisions.

APPENDIX A. THE sVGP

On September 10, 2014, EPA issued notice of final permit issuance of the sVGP for small vessels. This general permit was scheduled to become effective on December 19, 2014, unless the existing temporary permit moratorium were extended by congressional action—as in S. 2444, passed by Congress on December 10 (see "Small Vessel Permit Moratorium"). By announcing this final permit more than three months ahead of the effective date, EPA intended to give affected vessel owners and operators sufficient lead time to meet the permit's requirements.

The 2014 sVGP was to apply to non-military, non-recreational vessels operating in a capacity of transportation that are less than 79 feet in length. EPA estimated that approximately 115,000 to 138,000 domestic and 156 foreign vessels were potentially subject to the sVGP. They include various types of commercial fishing vessels, tugs and

[41] Executive Office of the President, Office of Management and Budget, *Statement of Administration Policy, H.R. 4909, National Defense Authorization Act for Fiscal Year 2017*, May 16, 2016.

towing vessels, water taxis and small ferries, tour boats, and various other types of vessels used for non-recreational purposes. Approximately 68,000 are commercial fishing vessels, comprising the largest category, which includes vessels involved in fish catching, fish processing, and charter fishing. The second-largest category is "unspecified" vessels (totaling 27,000), followed by passenger vessels (21,000), such as charter fishing vessels and harbor cruise vessels.[42]

These vessels were excluded from the 2008 VGP by the initial moratorium in P.L. 110-299, which Congress subsequently extended—most recently to December 18, 2017, in S. 2444.

The sVGP would have regulated several categories of discharges, including fuel management, engine and oil control, solid and liquid waste management, vessel hull maintenance, graywater, fish hold effluent, and ballast water. It prescribed BMPs such as preventive maintenance of engines and fuel tanks to minimize the occurrence of leaks and spills that could release fuel or oil to receiving waters, and the minimization of graywater discharges that may contain soaps and detergents or nutrients into sensitive water bodies and confined waters. Most of the practices are already widely implemented by vessels subject to the draft sVGP, according to EPA.

EPA concluded that few vessels covered by the sVGP are affected by ballast water management requirements, because vessels less than 100 feet long typically do not load and discharge ballast or rely on ballast for stability. However, for vessels less than 79 feet long that do use ballast as a stability enhancer, the sVGP prescribed BMPs, because in EPA's view, no existing treatment systems are believed to have been developed for vessels with these small amounts of ballast water. Appropriate ballast water management BMPs include avoiding or minimizing ballast water uptake in areas with a high potential to contain harmful organisms and only discharging the minimal amounts of ballast water necessary in U.S. coastal and inland waters.

Because of the large universe of vessels covered by the sVGP, EPA determined that requiring all of these vessels to submit an NOI would be an extremely large administrative burden. Further, requiring an NOI for these vessels would be of little value, because of the limited range of discharge types and the reduced likelihood that they will introduce significant quantities of toxic and conventional pollutants to waterways.[43] However, like the VGP, EPA would have required all vessel operators covered by the permit to comply with its requirements, including signing an sVGP Permit Authorization and Record of Inspection form and maintaining that form onboard at all times. The terms of the sVGP would expire five years after the permit's effective date.

EPA made several changes in the 2014 sVGP, compared with the 2011 draft permit. For example, the final permit added a condition that accumulated bilgewater must be

---

[42] As previously described, Congress permanently exempted recreational vessels of all size from CWA permit requirements in P.L. 110-288 (see "The 2008 Vessel General Permit and 2011 Draft Vessel General Permits").

[43] U.S. Environmental Protection Agency, *2014 Final Issuance of National Pollutant Discharge Eliminations System (NPDES) Small Vessel General Permit (sVGP) for Discharges Incidental to the Normal Operation of Vessels Less than 79 Feet (sVGP) Fact Sheet*, p. 14.

removed, to the extent practicable, prior to transporting a vessel from one waterbody to another over land.

Overall, EPA estimated that the sVGP requirements could result in total annual costs for domestic vessels ranging between $7.1 million and $16.9 million (in 2010 dollars), in the aggregate.[44] Approximately 35% of these costs were associated with vessel hull maintenance, 25% with recordkeeping and inspection, and 25% with engine and oil control BMPs. The average cost per vessel was estimated to range from $17 per year (for vessels that already implement control practices) to $133 per year. The estimated range depended on the number of applicable discharge categories and baseline practices.

EPA lacks data to quantify the environmental benefits of the sVGP, but qualitatively, the agency expected that reducing discharges incidental to the operation of small non-recreational vessels would have two broad categories of benefits: enhanced environmental quality from reduced loads of pollutants, and reduced risk of introducing and spreading invasive species.

As it did with the VGP, EPA acknowledged uncertainties about impacts of the sVGP, largely due to limitations of data regarding financial and operational characteristics of affected firms and compliance costs that firms may incur. Particularly for the smaller vessels covered by the sVGP, EPA said that uncertainty exists for the revenue data for firms and also on the number of firms that have vessels that could incur cost impacts. For example, EPA assumed the same range of cost per vessel for all industry sectors, based on the best and worst case scenarios, but the agency recognized that this simplifying assumption may be inaccurate, because some vessels may already be implementing discharge control practices and would therefore not incur additional costs. Despite uncertainties, EPA concluded that it is "unlikely that a significant number of firms in the commercial fishing industry incurring material economic impacts as a result of complying with the sVGP." The permit was expected to have some effect on small firms, because of the very large number of small operators, but "the exact impacts on the profitability of these small businesses are difficult to quantify ... due to limitations of the data."[45]

APPENDIX B. DISCHARGES INCIDENTAL TO THE NORMAL OPERATION OF MILITARY VESSELS

As noted in the text of this chapter, discharges incidental to the normal operation of military vessels are not subject to EPA's VGP or the 2012 Coast Guard rule concerning ballast water discharges. Instead, such discharges are subject to separate regulatory

---

[44] EPA did not estimate the compliance costs for foreign vessels, because their costs are not expected to have a direct impact on U.S. firms. U.S. Environmental Protection Agency, *Economic and Benefits Analysis of the Final 2014 Small Vessel General Permit (sVGP)*, August 2014, p. 17.

[45] Ibid., pp. 84-85.

requirements, pursuant to CWA Section 312(n), which Congress added to the act in 1996 (Section 325 of P.L. 104-106). Section 312(n) requires EPA and the Department of Defense (DOD) jointly to promulgate uniform national discharges standards (UNDS) for certain discharges incidental to the normal operation of a vessel of the Armed Forces, unless the Secretary of Defense finds that compliance with UNDS would not be in the national security interests of the United States. Once finalized, the standards will be applicable to discharges from U.S. Armed Forces vessels operating in the navigable waters of the United States, the territorial seas, and the contiguous zone.

Since enactment of these provisions, EPA and DOD have been working through a multi-stage rulemaking process to develop the required standards. This process has already been underway for 20 years and is likely to extend for an unknown number of additional years. First, in Phase I, the agencies promulgated regulations in 1999 identifying 25 discharges for which it is "reasonable and practicable" to require a marine pollution control device (MPCD). Phase I also identified 14 discharges as not requiring control with a MPCD. These 1999 regulations are codified at 40 C.F.R. Part 1700.

During Phase II, EPA and DOD are proposing discharge performance standards for the 25 categories of discharges identified in the 1999 rule. Phase II is occurring in three separate rulemakings. The first set of performance standards, which addressed 11 discharges, was published in 2014,[46] and the second set of performance standards, addressing another 11 discharges, was published in October 2016.[47] The third set of performance standards, which will address the final three discharges identified in the 1999 rule—all of which involve ballast discharges—will be published separately in a future rule.

Phase III of the UNDS rulemaking will be a DOD-only rule. It will require DOD, in consultation with EPA and the Coast Guard, to promulgate regulations governing the design, construction, installation, and use of MPCDs necessary to meet the performance standards covered by Phase II. Similar to Phase II, Phase III will be promulgated in three batches. CWA Section 312(n) requires DOD to promulgate the Phase III standards within one year of finalization of the Phase II standards, and these regulations are to become effective upon promulgation, unless DOD specifies another effective date.

EPA and DOD estimate that the universe of Armed Forces vessels affected by the UNDS rulemaking is about 6,230 ships of the U.S. Navy, Army, Marine Corps, Air Force, and Military Sealift Command. Approximately 18% of the total (1,090 vessels, including aircraft carriers, auxiliary ships, patrol ships, and submarines) are larger than 79 feet in length. The remaining 5,140 ships, including a variety of self-propelled boats, are less than 79 feet in length. According to EPA and DOD, in developing the Phase II

---

[46] Environmental Protection Agency and Department of Defense, "Uniform National Discharge Standards for Vessels of the Armed Forces—Phase II, Proposed rule," 79 *Federal Register* 6117-6135, February 3, 2014. These proposed standards have not been finalized.

[47] Environmental Protection Agency and Department of Defense, "Uniform National Discharge Standards for Vessels of the Armed Forced—Phase II Batch Two, Proposed rule," 81 *Federal Register* 69753-69772, October 7, 2016. The public comment on this proposal closes on December 6, 2016.

discharge performance standards, the agencies are referencing the 2013 VGP and the 2014 sVGP as the baseline for each comparable discharge incidental to the normal operation of a vessel of the Armed Forces.[48]

[48] Ibid., pp. 69755-69756.

In: Environmental Protection Agency (EPA)
Editor: Coby Schaefer
ISBN: 978-1-53614-933-3

Chapter 5

IMPLEMENTING EPA'S 2015 OZONE AIR QUALITY STANDARDS*

James E. McCarthy

SUMMARY

Implementation of revised ozone standards by the U.S. Environmental Protection Agency (EPA) is now moving forward, after the agency designated 52 areas with just over 200 counties or partial counties and two tribal areas as "nonattainment" for the standards. The standards— formally known as National Ambient Air Quality Standards (NAAQS) for ground-level ozone— are standards for outdoor (ambient) air. In 2015, EPA tightened both the primary (health-based) and secondary (welfare-based) standards from 75 parts per billion (ppb) to 70 ppb after concluding that protecting public health and welfare requires lower concentrations of ozone than were previously judged to be safe. Ozone aggravates heart and lung diseases and may contribute to premature death; the primary standard addresses these concerns. Ozone can also have negative effects on forests and crop yields, which the secondary NAAQS is intended to protect.

The designated nonattainment areas include counties in 22 states and the District of Columbia. Most of these areas have had previous experience as nonattainment for earlier versions of the NAAQS. Designation as nonattainment imposes more stringent permitting and pollution control requirement for new and modified stationary sources of emissions as compared with the requirements in areas that are in attainment of the NAAQS, and requires the development of State Implementation Plans demonstrating how emissions will be reduced sufficiently to reach attainment.

EPA estimates the cost of meeting the 70 ppb standard in all states except California at $1.4 billion annually in 2025. Because most California areas would have until the 2030s to reach attainment, EPA provided separate cost estimates for California ($0.80 billion annually, post-2025). These cost estimates are substantially less than those from the National Association of Manufacturers and other industry sources, which have been widely cited. The benefits of reducing ozone concentrations were estimated by EPA at $2.9-$5.9 billion annually by 2025. The dollar value of avoided premature deaths accounts for 94% to 98% of this estimate.

---

* This is an edited, reformatted and augmented version of Congressional Research Service Publication No. R43092, dated August 16, 2018.

The agency projects that most areas will be able to reach attainment of the new standards by 2025 as a result of already promulgated regulations for gasoline, autos, power plants, and other emission sources. These regulations are being implemented independently of the 2015 NAAQS revision.

Members of Congress have shown particular interest in whether the expected benefits of the standards justify their projected costs. There is controversy over the methods used to estimate both costs and benefits. As the Clean Air Act is currently written, however, the agency is prohibited from weighing costs against benefits in setting NAAQS standards. The statute simply directs EPA to set the primary standard at a level requisite to protect public health, allowing an adequate margin of safety.

Various interest groups have lobbied against strengthening the standards. In the 115th Congress, the House has passed H.R. 806 to delay implementation of the 2015 NAAQS until the mid-2020s and to make changes to the process of future NAAQS revisions. The House Appropriations Committee reported a similar delay as a rider to EPA's 2018 appropriation (in Section 432 of H.R. 3354), but the final version of the appropriation (in H.R. 1625/P.L. 115-141) did not include the rider.

INTRODUCTION

On October 1, 2015, under the authority of Sections 108 and 109 of the Clean Air Act (CAA) as amended,[1] the Environmental Protection Agency (EPA) finalized revisions[2] to the National Ambient Air Quality Standards (NAAQS) for ground-level ozone[3] from the then-current level of 75 parts per billion (ppb) to 70 ppb.

As described in detail below (see "Implementation of the 2015 Standards"), under the CAA, revision of the standards sets in motion a process in which states and EPA identify areas that do not meet the standards ("nonattainment areas") and the states prepare implementation plans to demonstrate how emissions will be lowered sufficiently to reach attainment in those areas.

The revised standards are based on a review of more than 1,000 scientific studies by EPA staff.[4] The EPA science assessment was itself reviewed by a panel of 20 outside scientific experts, a group known as the Clean Air Scientific Advisory Committee (CASAC).[5] The EPA Administrator was under a court order to sign any changes or reaffirm the standards by October 1, 2015. A comment period, public hearings, and interagency review had followed the publication of a proposed range of standards in December 2014.

---

[1] The Clean Air Act, codified as 42 U.S.C. 7401 *et seq*.

[2] EPA, "National Ambient Air Quality Standards for Ozone: Final Rule," October 1, 2015, http://www3.epa.gov/ozonepollution/actions.html. The rule appeared in the Federal Register October 26, 2015, at 80 *Federal Register* 65292.

[3] Ground-level (tropospheric) ozone can be a lung irritant with serious adverse health effects. In the stratosphere, however, ozone protects the Earth from harmful ultraviolet rays of the sun. For more information, see EPA, "Ozone— Good Up High Bad Nearby," at https://cfpub. epa.gov/airnow/index.cfm?action=gooduphigh.index.

[4] See EPA, "Final Report: Integrated Science Assessment of Ozone and Related Photochemical Oxidants," U.S. EPA, Washington, DC, EPA/600/R-10/076F, 2013, http://cfpub.epa.gov/ncea/isa/recordisplay.cfm?deid=247492.

[5] CASAC's review of the ISA, dated November 14, 2012, can be found at http://yosemite.epa.gov/sab/sabproduct.nsf/60C2732674A5EEF385257AB6007274B9/$File/EPA-CASAC-13-001+unsigned.pdf.

Ground-level ozone (often referred to as "smog") is associated with potentially serious health effects when present in high enough concentrations. These health effects include aggravated asthma, chronic bronchitis, and heart attacks, and in some cases premature death. In the Regulatory Impact Analysis (RIA) accompanying the 2015 rule, EPA stated that reducing concentrations to 70 ppb versus the previous standard of 75 ppb[6] would avoid 320 to 660 premature deaths annually in 2025 nationwide (excluding California).[7] California was excluded from EPA's estimate of both costs and benefits of the nationwide standard, because most areas of the state will have until the 2030s to reach attainment of the NAAQS.[8] California costs and benefits are presented separately in the "Costs and Benefits of Control" section of this chapter.

High ozone concentrations also affect the growth of plants, causing damage to both forests and field crops. In 2006, the U.S. Forest Service examined 380 monitoring sites in the Mid-Atlantic and Southeastern states and found visible injury to forest plants from ozone at 121 (32%) of them. At 20 of the sites, the damage was described as "severe."[9] In addition, EPA found that "several economically important crop species are sensitive to ozone levels typical of those found in the United States," and estimated that crop losses could be reduced by $400 million to $620 million annually by implementation of a more stringent ozone standard.[10]

While EPA's analysis has found that there would be substantial health and welfare benefits to reducing ozone, the agency also has concluded that there could be substantial costs. In the RIA, EPA estimated the economic value of the benefits of reducing ozone concentrations to 70 ppb (for areas other than California) at $2.9-$5.9 billion annually by 2025, with costs at $1.4 billion annually in 2025.[11]

---

[6] The ozone NAAQS is also frequently expressed in terms of parts per *million* rather than parts per *billion*. A standard of 0.075 parts per million (ppm) would be the same as a standard of 75 parts per billion (ppb), and is often referred to as such. This report uses parts per billion in most cases, as does EPA in most of the explanatory materials accompanying the standard. Until 2008, the standard was expressed in parts per million with two decimal places (e.g., 0.08 ppm from 1997 to 2008). Without a third decimal place, concentrations as high as 0.084 ppm (or 84 ppb) could be rounded to 0.08 and considered to be in attainment of the standard. The 2008 standard added greater precision to the standard by adding a third decimal place (0.075 ppm).

[7] EPA, *Regulatory Impact Analysis of the Final Revisions to the National Ambient Air Quality Standards for Ground-Level Ozone*, September 2015, pp. ES-14 to ES-19, at https://www3.epa.gov /ttn/naaqs/standards/ozone/data/20151001ria.pdf. Hereinafter, "EPA, 2015 RIA." A separate estimate for California shows a 70 ppb NAAQS resulting in 120-220 premature deaths avoided annually post-2025 (p. ES-18).

[8] Most of California's population lives in areas classified as Severe or Extreme for their nonattainment, under the Clean Air Act's statutory categorization scheme. These areas have more stringent emission control requirements and more time to reach attainment, so EPA provided separate cost-benefit estimates for the state.

[9] EPA, *Regulatory Impact Analysis, Final National Ambient Air Quality Standard for Ozone*, July 2011, p. 64, at https://www.nrc.gov/docs/ML1224/ML12240A237.pdf.

[10] Ibid., p. 74.

[11] EPA, 2015 RIA, p. ES-15. Most of California's population lives in areas classified as Severe or Extreme for their nonattainment, under the Clean Air Act's statutory categorization scheme. These areas have more stringent emission control requirements and more time to reach attainment, so EPA provided separate cost-benefit estimates for the state.

Source: U.S. EPA Green Book, https://www3.epa.gov/airquality/greenbook/map/map8hr_ 2008.pdf. Map shows areas designated nonattainment with respect to the 2008 ozone standard by EPA as of June 30, 2018.

Notes: Nonattainment designations were based on 2008-2010 monitoring data in most cases. Eighteen of the 38 areas shown now have monitoring data indicating attainment of the standard, but, as of June 2018, had not completed administrative requirements to be reclassified to "attainment."

Figure 1. Ozone Nonattainment Areas (2008 Standard, 0.075 ppm) as of June 2018.

The ozone NAAQS are among EPA's most far-reaching standards. At the level set in March 2008 (75 ppb), 107 million people lived in areas that were designated nonattainment for the standards (see *Figure 1*), as of June 2018.[12] These 38 areas (referred to as "nonattainment areas") include 168 counties in 17 states and the District of Columbia.[13] The areas generally coincide with metropolitan areas, but may be larger or smaller. Of the 38 areas, 18 have ozone readings that indicate they have attained the standard, which would leave 20 areas with a combined population of nearly 78 million not attaining the standard.[14]

When it promulgated the 2015 NAAQS revision, EPA stated that if the most recent monitoring data then available (for the years 2012-2014) were used to identify nonattainment areas for the revised standards, the number of counties showing nonattainment would be 241 in 33 states and the District of Columbia.[15] The actual

[12] For specifics, see EPA's "Green Book," at https://www.epa.gov/green-book.

[13] EPA, *Green Book 8-Hour Ozone (2008) Area Information*, Designated Area/State Information, June 30, 2018, https://www3.epa.gov/airquality/greenbook/hbtc.html.

[14] EPA may issue a determination of attainment, sometimes referred to as a "Clean Data Determination," after notice and comment rulemaking determining that a specific area is attaining the relevant standard. This is not the same as redesignation to attainment, however. EPA does not formally redesignate areas as attainment until the state meets statutory criteria, including submission of a certified monitoring data and obtaining EPA approval for a Maintenance Plan. The Maintenance Plan shows how the nonattainment area will maintain compliance with the NAAQS for a 10- year period. For additional discussion see EPA, "Determinations of Attainment by the Attainment Date, Extensions of the Attainment Date, and Reclassification of Several Areas for the 2008 Ozone National Ambient Air Quality Standards," 86 *Federal Register* 26697, 26701, May 4, 2016. For list of areas with Clean Data Determinations/Attainment Determinations that are designated as nonattainment with respect to the 2008 ozone standard (as of June 30, 2008), see EPA Green Book, https://www3.epa.gov/airquality/greenbook/hfr2rpt5.html.

[15] See EPA, "County-level Design Values for the 2015 Ozone Standards," at https:// www.epa.gov/ozone-pollution/ozone-data-tables-1997-2008-and-2015-standards.

nonattainment designations promulgated in 2018 are in most cases based on 2014-2016 monitoring data and include just over 200 counties or partial counties in 22 states and the District of Columbia, plus two tribal areas. (See *Figure 2.*) In the intervening years, the emissions that contribute to ozone formation have declined, in response to already promulgated standards for motor vehicles, gasoline, power plants, and other sources of emissions. Thus, the number of counties showing nonattainment is lower than the EPA estimate released in 2015.[16]

The potential economic, health, and environmental impacts of a change in the ozone NAAQS have led to great interest in EPA's review and implementation of the standards. To assist Members and staff in evaluating the process and its potential impacts, this chapter provides background on NAAQS, the process used to establish them, the 2008 and 2015 ozone standards, and issues that have been raised as EPA implements the final rule.

WHAT ARE NAAQS?

NAAQS are federal standards that apply to ambient (outdoor) air. Section 109 of the Clean Air Act directs EPA to set both primary NAAQS, which are standards, "the attainment and maintenance of which in the judgment of the [EPA] Administrator ... are requisite to protect the public health," with "an adequate margin of safety," and secondary NAAQS, which are standards necessary to protect public welfare, a broad term that includes damage to crops, vegetation, property, building materials, climate, etc.[17]

The pollutants for which NAAQS have been set are generally referred to as "criteria" pollutants. The act defines them as pollutants that "endanger public health or welfare," and whose presence in ambient air "results from numerous or diverse mobile or stationary sources."[18] Six pollutants (or categories of pollutants) are currently identified as criteria pollutants: ozone, particulates, carbon monoxide, sulfur dioxide, nitrogen oxides, and lead. The EPA Administrator can add to this list if he determines that additional pollutants meet the definition, or delete pollutants from the list if they no longer meet the definition. There have been no changes to the list, however, since the late 1970s.

NAAQS are at the core of the Clean Air Act even though they do not directly regulate emissions. In essence, they are standards that define what EPA considers to be

[16] When the agency promulgated a revision to the ozone NAAQS in March 2008, it identified 345 counties with monitors showing nonattainment of the new standard. By the time the agency designated nonattainment areas in May 2012, the number of nonattaining counties had declined to 232.

[17] The CAA's definition of welfare is found in Section 302(h) of the act (42 U.S.C. 7602(h)).

[18] Authority to establish NAAQS comes from both Sections 108 and 109 of the act; this definition of criteria pollutants is found in Section 108. The authority and procedures for controlling the sources of criteria pollutants are found throughout Titles I, II, and IV of the act. Many pollutants that are less widely emitted are classified as "hazardous air pollutants" and are regulated under a different section of the act (Section 112). That section lists 187 pollutants or groups of pollutants as hazardous and establishes different authorities and requirements for controlling their emissions.

clean air for the pollutant in question. Once a NAAQS has been set, the agency, using monitoring data and other information submitted by the states, identifies areas that exceed the standards and must reduce pollutant concentrations; these areas are referred to as "nonattainment areas." The designation process is often delayed by litigation over the standards, by EPA's agreement to reconsider aspects of them, or by consultations with the states regarding the specifics of the areas to be designated. Designation of nonattainment areas for the 1997 ozone NAAQS, for example, took seven years. Designations under the 2008 standards took four years.

After nonattainment areas are designated, state and local governments have up to three years to produce State Implementation Plans (SIPs), which outline the measures that will reduce emission levels and attain the standards. Finalizing SIPs, through EPA review and approval, often takes longer.

Under the CAA, actual attainment of the ozone NAAQS is allowed to stretch over a three-year to 20-year period, depending on the severity of the area's pollution. Ozone nonattainment areas are designated as Marginal, Moderate, Serious, Severe, or Extreme, depending on the level of pollution. Each of these classifications comes with required pollution control measures: the more severe the pollution, the more stringent are the required controls, and the longer the area is allowed before it must demonstrate attainment.[19]

Thus, establishment or revision of a NAAQS is not an event that requires immediate compliance with an air quality standard; rather, it sets in motion a long and complex implementation process. That process may ultimately have far-reaching impacts for public health and welfare, for sources of pollution in numerous economic sectors, and for states and local governments.

EPA has several tools available to encourage nonattainment areas to comply with a NAAQS. The most frequently mentioned of these is the potential for highway-fund sanctions: failure to submit or implement a SIP adequate to attain or maintain compliance with the NAAQS can lead to the temporary suspension of federal highway funds for new nonsafety-related projects. Ultimately, EPA can impose a federal implementation or maintenance plan (a FIP) in an area that does not have an approved SIP. Imposition of sanctions or FIPs is relatively rare, however: generally the states avoid sanctions by submitting plans that require sufficient emission reductions to be deemed adequate by EPA.[20]

In addition to state and local actions to address ambient concentrations of NAAQS pollutants, EPA itself acts to control emissions and concentrations of criteria pollutants through national standards. These national standards apply to products that contribute to ozone and other criteria pollution (particularly mobile sources, such as automobiles) and to stationary emission sources (such as power plants). These standards lead to emission

---

[19] For a more detailed discussion, see CRS Report RL30853, *Clean Air Act: A Summary of the Act and Its Major Requirements.*

[20] See CRS Report R44050, *Transportation Conformity Under the Clean Air Act.*

reductions that states can factor into their implementation plans, reducing the need for local air pollution control measures.

REVIEWING THE OZONE NAAQS

Section 109(d) of the Clean Air Act requires the agency to review each NAAQS every five years. That schedule is rarely met, but it often triggers lawsuits that force the agency to undertake or complete a review.[21] In June 2013, the Sierra Club and three other groups filed suit over EPA's failure to complete the required review of the 2008 ozone NAAQS by the March 2013 deadline, and a court subsequently ordered EPA to propose any changes to the standards by December 1, 2014, and complete the review, with promulgation of any revisions by October 1, 2015.[22]

A historical review of the ozone NAAQS and their revisions is presented in *Table 1.*

The NAAQS Review Process

Reviewing an existing NAAQS is generally a long process. To begin the process, EPA scientists compile the scientific literature published since the last NAAQS revision, and summarize it in a report known as a Criteria Document or Integrated Science Assessment (ISA). The ISA for ozone, completed in February 2013, reviewed more than 1,000 scientific studies.[23]

Ozone ISAs cover topics as wide-ranging as the physics and chemistry of ozone in the atmosphere; environmental concentrations, patterns, and exposure; dosimetry; animal-to-human extrapolation; toxicology; interactions with co-occurring pollutants; controlled human exposure studies; epidemiology; effects on vegetation and ecosystems; effects on UVB (ultraviolet light) exposures and climate; and effects on man made materials.

Following completion of the ISA, EPA prepares a Risk and Exposure Assessment to identify exposure pathways, at-risk populations, and health endpoints. This document was completed for the 2015 review in August 2014.[24]

A final document prepared by EPA staff, the Staff Paper or Policy Assessment, summarizes the information compiled in the ISA and Risk Assessment and provides the

---

[21] CRS Report R41563, Clean Air Issues in the 112th Congress, summarizes EPA's recent efforts to review the NAAQS and implement revisions, including the next steps as of 2012 for each of the six criteria pollutants. Reviews of all six pollutants (ozone, PM, lead, NO2, CO, and SO2) have been completed since 2006, with the standards being made more stringent for five of the six.

[22] Sierra Club v. EPA, No. 13-2809 (N.D. Cal., Apr. 30, 2014) (unpublished).

[23] EPA, Office of Research and Development, *Integrated Science Assessment of Ozone and Related Photochemical Oxidants*, Final Report, February 2013, at http://www.epa.gov/ ncea/isa/ozone.htm. For access to references cited in the first, second, and third external review draft and final ozone ISAs, see EPA's "Health and Environmental Research Online" page at http://hero.epa.gov/index.cfm/project/page/project_id/1628.

[24] EPA, *Ozone (O3) Standards - Documents from Current Review - Risk and Exposure Assessments*, at https://www.epa.gov/naaqs/ozone-o3-standards-risk-and-exposure-assess ments-current-review.

Administrator with options regarding the indicators, averaging times, statistical form, and numerical level (concentration) of the NAAQS. A Policy Assessment for the 2015 NAAQS revision was completed and publicly released on August 29, 2014.[25]

Table 1. History of the National Ambient Air Quality Standards for Ozone (1971-2015)

Year	Final Rule/Decision	Primary/ Secondary	Indicator	Averaging Time	Level	Form
1971	36 *FR* 8186 Apr 30, 1971	Primary and Secondary	Total photochemical oxidants	1-hour	0.08 ppm	Not to be exceeded more than one hour per year.
1979	44 *FR* 8202 Feb 8, 1979	Primary and Secondary	Ozone	1-hour	0.12 ppm	Attainment is reached when the number of days per calendar year, with maximum hourly average concentration greater than 0.12 ppm, is equal to or less than one.
1993	58 *FR* 13008 Mar 9, 1993	EPA decided that revisions to the standards were not warranted at the time.				
1997	62 *FR* 38856 Jul 18, 1997	Primary and Secondary	Ozone	8-hour	0.08 ppm	Annual fourth highest Daily maximum 8-hr concentration, averaged over 3 years.
2008	73 *FR* 16483 Mar 27, 2008	Primary and Secondary	Ozone	8-hour	0.075 ppm	Annual fourth highest Daily maximum 8-hr concentration, averaged over 3 years.
2015	80 *FR* 65292 Oct 26, 2015	Primary and Secondary	Ozone	8-hour	0.070 ppm	Annual fourth highest Daily maximum 8-hr concentration, averaged over 3 years.

Source: EPA, http://epa.gov/ttn/naaqs/standards/ozone/s_o3_history.html.

[25] EPA, Office of Air and Radiation, *Policy Assessment for the Review of the Ozone National Ambient Air Quality Standards*, August 2014, at http://www.epa.gov/ttn/naaqs/ standards/ozone/data/20140829pa.pdf. Hereinafter "EPA, 2014 Policy Assessment."

To ensure that NAAQS reviews meet the highest scientific standards, the 1977 amendments to the Clean Air Act required the Administrator to appoint an independent Clean Air Scientific Advisory Committee (CASAC). CASAC has seven members, largely from academia, state air pollution agencies, and private research institutions. In conducting NAAQS reviews, their expertise is supplemented by panels of the nation's leading experts on the health and environmental effects of the specific pollutants that are under review. These panels can be rather large. The panel for the 2015 ozone review, for example, had 20 members. CASAC and the public make suggestions regarding the membership of the panels on specific pollutants, with the final selections made by EPA. The panels evaluate the agency's work during NAAQS-setting and NAAQS-revision, rather than conducting their own independent review of the standards.

Recent Reviews of the Ozone Standard

Before the 2015 review, EPA last changed the NAAQS for ozone in March 2008, from 0.08 ppm (effectively 84 ppb) to 0.075 ppm (75 ppb). Although the standard was strengthened, the level chosen at that time was subject to controversy. A 23-member CASAC Review Panel unanimously recommended a range of standards more stringent than what the Administrator chose.[26]

In September 2009, EPA agreed to reconsider the 2008 standard. As a result, a more stringent primary standard and a different version of the secondary standard were proposed in January 2010. After a year-and-a-half of public comment and review, EPA sent what it considered a final set of standards to the Office of Management and Budget (OMB) for interagency review. The process was short-circuited, however, by a presidential decision to await conclusion of the next regular review—the review completed in 2015—before promulgating any change.

The Primary Standard

The review completed in 2008 found evidence of health effects, including mortality, at levels of exposure below the then-current 0.08 ppm standard. As a result, both EPA staff and CASAC recommended strengthening the standard. CASAC stated, "There is no

---

[26] In a letter to the Administrator sent after promulgation of the NAAQS, the panel's chair stated the following:

> Nevertheless, the members of the CASAC Ozone Review Panel do not endorse the new primary ozone standard as being sufficiently protective of public health. The CASAC—as the Agency's statutorily-established science advisory committee for advising you on the national ambient air quality standards—unanimously recommended decreasing the primary standard to within the range of 0.060–0.070 ppm. It is the Committee's consensus scientific opinion that your decision to set the primary ozone standard above this range fails to satisfy the explicit stipulations of the Clean Air Act that you ensure an adequate margin of safety for all individuals, including sensitive populations.

Letter of Rogene F. Henderson, Chair, Clean Air Scientific Advisory Committee, to EPA Administrator Stephen L. Johnson, April 7, 2008, at http://yosemite.epa.gov/sab/sabproduct.nsf/4AF8764324331288852574250069E494/$File/EPA-CASAC-08-009-unsigned.pdf.

scientific justification for retaining the current [0.08 ppm] primary 8-hr NAAQS."[27] The panel unanimously recommended a range of 0.060 to 0.070 ppm (60 to 70 parts per billion) for the primary (health-based) 8-hour standard.

EPA staff also recommended strengthening the primary standard. They recommended "considering a standard level within the range of somewhat below 0.080 parts per million (ppm) to 0.060 ppm."[28]

Based on these recommendations, and his own judgment regarding the strength of the science, Stephen Johnson, the Bush Administration's last EPA Administrator, chose to finalize the standard at 0.075 ppm (75 parts per billion).[29] That revision led to designation of nonattainment areas in April and May 2012, as shown above in *Figure 1*.[30]

The Regulatory Impact Analysis that accompanied the final 2008 standard identified 345 counties in 36 states and the District of Columbia in exceedance of the 0.075 ppm standard, using data for 2004-2006 (the most recent available at the time). By May 2012, when the nonattainment areas were actually designated, the number of counties in nonattainment had fallen to 232 in 26 states and the District of Columbia, based mostly on data for 2008-2010.[31] In the intervening years, emissions declined in most areas as more stringent standards for both mobile and stationary sources took effect. The 2008-2009 recession and other economic factors also contributed to the lower numbers. When the economy is operating well below capacity, emissions generally decline; and changes in the economy (e.g., fewer vehicle miles traveled and a switch to cleaner fuels) have also resulted in lower emissions. As of April 2016, only 115 counties in 12 states exceeded the 2008 ozone NAAQS, although about double that number remained classified as nonattainment pending completion of administrative requirements.

As noted, EPA began a process to reconsider the 2008 ozone NAAQS in September 2009, and proposed a more stringent primary NAAQS in January 2010. The reconsideration process, which generally relied on the same data as that used to set the 2008 standard, led EPA to recommend a primary NAAQS of 0.070 ppm (70 ppb), within the range recommended by the CASAC Ozone Review Panel in 2008. A draft final standard was prepared and sent to OMB for final interagency review in the summer of 2011, but was withdrawn at the President's request in September 2011.

---

[27] Letter of Rogene Henderson, Chair, Clean Air Scientific Advisory Committee, to Hon. Stephen L. Johnson, EPA Administrator, October 24, 2006, available at http://yosemite. epa.gov/sab/ sabproduct.nsf/AB290E0DB8B72A33852572120055858F/$File/casac-07-001.pdf.

[28] "Review of National Ambient Air Quality Standards for Ozone Final Staff Paper, Human Exposure and Risk Assessments and Environmental Report," Fact Sheet, at http://www.epa.gov/ttn/naaqs/standards/ozone/data/ 2007_01_finalsp_factsheet.pdf.

[29] All of EPA's references to the 2008 standard are expressed as parts per million (e.g., 0.075 ppm), but many references in the press convert this to a more readable parts per billion (i.e., 75 parts per billion).

[30] Detailed information on the designations, including links to *Federal Register* notices, can be found on EPA's website at https://www.epa.gov/green-book.

[31] EPA, "Nonattainment Designations for the 2008 Ozone Standards—Counties by State, April 30, 2012 and May 31, 2012," at https://archive.epa.gov/ozonedesignations/ web/html/ finaldes.html. A few states had certified monitoring data for 2009-2011, and their designations were based on that three-year period. As of October 2015, the number is 224 counties in 25 states and the District of Columbia. For a listing, see U.S. EPA Green Book, "8-Hr Ozone (2008) Nonattainment Areas," at https://www.epa.gov/green-book.

The 2015 Review

The 2015 review, after assessing more than 1,000 new studies, reached conclusions similar to those of the 2008 process:

> The available scientific evidence and exposure/risk information provide strong support for considering a primary O3 [ozone] standard with a revised level in order to increase public health protection, including for at-risk populations and lifestages. Staff concludes that it is appropriate in this review to consider a revised primary O3 standard level within the range of 70 ppb to 60 ppb. A standard set within this range would result in important improvements in public protection, compared to the current [2008] standard, and could reasonably be judged to provide an appropriate degree of public health protection, including for at-risk populations and lifestages. In its advice to the Administrator, CASAC also concluded that the scientific evidence and exposure/risk information support consideration of standard levels from 70 to 60 ppb. Within this range, CASAC concluded that a level of 70 ppb would provide little margin of safety and, therefore, provided the policy advice that the level of the O_3 standard should be set below 70 ppb.[32]

Based on these recommendations, the EPA Administrator proposed revising the primary NAAQS on November 26, 2014, to a level somewhere in the range of 65 to 70 ppb. She also asked for comments on retaining the 2008 75 ppb standard or promulgating a standard of 60 ppb. In October 2015, she completed the review process, setting the primary ozone NAAQS at 70 ppb.

The Secondary Standard

As part of the review completed in 2008, the 2009-2011 reconsideration process, and the review completed in 2015, EPA also assessed the secondary NAAQS for ozone. As explained above, secondary NAAQS are standards necessary to protect public welfare, a broad term that includes damage to crops, vegetation, property, building materials, climate, etc.[33] Prior to 2008, the secondary standard was identical to the primary standard—0.08 ppm beginning in 1997.

Ozone affects both tree growth and crop yields, and the damage from exposure is cumulative over the growing season. In order to address this damage, EPA staff recommended in the 2008 review that the Administrator establish a new form for the secondary standard: a seasonal (three-month) average that would cumulate hourly ozone exposures for the daily 12-hour daylight window (termed a "W126 index").[34] The staff initially recommended a standard in a range of 7-21 parts per million-hours (ppm-hours). CASAC's ozone panel agreed unanimously that the form of the secondary standard should be changed as the staff suggested, but it did not agree that the upper bound of the

---

[32] EPA, 2014 Policy Assessment, p. ES-5.

[33] The Clean Air Act's definition of welfare is found in Section 302(h) of the act (42 U.S.C. 7602(h)).

[34] The index gives greater weight to higher concentrations of ozone. For a detailed explanation of how it is calculated, see EPA, "Ozone W126 Index," at https://www.epa.gov/air-quality-analysis/ozone-w126-index.

range should be as high as 21 ppm-hours, suggesting that the upper bound be no higher than 15 ppm-hours.[35]

The Administrator's June 2007 proposal was in line with the staff recommendation, 7-21 ppm hours, but his final March 2008 choice was to duplicate the primary standard he promulgated at that time. He set a secondary standard at 0.075 ppm averaged over 8 hours, rejecting the advice of both CASAC and his staff.

The secondary standard carries no deadline for attainment and has never been the subject of penalties or sanctions for areas that failed to meet it, but there was substantial disagreement between the Bush Administration EPA and the Office of Management and Budget over the form in which this standard should be set. OMB maintained that EPA had failed to consider or evaluate the effects of a W126 standard on "economic values, personal comfort, and well-being"—terms that are also included in the Clean Air Act's definition of welfare—and thus did not provide a balanced consideration of welfare effects, as required by the act. OMB also maintained that EPA had not adequately demonstrated that the proposed secondary standard would be more protective than one set equal to the primary standard.[36] Ultimately, OMB prevailed.[37]

Upon reconsideration, the 2011 draft final standards would also have adopted the W126 index and would have set the secondary standard at 13 ppm-hours, in line with CASAC's recommendations. With the President's request to withdraw the draft standard and await completion of the ongoing five-year review, a seasonal standard was not implemented.

EPA's August 2014 Policy Assessment renewed this debate, this time with an additional thumb on the scale. The agency's staff again recommended that the Administrator set a secondary standard using the W126 index. The staff recommended a standard somewhere in the range of 7 to 17 ppm-hours, similar to CASAC's recommended range of 7 to 15 ppm-hours.[38] In the interim, the D.C. Circuit Court of Appeals also weighed in. In a July 23, 2013, decision, *Mississippi v. EPA*, the court remanded the 2008 secondary standard to EPA for further explanation or reconsideration: the court found that "EPA must expressly 'determine what level of ... protection is requisite to protect the public welfare,' [citation omitted] and explain why this is so."[39]

The 2015 final rule is something of a hybrid: it describes the protection offered by the revised secondary standard in terms of a cumulative seasonal approach, but retains the

[35] Letter of Rogene Henderson, Chair, Clean Air Scientific Advisory Committee, to Hon. Stephen L. Johnson, EPA Administrator, March 26, 2007, p. 3, available at http://yosemite.epa.gov/sab/sabproduct.nsf/FE915E916333D776852572AC007397B5/$File/casac-07-002.pdf.

[36] See "Ozone Secondary Standard," Memorandum of Marcus Peacock, EPA Deputy Administrator, to Susan Dudley, OMB, March 7, 2008, p. 2, at http://www.reginfo.gov/public/postreview/Steve_Johnson_Letter_on_NAAQs_final_3- 13-08_2.pdf. For additional discussion, also see CRS Report R41062, *Ozone Air Quality Standards: EPA's Proposed Revisions*.

[37] Letter of Susan E. Dudley, Administrator, Office of Information and Regulatory Affairs, OMB, to Stephen L. Johnson, EPA Administrator, March 12, 2008, at http://www.reginfo.gov/public/postreview/Steve_Johnson_Letter_on_NAAQs_final_3-13-08_2.pdf.

[38] See EPA, 2014 Policy Assessment, pp. ES-9 to ES-12.

[39] Mississippi v. EPA, 723 F.3d 246, 272, 273 (D.C. Cir. 2013) (ellipses in original).

practice of making the standard identical to the primary standard. The overview of the agency's rule states the following:

> EPA also is strengthening the secondary standard to improve protection for trees, plants and ecosystems. Like the primary, an area will meet the standard if the fourth-highest maximum daily 8-hour ozone concentration per year, averaged over three years, is equal to or less than 70 ppb. [...]
>
> The agency has assessed ozone exposure to vegetation using a seasonal index known as a "W126 index." A W126 index, named after portions of the equation used to calculate it, is a weighted index designed to reflect the cumulative exposures that can damage plants and trees during the consecutive three months in the growing season when daytime ozone concentrations are the highest and plant growth is most affected.
>
> EPA determined that a W126 index level of 17 parts per million-hours (ppm-hours) is sufficient to protect the public welfare based on the latest science.
>
> Analyses of data from air quality monitors show that an 8-hour standard of 70 ppb will limit cumulative, seasonal exposures above a W126 index level of 17 ppm-hours, averaged over three years.[40]

Whether this explanation is sufficient to address the D.C. Circuit decision in the *Mississippi* case is an open question. CASAC, in its review of the EPA staff recommendations, made two points that might argue against it being sufficient.

First, they stated the following:

> The CASAC does not support a level higher than 15 ppm-hrs. For example, at 17 ppm-hrs, the median tree species has 6% relative biomass loss, and the median crop species has over 5% yield loss. These levels are unacceptably high.... Furthermore, there are specific economically significant crops, such as soybeans, that may not be protected at 15 ppm-hrs but would be protected at lower levels. A level below 10 ppm-hrs is required to reduce foliar injury. A level of 7 ppm-hrs is protective of relative biomass loss for trees and offers additional protection against crop yield loss and foliar injury.[41]

Second, CASAC specifically did not recommend a three-year average for the secondary standard, stating that if the agency chose a three-year average, "then the level of the standard should be revised downward such that the level for the highest three-month summation in any given year of the three-year period would not exceed the scientifically recommended range of 7 ppm-hrs to 15 ppm-hrs."[42]

---

[40] EPA, "Overview of EPA's Updates to the Air Quality Standards for Ground-Level Ozone," October 1, 2015, pp. 4-5, at https://www.epa.gov/sites/production/files/2015-10/documents/ overview_of_2015_rule.pdf.

[41] Letter of Dr. H. Christopher Frey, Chair, Clean Air Scientific Advisory Committee, to Hon. Gina McCarthy, Administrator, U.S. EPA, June 26, 2014, p. iii, at http://yosemite. epa.gov/sab/sabproduct.nsf/ 5EFA320CCAD326E885257D030071531C/$File/EPA-CASAC-14-004+unsigned.pdf.

[42] Ibid., pp. iii-iv.

Implementation of the 2015 Standards

Promulgation of the NAAQS Revisions in October 2015 set in motion a long implementation process in which EPA works with state, local, and tribal air quality agencies to identify areas that do not meet the revised ozone standards and then implements measures designed to reach attainment.[43]

A first major step occurred in October 2016, when the states (and six authorized Indian tribes), using monitoring data and other information, submitted to EPA lists showing recommended attainment/nonattainment area designations. As shown in *Table 2*, the states identified 213 counties in 55 potential nonattainment areas. The list included areas in 21 states and the District of Columbia. Indian tribes with designated Clean Air Act authority identified two additional areas as nonattainment.

The list contained approximately the same number of states and counties as were designated nonattainment for the previous (2008) ozone NAAQS at the time, and most of the counties identified by the states are currently in nonattainment for the 2008 standard or are covered by a maintenance plan, having attained that or an earlier ozone NAAQS.

The second step of the implementation process, in which EPA responds to the state and tribal recommendations, is now complete. In this step, EPA reviews the recommendations and finalizes designations after what is often significant dialogue with the states.

Negotiations between EPA and the states during the designation process tend to focus on the quality of the monitoring data, whether exceptional events[44] or other factors outside the state's control affected air quality during the monitoring period, and what the boundaries of identified nonattainment areas should be.[45] The boundary question may lead to the inclusion of additional counties or partial counties where emissions contribute to nonattainment downwind, even though the county itself has monitored data demonstrating attainment.[46]

---

[43] The implementation process is outlined in "Implementing the 2015 Ozone NAAQS," Memorandum of Janet G. McCabe, Acting Assistant Administrator, Office of Air and Radiation, EPA, to Regional Administrators, Regions 1-10, October 1, 2015, at https://www.epa.gov/sites/production/files/2015-10/documents/implementation_memo.pdf.

[44] EPA regulations can allow the states to exclude monitoring data for periods affected by exceptional events (forest fires, unusual weather conditions, volcanic or seismic activities, etc.) from the data used to make designation decisions. For additional information, see U.S. EPA, "Treatment of Data Influenced by Exceptional Events," at https://www.epa.gov/air-quality-analysis/treatment-data-influenced-exceptional-events.

[45] EPA provided guidance to states and tribes to consider when making their recommendations for initial designations. In particular, EPA recommended states and tribes evaluate five factors: (1) air quality data, (2) emissions and emissions-related data, (3) meteorology, (4) geography/topography, and (5) jurisdictional boundaries. See Janet G. McCabe, Acting Assistant Administrator, Office of Air and Radiation, EPA, "Area Designations for the 2015 Ozone National Ambient Air Quality Standards," Memorandum to Regional Administrators, Regions 1-10, February 26, 2016, at https://www.epa.gov/sites/production/ files/2016-02/documents/ozone-designations-guidance-2015.pdf.

[46] For example, EPA disagreed with Nevada's recommended boundary for the portion of Clark County to be designated nonattainment, in part because the state's recommended boundary excluded a monitor showing ozone violations. See EPA, Las Vegas Nonattainment Area, Intended Designations for the 2015 Ozone National Ambient Air Quality Standards, Technical Support Document, December 2017, p. 30, https://www.epa.gov/sites/ production/files/2017-12/documents/nv_120d_tsd_ combined_final.pdf.

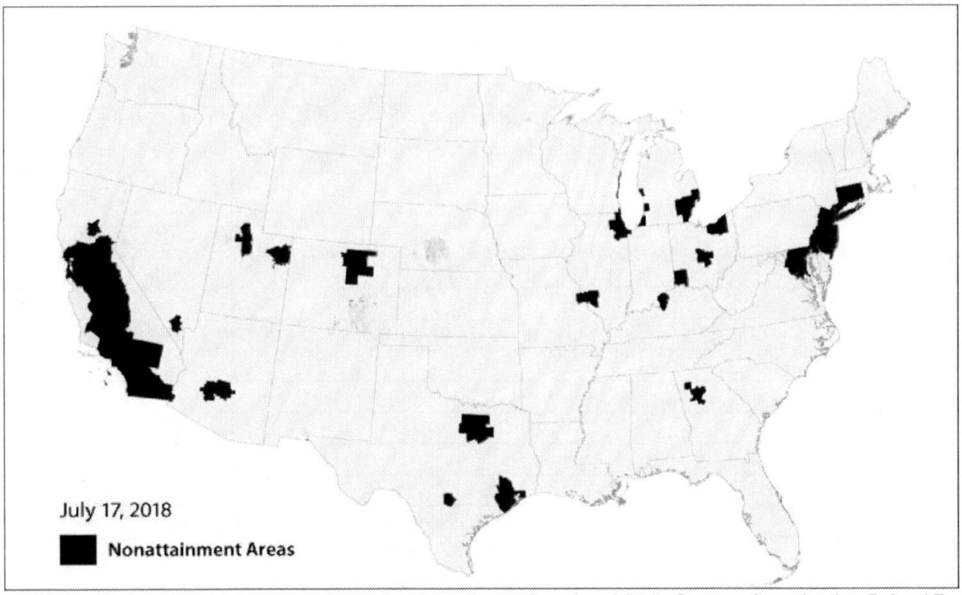

Source: U.S. EPA, Nonattainment Area Designations for the 2015 Ozone Standards, July 17, 2018, available at https://www.epa.gov/sites/production/files/2018-07/ documents/ozone_designations _national_map_with_san_antonio_added.pdf.

Notes: Alaska and Hawai'i are not shown: no areas in Alaska and Hawai'i exceeded the 70 ppb standard, based on 2013-2015 monitoring data.

Figure 2. Nonattainment Areas for 2015 Ozone NAAQS (0.070 ppm) (as of July 17, 2018)

The states often provide more recent monitoring data during this process, which may result in an area demonstrating that it has attained the standard or may change the area's classification (e.g., as Marginal or Moderate nonattainment).[47] This second step resulted in three separate actions designating areas under the 2015 ozone NAAQS

- on November 16, 2017, EPA designated 2,646 counties as in attainment with the 2015 ozone standards;[48]
- on April 30, 2018, the agency identified 51 areas with 159 entire counties and 41 partial counties as nonattainment.[49] All other areas, with the exception described in the next bullet, were designated attainment/unclassifiable or unclassifiable; and

[47] The 2016 state recommendations were based, in general, on 2013-2015 monitoring data. In 2017, some states submitted 2014-2016 data to EPA for consideration. For example, in April 2017, Pennsylvania submitted 2014-2016 monitoring data to EPA. See Cosmo Servidio, Regional Administrator, EPA Region 3, letter to The Honorable Tom Wolfe, Governor of Pennsylvania, December 20, 2017, https://www.epa.gov/sites/production/ files/2018-01/documents/pa-epa-response-ozone2.pdf.

[48] EPA, "Air Quality Designations for the 2015 Ozone National Ambient Air Quality Standards (NAAQS)," 82 *Federal Register* 54232, November 16, 2017.

[49] EPA, "Additional Air Quality Designations for the 2015 Ozone National Ambient Air Quality Standards," 83 *Federal Register* 25776, June 4, 2018.

Table 2. List of Nonattainment Areas for 2015 Ozone NAAQS

State or Tribe	State and Tribe Recommendations (2016)		EPA Nonattainment Designations (2018)	
	Nonattainment Areas	Counties and Partial Counties in the Identified Areas	Nonattainment Areas	Counties and Partial Counties in the Identified Areas[a]
Alabama	0	0	0[b]	0[b]
Alaska	0	0	0	0
Arizona	2	4	2	4
Arkansas	0	0	0[b]	0[b]
California	19	37	19	37
Colorado	1	9	1	9
Connecticut	2	8	2	8
Delaware	1	1	1	1
District of Columbia	1	1	1	1
Florida	0	0	0	0
Fond du Lac (MN)	0	0	0	0
Forest County Potawatomi (WI)	0	0	0	0
Georgia	1	8	1	7
Gila River (AZ)	0	0	0	0
Hawai'i	0	0	0[b]	0[b]
Idaho	0	0	0[b]	0[b]
Illinois	2	11	2	9
Indiana	0	0	2	3
Iowa	0	0	0[b]	0[b]
Kansas	0	0	0[b]	0b
Kentucky	1	3	2	6
Louisiana	1	5	0	0
Maine	0	0	0[b]	0[b]
Maryland	3	12	3	12
Massachusetts	0	0	0	0
Michigan	5	11	4	10
Minnesota	0	0	0[b]	0[b]
Mississippi	0	0	0[b]	0[b]
Missouri	1	5	1	4
Montana	0	0	0[b]	0[b]
Morongo Band (CA)	1	1	1	1
Nebraska	0	0	0[b]	0[b]
Nevada	1	1	1	1
New Hampshire	0	0	0[b]	0[b]
New Jersey	1	21	2	21
New Mexico	1	1	1	1
New York	1	9	1	9
North Carolina	0	0	0[b]	0[b]
North Dakota	0	0	0[b]	0[b]
Ohio	3	15	3	15
Oklahoma	0	0	0	0
Oregon	0	0	0	0
Pechanga (CA)	1	2	1	2

Table 2. (Continued)

State or Tribe	State and Tribe Recommendations (2016)		EPA Nonattainment Designations (2018)	
	Nonattainment Areas	Counties and Partial Counties in the Identified Areas	Nonattainment Areas	Counties and Partial Counties in the Identified Areas[a]
Pennsylvania	4	14	1	5
Rhode Island	0	0	0	0
South Carolina	0	0	0[b]	0[b]
South Dakota	0	0	0[b]	0[b]
Tennessee	0	0	0[b]	0[b]
Texas	4	21	3	16
Utah	3	7	3	7
Ute (UT)	0	0	0	0
Vermont	0	0	0[b]	0[b]
Virginia	1	9	1	9
Washington	0	0	0	0
West Virginia	0	0	0	0
Wisconsin	0	0	5	6
Wyoming	0	0	0	0
Total [c]	55	213	52	201[c]

Source: U.S. EPA, *2015 Ozone Standards—State Recommendations, EPA Responses, and Technical Support Documents and 2015 Ozone Standards—Tribal Recommendations and EPA Responses*, accessed April 3, 2018, available at https://www.epa.gov/ozone-designations/2015-ozone-standards-state-recommendations. *U.S. EPA, Ozone Designations Regulatory Actions*, accessed August 8, 2018, available at https://www.epa.gov/ozone-designations/ozone-designations-regulatory-actions.

Notes: Most states submitted initial recommendations to EPA on or about October 1, 2016, using monitoring data for 2013-2015 to make their recommendations. In some cases, the submission letter included preliminary information for 2016, through August 31 of the year. EPA notified states that they could submit more recent data (e.g., for years 2015-2017) by February 28, 2018. Final designations were based on 2014-2016 or 2015-2017 data.

[a]The term "county" includes noncounty administrative or statistical areas that are comparable to counties. Louisiana parishes; the organized boroughs of Alaska; the District of Columbia; and the independent cities of the states of Virginia, Maryland, Missouri, and Nevada are equivalent to counties for administrative purposes. Alaska's Unorganized Borough is divided into 10 census areas that are statistically equivalent to counties. Using this definition, as of 2017, there are 3,142 counties and county-equivalents in the United States.

[b]EPA designated the entire state as attainment/unclassifiable. See EPA, "Air Quality Designations for the 2015 Ozone National Ambient Air Quality Standards (NAAQS)," 82 *Federal Register* 54232, November 16, 2017.

[c]Several areas include counties in more than one state, including areas in Kentucky and Ohio (Cincinnati nonattainment area); Connecticut, New Jersey, and New York (metropolitan New York City nonattainment area); Delaware, Maryland, New Jersey, and Pennsylvania (Philadelphia nonattainment area); Illinois and Missouri (St. Louis nonattainment area); and the District of Columbia, Maryland, and Virginia (Washington, DC, nonattainment area). To avoid double-counting the number of nonattainment areas, the total counts each of these areas only once.

[d]The number of counties includes 160 whole nonattainment counties and county equivalents and 41 partial nonattainment counties. Parts of five counties are included in multiple areas but are only counted once in this total. In addition, EPA designated nonattainment areas in two tribal areas. Source: EPA, *Nonattainment and Unclassifiable Area Designations for the 2015 Ozone Standards*—April 30, 2018, May 2018, p. 5, https://www.epa.gov/sites/production/files/2018-04/ documents/placeholder_1.pdf.

- on July 17, 2018, the agency made final designation decisions for an eight-county area that includes San Antonio, TX, designating one county as nonattainment and the remaining seven counties as attainment. This action completed the area designations with respect to the 2015 ozone standard.[53]

Once EPA designates nonattainment areas, it classifies the areas into categories based on the severity of nonattainment. The classifications are, in increasing order of severity: Marginal, Moderate, Serious, Severe, and Extreme. EPA also sets a schedule for states to submit State Implementation Plans (SIPs), which outline the measures that will reduce emission levels. In a Fact Sheet accompanying the designations, EPA states that "Areas classified as Marginal are not required to submit plans demonstrating how they will meet the ozone standards. For all other nonattainment areas, states have three to four years after the effective date of final designations to develop and submit required SIP elements to EPA."[54] Finalizing SIPs, through EPA review and approval, often takes longer than the three- to four-year period.

Finally, under the CAA, actual attainment of the standards is allowed to stretch over a three-year to 20-year period, depending on the severity of the area's pollution.[55]

Controlling Ozone Pollution

Controlling ozone pollution is more complicated than controlling many other pollutants, because ozone generally is not emitted directly by pollution sources. Rather, it forms in the atmosphere when volatile organic compounds (VOCs) and nitrogen oxides (NOx) react in the presence of sunlight. The ozone concentration is as dependent on the temperature and amount of sunshine as it is on the presence of the precursor gases. Thus, in general, ozone is a summertime pollutant. Other factors being equal, a cool, cloudy summer will produce fewer high ozone readings than a warm, sunny summer.

There are also complicated reactions that affect ozone formation. In general, lower emissions of precursor gases (particularly lower emissions of VOCs) lead to less ozone. But under some conditions, *higher* emissions of NOx lead to lower ozone readings. This makes modeling ozone air quality and predicting attainment more difficult and contentious than the modeling of other air pollutants, and can affect consultations between EPA and the states to determine the boundaries of nonattainment areas and the adequacy of SIPs.

Most stationary and mobile sources of air pollution are considered to be contributors to ozone pollution. Thus, there are literally hundreds of millions of sources of the

---

[53] EPA, "Additional Air Quality Designations for the 2015 Ozone National Ambient Air Quality Standards—San Antonio, Texas Area," 83 *Federal Register* 35136, July 25, 2018.

[54] U.S. Environmental Protection Agency, *Fact Sheet: Final Area Designations for the National Ambient Air Quality Standards for Ozone Established in 2015*, May 2018, p. 1, https://www.epa.gov/sites/production/files/2018-04/documents/placeholder_0.pdf.

[55] U.S. Environmental Protection Agency, "Implementation of the 2015 National Ambient Air Quality Standards for Ozone: Nonattainment Area Classifications Approach," 83 *Federal Register* 10380, March 9, 2018.

pollutants of concern, and control strategies require implementation of a wide array of measures. Among the sources of VOCs are motor vehicles (about one-fourth of total emissions), industrial processes, particularly the chemical and petroleum industries, and any use of paints, coatings, and solvents (about 40% for these sources combined). Service stations, pesticide application, dry cleaning, fuel combustion, and open burning are other significant sources of VOCs. Nitrogen oxides come overwhelmingly from fuel combustion by motor vehicles, electric utilities, and other industrial sources.

WINTERTIME OZONE

An emerging set of issues has arisen in regard to wintertime ozone pollution in rural areas of the western United States. Ozone is generally considered a summertime pollutant, but winter exceedances of the ozone NAAQS have recently been found to occur near oil and gas fields in the West.[56] At times, ozone concentrations as high as those in Los Angeles, the nation's smoggiest city, have been found in these areas—principally the Upper Green River Basin of Wyoming, the Uintah Basin of Utah, and a nearby area of Colorado.

The mechanism of ozone formation in the areas is still being studied, but recent assessments have found that the thousands of oil and gas wells in the two basins release VOCs that react with NOx from oil and gas operations and coal-fired power plants to create ozone. A study of the Uintah Basin by the National Oceanic and Atmospheric Administration, EPA, the Bureau of Land Management, the Western Energy Alliance, the Utah Department of Environmental Quality, and seven universities found that sources external to the basin are not major sources of the ozone found within it, and that among inventoried sources within the basin, 98% to 99% of the VOCs and 57% to 61% of the NOx come from oil and gas operations.[57] The sunlight necessary for ozone to be created is magnified when it is reflected off of heavy snow cover. Snow cover also helps create temperature inversions that trap polluted air in the basins. In winters with little snow, there have been few exceedances of the standards.[58]

In 2012, EPA promulgated standards requiring the reduction of VOC emissions from new and modified oil and gas production and transmission operations, including a requirement to use "green completions" on hydraulically fractured onshore natural gas

[56] Wintertime ozone levels are occasionally elevated in urban areas, as well, when concentrations of pollution become trapped in cold stagnant air near the Earth's surface. What is unusual about the rural areas discussed in this section of the report is that they do not experience high ozone concentrations in warm weather. In addition, elevated ozone in western rural areas is a newly observed phenomenon, whereas pollution episodes associated with temperature inversions in urban areas have been observed for decades.

[57] *2012 Uintah Basin Winter Ozone & Air Quality Study*, Final Report, February 1, 2013, p. 2, at http://rd.usu.edu/files/uploads/ubos_2011-12_final_report.pdf. The remaining NOx comes primarily from the Bonanza Power Plant.

[58] Ibid. In the Uintah Basin, for example, winter 2011-2012 measurements indicated no exceedance of the 75 ppb NAAQS for ozone: the highest 8-hour average reading was 63 ppb. In 2010-2011, by contrast, there were 25 winter days with ozone levels exceeding the 75 ppb standard, with the highest 8-hour reading being 139 ppb.

wells. In 2016, these standards were extended to cover hydraulically fractured oil wells. Further, in 2015, EPA proposed Control Techniques Guidelines intended to provide state, local, and tribal air agencies with information to assist them in determining "reasonably available control technology" for reducing VOC emissions from existing oil and natural gas sources in ozone nonattainment areas and states in the Ozone Transport Region.[59] The impact of these regulations on wintertime ozone concentrations is yet to be determined, but interestingly the Upper Green River Basin, which was designated nonattainment under the 2008 ozone NAAQS, has been designated attainment for the 2015 standard.

COSTS AND BENEFITS OF CONTROL

As noted elsewhere in this chapter ("The Role of Cost," below), EPA is prohibited by statute from taking cost into account in setting NAAQS. Despite that prohibition, in order to comply with an executive order (E.O. 12866),[60] the agency produces a Regulatory Impact Analysis (RIA) analyzing in detail the costs and benefits of new or revised NAAQS standards when the NAAQS are proposed and promulgated.

EPA' s Cost Estimates

Although EPA chose a standard of 70 ppb in the 2015 final rule, the RIA for the rule shows estimates for two options: 70 ppb (the high end of the range for the proposed NAAQS) and 65 ppb (the low end of the range for the proposed NAAQS). EPA's estimates of the nationwide benefits and costs of the two options for all areas except California are summarized in *Table 3.*

Table 3. Estimated Annual Costs and Health Benefits of Ozone NAAQS Options, Nationwide Except California, in 2025 (in billions of 2011 dollars)

Option	Costs	Health Benefits
70 ppb	$1.4	$2.9 to $5.9
65 ppb	$16.0	$15.0 to $30.0

Source: EPA, "Regulatory Impact Analysis" of the Final Revisions to the NAAQS for Ground-Level Ozone, September 2015.

Notes: The data reflect annualized costs and annual monetized benefits of achieving the standard (or a more stringent 65 ppb option considered, but not adopted) in 2025 for all areas of the United States except California. Because of more severe pollution, the Clean Air Act will give most areas of California until the 2030s to reach attainment. The estimates assume a 7% discount rate.

---

[59] EPA, "Control Techniques Guidelines for the Oil and Natural Gas Industry (Draft)," EPA-453/P-15-001, August 2015, https://www.epa.gov/sites/production/files/2016-09/docu ments/og_ctg_draft_081815.pdf.

[60] Executive Office of the President, "Regulatory Planning and Review, Executive Order 12866 of September 30, 1993," 58 *Federal Register* 51735, October 4, 1993, http://www. archives.gov/federal-register/executive-orders/pdf/ 12866.pdf.

The estimated annual nationwide cost (excluding the cost in California) is $1.4 billion in 2025 for a 70 ppb standard.[61] Although this is a large sum, it is substantially less than the cost estimates EPA provided for the same range of standards in 2008 and 2011. At that time, EPA projected costs of $19 billion to $25 billion to attain a 70 ppb standard. Two factors account for the reduction in cost:

1. The baseline from which additional costs are projected is now set at 75 ppb (the 2008 standard). In 2011, EPA projected $7.6 billion to $8.8 billion in costs to reach what is now that baseline.
2. Other rules promulgated since 2011 (notably the Tier 3 auto emission and gasoline standards and two rules affecting power plants) are expected to reduce ozone precursors whether or not EPA revises the NAAQS. As shown in *Figure 3*, the 2015 RIA projects that, by 2025, these other (already promulgated) rules will bring monitored ozone levels to 70 ppb or below in all but 14 counties (excluding California) of the 241 counties that showed nonattainment with the 70 ppb level, based on 2012-2014 monitoring data.[62]

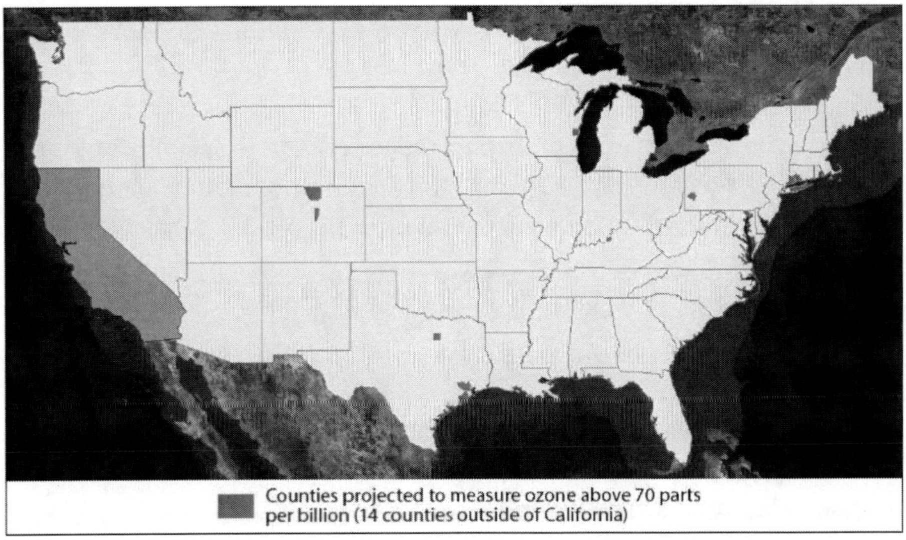

Source: EPA, "Ozone Maps" accompanying the October 2015 release of the final ozone NAAQS rule, https://ozoneairqualitystandards.epa.gov/OAR_OAQPS/ OzoneSliderApp/index.html#.

Notes: EPA did not include areas in California. Because of more severe pollution, the Clean Air Act will give many areas in California until the 2030s to reach attainment. No areas in Alaska and Hawaii are projected to measure ozone above 70 ppb.

Figure 3. EPA Projection of Counties That Will Not Meet the Revised Ozone Standards in 2025 Without Promulgation of Additional Emission Controls.

---

[61] EPA, 2015 RIA, p. ES-15.

[62] For documentation of 241 counties showing nonattainment, see EPA, "County-level Design Values for the 2015 Ozone Standards," at https://www.epa.gov/ozone-pollution/ozone-data-tables-1997-2008-and-2015-standards.

EPA's Benefit Estimates

As presented in its 2015 RIA, EPA's estimate of monetized nationwide benefits (excluding those in California) ranges from $2.9 billion to $5.9 billion annually in 2025 for a 70 ppb standard. The public health benefits of setting a more stringent ozone standard are the monetized value of such effects as fewer premature deaths, fewer hospital admissions, fewer emergency room visits, fewer asthma attacks, less time lost at work and school, and fewer restricted activity days.[63] The RIA for the final standard states that the nationwide benefits of the revised standard (excluding California) would include the avoidance of 320 to 660 premature deaths annually in 2025 (at 70 ppb) or 1,590 to 3,320 deaths annually (at 65 ppb).[64] The dollar value of the avoided premature deaths accounts for 94% to 98% of EPA's total monetized benefits.[65]

The benefit estimates include benefits of reduced fine particle (PM2.5) concentrations associated with ozone controls, in addition to the benefits of reduced ozone itself. The RIA states that "PM2.5 co-benefits account for approximately half to three-quarters of the estimated benefits, depending on the standard analyzed and on the choice of ozone and PM mortality functions used."[66] Including these co-benefits is consistent with the methodology EPA has used in valuing benefits of many other proposed and promulgated standards, but some observers are critical of this approach, noting that including these co-benefits results in a net benefit, whereas an analysis that considered the costs and benefits of ozone reductions in isolation would not.[67] The control technologies used to capture ozone precursors do capture particles and their precursors, however. Since they do so at no additional cost, EPA has considered this a benefit of the controls.

Other EPA projections of benefits associated with a 70 ppb standard in 2025 include preventing the following, annually:

- 28-260 nonfatal heart attacks,
- 960 hospital admissions and emergency room visits,
- 340 cases of acute bronchitis,

---

[63] For a full discussion of these variables and their monetized values, see Chapter 6 of the 2008 RIA at http://www.epa.gov/ttn/ecas/regdata/RIAs/452_R_08_003.pdf.

[64] EPA, 2015 RIA, p. ES-16.

[65] EPA's estimate uses an approach called the Value of a Statistical Life (VSL). See EPA, 2015 RIA, pp. 6-55. For additional information on VSL, see archived CRS Report R41140, *How Agencies Monetize "Statistical Lives" Expected to Be Saved By Regulations*, by Curtis W. Copeland. (The author of that report is no longer with CRS. Questions on it should be directed to Maeve Carey, Analyst in Government Organization and Management.)

[66] EPA, 2015 RIA, pp. 8-11.

[67] In 2017, EPA sought comment on an alternative way to account for human health co-benefits in its benefit-cost analysis of the proposal to repeal the Clean Power Plan. EPA's 2017 analysis reduced some estimates of the human health co-benefits by assuming no health benefits below specified thresholds for some air pollutants. EPA has not yet finalized the proposed repeal of the Clean Power Plan, nor is it clear whether and how the agency will apply this approach in other regulatory analyses. For more discussion about the treatment of human health co-benefits in the proposed repeal of the Clean Power Plan, see CRS Report R45119, *EPA's Proposal to Repeal the Clean Power Plan: Benefits and Costs*, by Kate C. Shouse.

- 11,000 cases of upper and lower respiratory symptoms,
- 230,000 cases of exacerbated asthma,
- 188,000 days when people miss work or school, and
- 620,000 minor restricted activity days.[68]

California and Other Regional Costs and Benefits

EPA provided separate cost and benefit estimates for California. Because of the statutory classification scheme for ozone nonattainment areas, under which more polluted areas have more stringent emission control requirements and more time to reach attainment, most areas of California will have until the 2030s to reach attainment. EPA's California estimates use projections for post-2025 (see *Table 4*). The agency concluded: "Because of the differences in the timing of achieving needed emissions reductions, incurring costs, and accruing benefits for California, the separate costs and benefits estimates for post-2025 should not be added to the primary estimates for 2025."[69]

EPA also estimated the costs of the revised NAAQS separately for the eastern United States (defined as Texas and the 36 states to its north and east) and the western United States (Montana, Wyoming, Colorado, New Mexico, and other states to the west, excluding California). These regional estimates show that 98% of the $1.4 billion estimated cost to achieve a 70 ppb standard is incurred by emission sources in the eastern United States.[70]

Table 4. Estimated Annual Costs and Health Benefits of Ozone NAAQS Options in California, Post-2025 (in billions of 2011 dollars)

Option	Costs	Health Benefits
70 ppb	$0.8	$1.2 to $2.1
65 ppb	$1.5	$2.3 to $4.2

Source: EPA, RIA of the Final Revisions to the NAAQS for Ground-Level Ozone, September 2015.

Notes: The data reflect annualized costs and annual monetized benefits of achieving the standard (or a more stringent, 65 ppb option considered, but not adopted) post-2025 in California. Because of more severe pollution, the Clean Air Act will give most areas of California until the 2030s to reach attainment. The agency states that "estimates of costs and benefits for California in a post-2025 time frame are likely to be relatively more uncertain than the national attainment estimates for 2025." In particular, the agency did not project emissions and air quality for any year other than 2025, although it assumed that emission controls and associated costs would occur through the beginning of 2038; benefits for California were modeled using population demographics in 2038. The estimates assume a 7% discount rate for future costs and benefits.

[68] EPA, 2015 RIA, p. ES-16.
[69] EPA, 2015 RIA, p. ES-17.
[70] EPA, 2015 RIA, p. ES-17.

Industry Estimates

Industry sources have generally estimated the future cost of emission controls necessary to attain a revised ozone NAAQS to be greater than EPA has. A 2015 study by the National Association of Manufacturers (NAM), for example, projected the cost of attaining a more stringent ozone NAAQS, as measured in reduced Gross Domestic Product, at up to $140 billion annually from 2017 to 2040.[71]

In reaching these conclusions, the NAM study made a number of assumptions different from those in EPA's 2015 RIA

- The study looked only at the most stringent option in the range proposed by EPA (65 ppb). EPA proposed a standard in the range of 65 to 70 ppb. In EPA's analysis, the cost of attaining a 65 ppb NAAQS—an option not ultimately chosen by EPA—would be over 10 times that of a 70 ppb NAAQS.
- The NAM study's baseline differs from that of EPA. It appears not to include some recently promulgated EPA regulations that will reduce NOx emissions (e.g., the Cross State Air Pollution Rule). It also does not include EPA's Clean Power Plan, which was finalized after the release of NAM's study, but before promulgation of the final ozone NAAQS. EPA projected that the Clean Power Plan would cause coal-fired electricity generation to decline 12% to 15% in 2025,[72] resulting in substantial reductions of NOx emissions.[73]
- The NAM study concluded that more than half of the emission reductions needed to reach attainment would have to come from "unknown controls,"[74] which, in an earlier version of its study, it estimated could cost as much as $500,000 per ton of emissions reduced.[75] EPA's modeling of the rule's costs found that unknown controls would play a much smaller role: both the number of tons to which such controls would apply and the cost per ton would be substantially less.
- NAM's analysis focused exclusively on emissions of NOx, without any consideration of VOC controls. VOC emissions from petroleum and related industries have increased in the past decade, according to EPA, while emissions

[71] NERA Economic Consulting for the National Association of Manufacturers, *Assessing Economic Impacts of a Stricter National Ambient Air Quality Standard for Ozone*, February 2015, p. S-1, at http://www.nam.org/Special/Media-Campaign/EPA-Overregulation/Ozone-Regulations.aspx. Hereinafter, "NAM Report."

[72] EPA, Office of Air and Radiation and Office of Air Quality Planning and Standards, *Regulatory Impact Analysis for the Clean Power Plan Final Rule*, August 2015, pp. 3-27, at https://www3.epa.gov/ttnecas1/docs/ria/utilities_ria_finalclean-power-plan-existing-units_2015-08.pdf.

[73] Although the Supreme Court has stayed implementation of the Clean Power Plan, and EPA has proposed its repeal, coal-fired generation had already declined more than EPA's 2025 projections by 2017 as a result of market forces, state regulations, and other factors.

[74] NAM Report, p.

[75] NAM Report, July 2014, pp. 12-19, at http://www.nam.org/Issues/Energy-and-Environment/Ozone-Regulations/NERA-NAM-Ozone-Full-Report-20140726/. The $500,000 figure does not appear in the February 2015 NAM report, but the February report continues to rely on scrappage programs to estimate the cost of unknown controls. NAM's present value of the cost of all necessary controls is more than six times that of EPA. See NAM Report, p. 10.

from most other sources have declined.[76] Recent EPA analyses suggest that there are low cost emission control options in the oil and gas sector.[77]

In 2015, the American Petroleum Institute (API), the American Chemistry Council, and NAM all released state maps showing potential nonattainment areas and the potential cost of the rule on a state-by-state basis.[78] A map of the entire country distributed by API identified 958 counties nationwide as potentially in violation of a 70 ppb ozone NAAQS.[79] The larger number of affected counties (by comparison, in 2015, EPA projected only 241 counties[80]) would also contribute to higher cost estimates.

As shown in *Table 2*, the number of nonattainment counties recommended by states (213) and designated by EPA (201) are less than one-fourth of the number assumed in industry projections.

At the state level, the contrast between industry projections and state data is particularly striking: API's map showed 59 of Oklahoma's 77 counties potentially in nonattainment; the state data (submitted one year later) indicated that all 77 counties have already attained the 2015 standard. Similarly, API's map showed Kansas with 75 counties in nonattainment; according to the state's 2016 submission, the standard has already been attained at every monitored location. In other states—Arizona, Colorado, Indiana, Michigan, Missouri, Texas, and Utah among them—the number of counties identified on the API map is substantially higher than the number identified by the state and ultimately designated nonattainment by EPA.

Issues

The 2015 ozone NAAQS review has raised issues regarding the cost of attainment, background ozone levels, and the schedule for NAAQS reviews. The adequacy of the ozone monitoring network and the role of federal versus state and local pollution control

---

[76] See EPA, National Emissions Inventory (NEI) Air Pollutant Emission Trends Data, at http://www.epa.gov/ttn/chief/ trends/. As shown in the NEI, VOC emissions from petroleum and related industries increased 85% between 2008 and 2014. This increase is due in part to improvements in the methodologies used to estimate emissions and the addition of subcategories of sources to the petroleum and related industries category, but available information suggests that the increase is also due in part to greater emissions. Personal communication, EPA Office of Air Quality Planning and Standards, September 6, 2017.

[77] See, for example, EPA, "Regulatory Impact Analysis, Final New Source Performance Standards and Amendments to the National Emissions Standards for Hazardous Air Pollutants for the Oil and Natural Gas Industry," April 2012, at http://www.epa.gov/ttnecas1/regdata/RIAs/oil_natural_gas_final_neshap_nsps_ria.pdf.

[78] See, for example, API, "New Ozone Rules Could Be Most Expensive Regulations Ever," Ozone State Map Series, at http://www.api.org/~/media/files/policy/environment/economic-impact-ozone/economic-impact-ozone-nm.pdf.

[79] See Mark Green, "EPA Fumbles on Unneeded Ozone Regs," *Energy Tomorrow*, October 2, 2015, at http://www.energytomorrow.org/blog/2015/10/02/epa-fumbles-on-unneeded-ozone-regs. *Energy Tomorrow* describes itself as "a project of the American Petroleum Institute ... speaking for the industry to the public, Congress and the Executive Branch, state governments and the media."

[80] NAM Report, p. 2.

measures are other important issues. More recently, EPA cited concerns about NAAQS review and implementation and identified "NAAQS reform" as one of four initiatives that the agency plans to undertake to reduce "unnecessary burdens on the development and use of domestic energy resources."[81] In particular, EPA's "Back-to-Basics" memorandum identified principles to guide forthcoming NAAQS reviews.[82] Among other things, EPA seeks to streamline the NAAQS review process and obtain Clean Air Scientific Advisory Committee advice regarding background pollution and potential adverse effects from NAAQS compliance strategies.[83] EPA has also formed the "Ozone Cooperative Compliance Task Force" to review issues related to the ozone NAAQS and that the task force "is reviewing administrative options to enable states to enter into cooperative agreements with EPA to provide regulatory relief and meaningfully improve ozone air quality."[84]

The Role of Cost

The issue of cost is a perennial one in NAAQS decisions, even though EPA is prohibited by the Clean Air Act from considering costs in setting the standards. The Clean Air Act's Section 109 has been so interpreted since the NAAQS provisions were added to the act in 1970; in 2001, this interpretation was affirmed in a unanimous Supreme Court decision, *Whitman v. American Trucking Associations*.[85] The Court pointed to numerous other CAA sections where Congress had explicitly allowed consideration of economic factors, concluding that if Congress had intended to allow such factors in the setting of a primary NAAQS, it would have been more forthright—particularly given the centrality of the NAAQS concept to the CAA's regulatory scheme. The court concluded that Section 109(b)(1) "unambiguously bars cost considerations from the NAAQS-setting process."[86]

This is not to say that cost considerations play no role in Clean Air Act decisions, including in implementation of a NAAQS. Cost-effectiveness is considered extensively by EPA and the states in selecting emission control options to meet the standards. Also, as discussed above, the agency prepares cost and benefit estimates at the time it proposes or promulgates a NAAQS, both for information purposes, and in order to comply with Executive Order 12866, under which the OMB requires cost-benefit analysis of

[81] EPA, Final Report on Review of Agency Actions that Potentially Burden the Safe, Efficient Development of Domestic Energy Resources Under Executive Order 13783, October 25, 2017, p. 2, https://www.epa.gov/sites/production/files/ 2017-10/documents/eo-13783-final-report-10-25-2017.pdf.

[82] E. Scott Pruitt, EPA Administrator, "Back-to-Basics Process for Reviewing National Ambient Air Quality Standards," memorandum to EPA Assistant Administrators, May 9, 2018, https://www.epa.gov/sites/production/files/2018-05/documents/image2018-05-09-173219.pdf.

[83] Pruitt, "Back-to-Basics Process."

[84] EPA, Final Report on Review of Agency Actions that Potentially Burden the Safe, Efficient Development of Domestic Energy Resources Under Executive Order 13783, October 25, 2017, p. 5, https://www.epa.gov/sites/production/files/2017-10/documents/eo-13783-final-report-10-25-2017.pdf.

[85] 531 U.S. 457 (2001).

[86] Ibid. at 471.

economically significant rules. But in deciding what level of ambient pollution poses a health threat, the statute bars consideration of costs.

Some Members of Congress would like to change this, by revising the CAA to require consideration of cost in NAAQS decisions. In the 112th Congress, the House twice passed legislation that would have done so: H.R. 2401 and H.R. 3409 would have required the EPA Administrator to take feasibility and costs into consideration in setting National Ambient Air Quality Standards. The Senate did not pass either bill. In the 114th Congress, H.R. 1388/S. 751 would have required EPA to take into consideration feasibility and cost in setting ozone NAAQS, as well as establishing several other conditions on the Administrator's ozone-NAAQS-setting authority. H.R. 4265 and H.R. 4775 would have allowed the Administrator to consider likely technological feasibility "as a secondary consideration" in establishing and revising NAAQS under certain circumstances. In the 115th Congress, S. 263/H.R. 806 contain similar provisions. The House passed H.R. 806 on July 18, 2017.

Background Ozone Levels

A number of states, particularly in the inter-mountain West, experience what are termed high "background" levels of ozone. As EPA explains

> Any ozone formed by processes other than the chemical conversion of local or regional ozone precursor emissions is generically referred to as "background" ozone. Background ozone can originate from natural sources of ozone and ozone precursors, as well as from manmade international emissions of ozone precursors. Natural sources of ozone precursor emissions such as wildfires, lightning, and vegetation can lead to ozone formation by chemical reactions with other natural sources. Another important component of background is ozone that is naturally formed in the stratosphere through interactions of ultraviolet light with molecular oxygen. Stratospheric ozone can mix down to the surface at high concentrations in discrete events called intrusions, especially at higher-altitude locations. The manmade portion of the background includes any ozone formed due to anthropogenic sources of ozone precursors emitted far away from the local area (e.g., international emissions). Finally, both biogenic and international anthropogenic emissions of methane, which can be chemically converted to ozone over relatively long time scales, can also contribute to global background ozone levels.[87]

In the final rule, EPA identified two definitions of background ozone[88]

- natural background (NB) (ozone "that would exist in the absence of any manmade precursor emissions"); and

[87] EPA, 2014 RIA, pp. 2-10 and 2-11.
[88] 80 *Federal Register* 65327, note 84.

- United States background (USB) (ozone "that would exist in the absence of any manmade emissions inside the U.S.").

What the definitions have in common is that they identify ozone levels that cannot be influenced by actions within the jurisdiction of concern.

EPA noted in the final rule that "[v]arious commenters maintained that it was inappropriate to revise the current NAAQS based on their view that natural background concentrations in several states are at or above O3 concentrations associated with meeting a NAAQS set at a level less than 75 ppb (presumably retaining the same indicator, form, and averaging time), making the NAAQS impossible for those states to attain and maintain, a result they claim is legally impermissible."[89]

The agency did not agree. In the final rule, EPA stated that it had "summarized a variety of modeling-based analyses of background O3 ... and conducted our own multi-model assessment of USB concentrations across the U.S.... The EPA analyses ... concluded that seasonal mean daily maximum 8-hour average concentrations of USB O3 range from 25-50 ppb, with the highest estimates located across the intermountain western U.S."

In response to the comments, the final rule stated that "EPA analyses indicate that there may be infrequent instances in a limited number of rural areas where background O3 would be appreciable but not the sole contributor to an exceedance of the revised NAAQS, but do not indicate U.S. background (USB) O3 concentrations will prevent attainment of a revised O_3 standard with a level of 70 ppb." Additionally, the rule noted that

> the courts have clearly established that "[a]ttainability and technological feasibility are not relevant considerations in the promulgation of [NAAQS]." *API v. EPA*, 665 F. 2d 1176, 1185 (D.C. Cir. 1981). Further, the courts have clarified that the EPA may consider proximity to background concentrations as a factor in the decision whether and how to revise the NAAQS only in the context of considering standard levels within the range of reasonable values supported by the air quality criteria and judgments of the Administrator. 79 *FR* 75242-43 (citing *ATA III*, 283 F. 3d at 379).[90]

Finally, the final rule outlined various forms of regulatory relief allowed under the statute and EPA implementing regulations, including "the exclusion of air quality monitoring data from design value calculations when there are exceedances caused by certain event-related U.S. background influences (e.g., wildfires or stratospheric intrusions)";[91] relief from more stringent requirements in areas designated as "rural transport areas"; and the application of CAA Section 179B, which allows EPA to approve

[89] 80 Federal Register 65327.
[90] 80 Federal Register 65328.
[91] Ibid.

state SIP submissions that demonstrate an area would have met the ozone NAAQS by the attainment date if not for international emissions contributing to the area.[92]

The invocation of these regulatory relief measures, as even EPA acknowledged, is not burden-free: each would require some level of assessment or demonstration by a state and/or EPA. States that have tried to invoke the "exceptional event" exceptions have expressed frustration with the lack of clarity and the burden involved in meeting EPA's data requirements and, thus, may not be confident in the agency's offer of regulatory relief.[93] But, as noted earlier (in *Figure 3*), EPA does not believe it will be necessary in most cases to invoke such measures. EPA modeling shows only 14 counties outside of California exceeding a 70 ppb standard without any emission control measures additional to those already promulgated as of this date.

In the 115th Congress, S. 263 and H.R. 806 would change the definition of an exceptional event in Section 319(b)(1) of the Clean Air Act to make it easier for states to exclude air quality monitoring data from the data used to determine nonattainment of a NAAQS. As noted earlier, the House passed H.R. 806 on July 18, 2017.

Schedule for Review

Section 109(d) of the Clean Air Act requires EPA to conduct reviews of the NAAQS and "make such revisions ... as may be appropriate" at five-year intervals. This required schedule is rarely adhered to: the 2015 ozone review took seven and a half years, and as shown in *Table 1*, previous ozone reviews have taken as long as 14 years. Once a review is complete, implementation of a revised standard (i.e., the designation of nonattainment areas and the submittal and review of SIPs) generally takes longer than five years; so if EPA were to meet the statutory NAAQS review schedule, it would be likely that the previous review would still be being implemented at the time the next revised NAAQS was being promulgated.

This has led many in Congress and in various affected industries to suggest that there be a longer time between mandatory reviews of the NAAQS: in the 114th Congress, at least eight bills were introduced that would have changed the required schedule for NAAQS reviews. In the 115th Congress, at least three bills would do so (see "Congressional Reaction" section, below).

Although a change in the mandated schedule might seem to be just a recognition that five years is an insufficient period of time for review, the five-year nondiscretionary deadline serves an important role in the minds of environment and public health groups: it allows them to file suit under the Clean Air Act's citizen suit provision to establish a judicial deadline for completion of a review. In the minds of plaintiffs, extending a

---

92 See EPA, "Tools for Addressing Background Ozone," https://www.epa.gov/sites/ production/files/2015-10/ documents/20151001_background_ozone.pdf.

93 See, for example, the comments of state and local officials on EPA's proposed 2012 Exceptional Events guidance, at http://www.regulations.gov/#!documentDetail;D=EPA-HQ-OAR-2011-0887-0051.

deadline to 10 years would mean a five-year delay before they could establish a court-ordered schedule for such reviews. Thus, there is substantial opposition to what might otherwise seem a mere recognition of reality.

Monitoring

The existing network of ozone monitors is concentrated in urban areas, because of the larger population potentially affected, and because most of the sources of ozone precursor emissions are located in such areas. But, as noted earlier, ozone is not generally emitted directly by pollution sources. It forms in the atmosphere, often downwind of emission sources. Thus, rural areas can have high ozone concentrations. The new form of the secondary NAAQS discussed by EPA suggested a need for additional monitoring in rural areas to detect impacts of ozone on forests and agricultural production. Both EPA and state monitoring budgets are constrained, however, raising questions as to how any additional monitoring requirement would be funded.

The agency, in a 2009 rulemaking separate from the NAAQS, proposed changing the minimum ozone monitoring requirements for both urban and nonurban areas.[94] That proposal would have required that each state operate at least three ozone monitors in nonurban areas. It would also have required at least one ozone monitor in each urban area with a population between 50,000 and 350,000. The requirements were not finalized.

In the final 2015 ozone NAAQS rule, EPA made some adjustments to its monitoring requirements,[95] including the following:

1. updating the Air Quality Index (AQI) based on the strengthened ozone health standard, to provide the public with the most up-to-date information about air quality where they live;
2. revising the monitoring season in 32 states and the District of Columbia, generally by extending the season to require monitoring in additional months;
3. requiring the modernization and streamlining of the Photochemical Assessment Monitoring Stations (PAMS) Network, which helps provide information on ozone formation and transport; and
4. changing the Federal Reference Method for monitoring to include an additional method that is based on advanced technology and monitoring methods.

Although EPA has finalized these changes, the agency is not requiring additional monitors in smaller urban or rural areas.

---

[94] 74 *Federal Register* 34525, July 16, 2009.

[95] For a summary of these actions, see EPA, "Updates to the Air Quality Index for Ozone and Ozone Monitoring Requirements," https://www.epa.gov/sites/production/files/2015-10/documents/20151001_air_quality_index_updates.pdf.

The absence of ozone monitors in rural areas is another factor that should help mitigate concerns of western states that background levels of ozone will cause rural areas in their states to be designated nonattainment. EPA's policy in designating ozone nonattainment areas has always relied on the submission of EPA-certified monitoring data. Areas without monitors (unless they are contributing to an area with monitored nonattainment) have generally been termed "unclassifiable." When nonregulatory monitors indicate exceedance of the ozone NAAQS, designation as nonattainment does not automatically follow. Rather, the state must first establish regulatory monitors[96] and collect three years of data before submitting a proposed designation to EPA.[97]

As of May 2015, approximately 814 counties had ozone monitors, according to EPA, while 2,330 counties did not.[98]

The Role of Federal Versus State and Local Pollution Control Measures

EPA has promulgated more stringent standards for most of the major sources of ozone precursors, including Tier 3 auto emission and fuel standards that began to take effect in 2017 and more stringent standards for power plants that began to take effect in 2015. These standards should make the task of demonstrating attainment with a more stringent ozone NAAQS substantially easier. As noted earlier, EPA estimates that nationwide (except in California) already promulgated EPA and state regulations would bring all but 14 counties into attainment of a 70 ppb ozone NAAQS by 2025. Some Members of Congress objected to the standards for motor vehicles, fuels, power plants, and other sources when they were under consideration, but at this date, the net effect of repealing them would be to shift the burden of attaining the ozone NAAQS more squarely in the direction of state and local governments.

A tighter ozone NAAQS has frequently raised the issue of how to control emission sources that contribute to pollution downwind, in other states. Ozone, which forms in the atmosphere from chemical reactions of precursor emissions, is the prime example of such downwind pollution. Under both the Bush and Obama Administrations, EPA has addressed this interstate air pollution issue—through the Clean Air Interstate Rule (CAIR, 2005), the Cross State Air Pollution Rule (CSAPR, 2011), and the CSAPR Update Rule (2016).

---

[96] An ambient air monitor used to determine whether an area is in attainment of a NAAQS (a "regulatory monitor") must meet three criteria: it must use methods specified in 40 C.F.R. Part 58, Appendix C; it must meet siting criteria specified in 40 C.F.R. Part 58, Appendix E; and it must meet quality assurance criteria specified in 40 C.F.R. Part 58, Appendix A.

[97] The Uintah Basin in Utah provides an example. Nonregulatory monitors indicated ozone levels that exceeded the NAAQS in December 2009, January-March 2010, and January-March 2011. The state began collecting regulatory monitoring data in April 2011. The area was not designated nonattainment until April 30, 2018.

[98] Office of Air Quality Planning and Standards, U.S. EPA, personal communication, May 1, 2015.

The D.C. Circuit Court of Appeals found fault with both the CAIR and CSAPR rules, although it allowed CAIR to take effect pending promulgation of an acceptable replacement.[99] Whether EPA, in the CSAPR rule, had correctly interpreted its authority to control emissions leading to downwind pollution ultimately reached the Supreme Court. On April 29, 2014, in a 6-2 decision (*EPA v. EME Homer City Generation, LLP*), the Court upheld the methodology at the heart of EPA's CSAPR standard-setting process.[100] The rule was remanded to the D.C. Circuit for consideration of additional issues, but it is now being implemented.

The complexity of establishing controls on air pollution that crosses state lines has suggested to many stakeholders that Congress should revisit the Clean Air Act sections that authorize or require such controls, and numerous bills have been introduced on the subject over the years. Since the last major CAA revision in 1990, however, none of those bills has made it to the floor of the House or Senate. The Supreme Court's decision in the *Homer City* case may have further reduced the odds of congressional action on the subject.

CONGRESSIONAL REACTION

Congress took an interest in what became the 2015 ozone standard even before a proposal was released. Three bills introduced in the 113th Congress would have affected the timing and/or EPA's authority to promulgate an ozone NAAQS. In the 114th Congress, which followed proposal of the standards, there were 15 bills or resolutions introduced to address the revision of the ozone (and, in some cases, other) NAAQS. Two of the bills, H.R. 4775 and H.R. 5538 (§438), passed the House, but were not acted on by the Senate.

In the 115th Congress, as of the August recess, six bills had been introduced that would affect the timing and/or EPA's authority to promulgate an ozone NAAQS[101]

- S. 263 (Senator Capito)/H.R. 806 (Representative Olson), the Ozone Standards Implementation Act, would delay the deadline for designation of nonattainment areas under the 2015 ozone NAAQS revision until October 2025 and the date for State Implementation Plan revisions under the standard until October 2026; would require future reviews of the NAAQS at 10-year rather than five-year intervals; would allow the Administrator to consider, as a secondary consideration, likely technological feasibility in establishing and revising primary

---

[99] For a discussion of the Bush and Obama Administration regulations and the D.C. Circuit decisions remanding them, see CRS Report R42895, *Clean Air Issues in the 113th Congress: An Overview*.

[100] EPA v. EME Homer City Generation, L.P., 134 S. Ct. 1584 (2014).

[101] In addition, an amendment to prohibit implementation and enforcement of the 2015 NAAQS was offered on 7/18/18—but not agreed to—in House deliberations about EPA's 2019 appropriation. See H. Amdt. 920 to H.R. 6147.

NAAQS in cases where the Administrator concludes that a range of levels of air quality for a pollutant are requisite to protect public health with an adequate margin of safety; would require the Administrator to publish regulations and guidance for implementing a new or revised NAAQS concurrently with the publication of the standard; would make changes to EPA's authority to exclude from air quality monitoring data used to determine attainment of NAAQS certain data influenced by "exceptional events"; would require a report to Congress on the impact of emissions originating outside the United States on attainment and maintenance of NAAQS; and other provisions. H.R. 806, but not S. 263, would require an EPA report to Congress on ozone formation, including the factors contributing to winter ozone formation, and effective control strategies, and would require EPA to incorporate the report's results into regulations and guidance implementing the 2015 ozone NAAQS. H.R. 806 was reported by the Energy and Commerce Committee on July 14, 2017; the bill passed the House on July 18, 2017.

- S. 452 (Senator Flake), the ORDEAL Act of 2017, would delay the enforcement and implementation of the 2015 ozone NAAQS until January 1, 2025, and would require future reviews at 10-year rather than five-year intervals.
- S. 1203 (Senator Hatch) would require the EPA Administrator to establish a program under which he shall defer the designation of an area as a nonattainment area for the eight-hour ozone NAAQS if the area achieves and maintains certain standards under a voluntary early action compact plan.
- H.R. 958 (Representative Sam Johnson), the Wasteful EPA Programs Elimination Act of 2017, would provide that no funds made available under any act may be used by EPA to implement any ozone standard promulgated after its date of enactment, among other provisions.
- H.R. 3354 (Representative Calvert), the Interior, Environment, and Related Agencies Appropriation Act, 2018, in Section 432 would delay the deadlines for designation of nonattainment areas under the 2015 ozone NAAQS revision until October 2025 and the date for State Implementation Plan revisions under the standard until October 2026. The bill was reported by the Appropriations Committee on July 21, 2017.

The ozone NAAQS has also been challenged in court. Twenty-six parties, including 10 states, have challenged the standards in the U.S. Court of Appeals for the District of Columbia (D.C. Circuit). A nearly equal number of parties, including six states and the District of Columbia, have sided with EPA. Oral argument in the case (*Murray Energy v. EPA*) was scheduled for April 19, 2017,[102] but 12 days before the argument, EPA requested a delay in the proceedings to review the standards. The court granted the

[102] Clerk's Order, Murray Energy Corp. v. EPA, No. 15-1385 (D.C. Cir. Dec. 19, 2016).

agency's request to delay the oral argument and held the case in abeyance pending further orders. EPA was required to file status reports with the D.C. Circuit every 90 days as it reviewed the rule. On August 1, 2018, EPA notified the court that it does not intend to revisit the 2015 standard.[103] The agency also reported that it has begun the next ozone NAAQS review, which will implement the principles from its May 2018 "Back-to-Basics" memorandum, and that it intends to complete this ozone NAAQS review by 2020.[104]

In addition, lawsuits have been filed challenging EPA's final designation of an area in Wisconsin as attainment—after proposing to designate it nonattainment—with respect to the 2015 standard.[105]

---

[103] Respondent EPA's Final Status Report at 1, 4, Murray Energy Corp. v. EPA, No. 15-1385 (D.C. Cir. Aug. 1, 2018).

[104] Respondent EPA's Final Status Report at 4, Murray Energy Corp. v. EPA, No. 15-1385 (D.C. Cir. Aug. 1, 2018).

[105] See Clerk's Order, Clean Wisconsin v. EPA, No. 18-1203 (D.C. Cir. Aug. 3, 2018) (consolidating cases challenging the attainment designation of an area of Wisconsin).

In: Environmental Protection Agency (EPA)
Editor: Coby Schaefer
ISBN: 978-1-53614-933-3

Chapter 6

EPA's Mid-Term Evaluation of Vehicle Greenhouse Gas Emissions Standards*

Richard K. Lattanzio

The One National Program

In 2009, the Obama Administration—through authorities provided to the National Highway Traffic Safety Administration (NHTSA) and the Environmental Protection Agency (EPA)—developed joint standards for fuel economy and greenhouse gas (GHG) emissions for new light-duty vehicles (defined generally as passenger cars and light trucks). The standards (referred to as the One National Program) were established in two phases: Phase 1 for vehicle model years (MY) 2012-2016, finalized on May 7, 2010; and Phase 2 for MY2017-2025, finalized on October 15, 2012. The agencies promulgated the joint rulemakings with the support of an array of stakeholders—including auto manufacturers, labor unions, the environmental community, the state of California, and other states—and intended them to provide industry with a single regulatory voice (to avoid conflicting regulations within the federal government and among the states) and a long-term regime (to provide greater certainty for product planning and engineering).

* This is an edited, reformatted and augmented version of CRS INSIGHT Publication No. IN10619, dated March 29, 2017.

THE MID-TERM EVALUATION

As part of the Phase 2 rulemaking, the agencies made a commitment to conduct a Mid-term Evaluation (MTE) for the second half of the standards: MY2022-2025 (40 C.F.R. 86.1818-12(h)). The agencies deemed an MTE appropriate given the long time frame at issue in setting the standards and given NHTSA's and California's competing statutory obligations. (That is, EPA, California, and some other states—through their authorities under the Clean Air Act [42 U.S.C. 7401 et seq.] and CA AB 1493, respectively—have finalized GHG emissions standards for MY2017-2025. Under the MTE, the agencies were to decide whether to revise them. NHTSA, through its authorities under the Energy Policy and Conservation Act [49 U.S.C. 32902(b)(3)(B)], has finalized standards for MY2017-2021, but requires de novo rulemaking for the period MY2022-2025).

Through the MTE, EPA was to determine whether their standards for MY2022-2025 were still appropriate given the latest available data and information. A final determination could result in strengthening, weakening, or retaining the current standards. If EPA determined that the standards were appropriate, the agency would "announce that final decision and the basis for that decision." If EPA determined that the standards should be changed, EPA and NHTSA would be required to "initiate a rulemaking to adopt standards that are appropriate." Throughout the process, the MY2022-2025 standards were to "remain in effect unless and until EPA changes them by rulemaking."

The Phase 2 rulemaking laid out several formal steps in the MTE process, including

- a Draft Technical Assessment Report (TAR) issued jointly by EPA, NHTSA, and the California Air Resources Board (CARB) with opportunity for public comment no later than November 15, 2017;
- a Proposed Determination on the MTE, with opportunity for public comment; and
- a Final Determination, no later than April 1, 2018.

THE DRAFT TECHNICAL ASSESSMENT REPORT

EPA, NHTSA, and CARB jointly issued the Draft TAR for public comment on July 27, 2016. The Draft TAR was a technical report, not a decision document, and examined a wide range of technology, marketplace, and economic issues relevant to the MY2022-2025 standards. The findings included that

- automakers are innovating in a time of record sales and fuel economy levels;

- the MY2022-2025 standards could be met largely with more efficient gasoline powered cars and with only modest penetration of hybrids and electric vehicles; and
- the standards preserve consumer choice, even as they protect the environment and reduce fuel consumption.

THE FINAL DETERMINATION

On November 30, 2016, EPA released a proposed determination stating that the MY2022-2025 standards remained appropriate and that a rulemaking to change them was not warranted. The agency based its findings on a Technical Support Document, the previously released Draft TAR, and input from the auto industry and other stakeholders. On January 12, 2017, the Administrator finalized the determination, stating "that the standards adopted in 2012 by the EPA remain feasible, practical and appropriate."

The final action arguably accelerated the time line for the MTE, and EPA announced it separately from any NHTSA or CARB process. EPA noted its "discretion" in issuing a final determination, saying that the agency "recognizes that longterm regulatory certainty and stability are important for the automotive industry and will contribute to the continued success of the national program."

On March 24, 2017, CARB passed a resolution to accept its staff's MTE of the state's Advanced Clean Car program— which includes MY2017-2025 vehicle GHG standards in line with EPA's final determination and the existing national rules. EPA had granted California a preemption waiver for its vehicle GHG standards on January 9, 2013, under Section 209(b) of the CAA (42 U.S.C. 7543(b)). Twelve other states have adopted California's standards under the provisions included in Section 177 of the CAA (42 U.S.C. 7507).

NHTSA has neither proposed nor promulgated standards for MY2022-2025.

REACTIONS

Reactions to EPA's final determination were swift. Critics of the standards reportedly vowed to work with the new Administration to revisit EPA's determination—citing a "rush to judgment" they argued contradicted the objectives of the One National Program. The avowed revisions include efforts to better harmonize the existing EPA/NHTSA/CARB standards, ease the MY2022-2025 standards, and/or eliminate them entirely.

Proponents of the standards reportedly suggested that the final determination would set up procedural hurdles against efforts to weaken the standards. They argued that any attempt to ease or overturn them would require new technical analysis and new rulemaking.

The Trump Administration

On March 15, 2017, EPA and NHTSA announced their joint intention to reconsider the Obama Administration's final determination and reopen the MTE process. The agencies reported that their aim was “to engage with stakeholders” and “complete the review in a transparent, data-driven manner.” EPA said that it would make a “new final determination ... no later than April 1, 2018.” The announcement did not address the relationship between federal and California standards.

In: Environmental Protection Agency (EPA)
Editor: Coby Schaefer
ISBN: 978-1-53614-933-3

Chapter 7

EPA'S METHANE REGULATIONS: LEGAL OVERVIEW*

Linda Tsang

SUMMARY

On March 28, 2017, President Trump signed Executive Order 13783, directing federal agencies to review existing regulations and policies that potentially burden the development or use of domestically produced energy resources. Acting pursuant to the order, the U.S. Environmental Protection Agency (EPA) is reviewing and reconsidering several regulations issued during the Obama Administration that address methane emissions from various industrial sectors. Methane is a short-lived greenhouse gas (GHG) with a Global Warming Potential of more than 25 times carbon dioxide that is emitted from various industrial activities.

President Trump's executive order specifically requires EPA to review the revised emission standards for new, modified, and reconstructed equipment, processes, and activities of the oil and natural gas sector issued by the Obama Administration in June 2016. EPA issued these standards for methane and volatile organic compound (VOC) emissions pursuant to Section 111 of the Clean Air Act (CAA). Based on its review and in response to several administrative petitions for reconsideration, EPA is now reconsidering certain emission requirements from the June 2016 rule, which remain in effect unless EPA finalizes a proposed two-year stay of these requirements or otherwise repeals those requirements.

In addition, EPA is reconsidering the emission standards and guidelines for new and existing municipal solid waste (MSW) landfills updated by the Obama Administration in August 2016. In those rules, EPA issued the updated and revised emission standards for MSW landfills built after 2014 to further reduce emissions, including methane emissions. The agency also revised emission guidelines established in 1996 for existing landfills operating prior to that date. At this time, the 2016 landfill rules are in effect. EPA has not

* This is an edited, reformatted and augmented version of Congressional Research Service Publication No. R44615, dated January 24, 2018.

formally proposed any revisions to the 2016 rules or initiated a public comment period for any issues under reconsideration.

EPA's review of the oil and natural gas sector and landfill methane rules has influenced the pending judicial challenges to the various 2016 rules. The U.S. Court of Appeals for the District of Columbia Circuit granted EPA's requests to pause the judicial challenges of both rules to allow EPA to complete its review of them. In addition, stakeholders have successfully challenged in court EPA's attempts to stay the various requirements that the agency is currently reconsidering. Judicial review of EPA's attempts to stay rules in effect could more broadly impact the Trump Administration's efforts to similarly stay other rules that are under reconsideration.

This chapter examines the statutory authority for issuing the methane regulations, legal challenges to the standards, and legal issues related to the reconsideration and stay of the regulations.

INTRODUCTION

On March 28, 2017, President Trump issued Executive Order 13783, which aims to promote the development or use of domestically produced energy resources.[1] Among its specific provisions, the order rescinded the Obama Administration's "Climate Action Plan" (CAP).[2] The CAP aimed to reduce emissions of carbon dioxide (CO2) and other greenhouse gases (GHGs), as well as to encourage adaptation to a changing climate.[3] One of the interagency initiatives within the CAP, the "Strategy to Reduce Methane Emissions" (Methane Strategy), focused on the control of methane emissions, a short-lived climate pollutant with a Global Warming Potential[4] of more than 25 times CO2.[5]

---

[1] *See* Exec. Order No. 13783, 82 Fed. Reg. 16, 093 (Mar. 31, 2017) (signed on Mar. 28, 2017). For additional information on the order, see CRS Legal Sidebar WSLG1789, *New Executive Order Directs Agencies to Revise or Rescind Climate Change Rules and Policies*, by Linda Tsang.

[2] Exec. Order No. 13783, § 3(b)(i)-(ii).

[3] EXEC. OFF. OF THE PRESIDENT, THE PRESIDENT'S CLIMATE ACTION PLAN (2013), https://obamawhitehouse.archives.gov/sites/default/files/image/president27sclimateactionplan.pdf. For additional background on the Climate Action Plan, see CRS Report R43120, *President Obama's Climate Action Plan*, coordinated by Jane A. Leggett.

[4] The Global Warming Potential (GWP) is a metric EPA has adopted to compare the climate impacts of different gases. *Understanding Global Warming Potentials*, ENVTL. PROT. AGENCY, https://www.epa.gov/ghgemissions/understandingglobal-warming-potentials (last visited Jan. 18, 2018). The GWP measures the total energy the emissions of 1 ton of a gas will absorb over a given period of time, relative to the emissions of 1 ton of CO2. *Id.*

[5] Methane (CH4) is estimated to have a GWP of 28–36 over 100 years. *Id.* EPA's *Inventory of U.S. Greenhouse Gas Emissions and Sinks* and Greenhouse Gas (GHG) Reporting Program follow the international GHG reporting standards under the United Nations Framework Convention on Climate Change, which requires the use of GWP values from the Intergovernmental Panel on Climate Change's (IPCC's) Fourth Assessment Report, published in 2007. *Id.* The report lists a GWP of 25 for methane. IPCC, *Fourth Assessment Report: Climate Change* (2007), https://www.ipcc.ch/ publications_and_data/ar4/wg1/en/ ch2s2-10-2.html. For additional information on methane and its GWP, see CRS Report R43860, *Methane: An Introduction to Emission Sources and Reduction Strategies*, coordinated by Richard K. Lattanzio; CRS Report R43860, *Methane: An Introduction to Emission Sources and Reduction Strategies*, coordinated by Richard K. Lattanzio; and CRS Report R42986, *Methane and Other Air Pollution Issues in Natural Gas Systems*, by Richard K. Lattanzio.

President Trump's order also requires agencies to review their existing regulations and "appropriately suspend, revise, or rescind those that unduly burden" domestic energy production and use.[6] The order specifically requires the U.S. Environmental Protection Agency (EPA) to review the revised volatile organic compounds (VOCs)[7] and GHG (namely methane) emission standards for new, modified, and reconstructed equipment, processes, and activities of the oil and natural gas sector issued by the Obama Administration in June 2016.[8] EPA issued these emission standards pursuant to Section 111 of the Clean Air Act (CAA) and the Methane Strategy.[9] Upon review of the standards and several administrative petitions for reconsideration, EPA has now decided to reconsider certain requirements in the emissions standards for the oil and natural gas sector.[10]

Although the methane emission standards and guidelines for new and existing municipal solid waste (MSW) landfills are not specifically mentioned in Executive Order 13783, EPA has also reviewed and is currently reconsidering several requirements of the MSW landfill emission standards and guidelines that the Obama Administration updated in August 2016.[11]

EPA's review of the oil and natural gas sector and landfill methane rules has influenced several judicial challenges to the Obama-era rules. The U.S. Court of Appeals for the District of Columbia Circuit (D.C. Circuit) granted EPA's requests to pause the judicial challenges to allow EPA to complete its review of the regulations.[12] In addition, EPA has attempted to stay the requirements that are being reconsidered by the agency, but courts have restricted the agency's ability to stay such requirements.[13]

This chapter examines the statutory authority for issuing the various methane regulations, legal challenges to the standards, and legal issues related to the reconsideration and stay of these regulations. The report begins by providing a general legal background of CAA Section 111 and Section 307 requirements for reconsidering rules, before addressing the oil and natural gas sector rules and the MSW landfill methane rules.

---

[6] Exec. Order No. 13783, § 1(c).

[7] EPA defines a volatile organic compound (VOC) as "any compound of carbon, excluding carbon monoxide, carbon dioxide, carbonic acid, metallic carbides or carbonates and ammonium carbonate, which participates in atmospheric photochemical reactions." *Technical Overview of Volatile Organic Compounds*, ENVTL. PROT. AGENCY, https://www.epa.gov/indoor-air-quality-iaq/technical-overview-volatile-organic-compounds#definition (last visited Jan. 18, 2018).

[8] *Id.* § 7.

[9] Oil and Natural Gas Sector: Emission Standards for New, Reconstructed, and Modified Sources, Final Rule, 81 Fed. Reg. 35,824 (June 3, 2016) (codified at 40 C.F.R. pt. 60, subpts. OOOO and OOOOa) [hereinafter *2016 Oil & Gas NSPSs*]; 42 U.S.C. § 7411.

[10] *See infra* "Reconsideration of the 2016 Oil and Gas NSPSs."

[11] *See infra* "Reconsideration of 2016 MSW Landfill NSPSs and Emission Guidelines."

[12] Per Curiam Order, North Dakota v. EPA, No. 16-1242 (D.C. Cir. May 18, 2017); Clerk's Order, North Dakota v. EPA, Nat'l Waste & Recycling Ass'n v. EPA, No. 16-1371 (D.C. Cir. June 14, 2017).

[13] *See infra* "EPA's Stays for the 2016 Oil and Gas NSPSs and Resulting Litigation" and "Stay of the MSW Landfill NSPSs and Emission Guidelines."

BACKGROUND

Regulations Under CAA Section 111

EPA issued the regulations addressing methane emissions from oil and natural gas sources and MSW landfills pursuant to CAA Section 111. Section 111 directs EPA to list categories of stationary sources that cause or contribute significantly to "air pollution which may reasonably be anticipated to endanger public health or welfare."[14] Once EPA lists a source category (e.g., "crude oil and natural gas production"), Section 111(b) requires EPA to establish "standards of performance" (known as new source performance standards, or NSPSs) for new, modified, and reconstructed sources within a listed source category.[15] A "standard of performance" is defined as an air pollution emission standard that reflects the "best system of emission reduction" (BSER) that has been "adequately demonstrated."[16] In determining the BSER, EPA must take into account the cost of achieving the emission reductions; non-air quality health and environmental impacts; and energy requirements of the regulated sources.[17]

At the same time or after the Section 111(b) standards are issued for *new, modified, and reconstructed* sources, EPA must issue a "procedure" requiring states to submit plans that establish standards of performance for *existing* sources in a source category, under certain conditions.[18] EPA refers to these Section 111(d) procedures as "emission guidelines" that states are required to follow when they develop plans to implement standards of performance for existing sources in their jurisdictions.[19] Congress established a single definition for standards of performance promulgated pursuant to Sections 111(b) and 111(d), requiring EPA to consider the same factors when

[14] 42 U.S.C. § 7411(b)(1)(A). For example, EPA added "municipal solid waste landfills" as a Section 111 source category "because, in the judgement of the [EPA] Administrator, it contributes significantly to air pollution which may reasonably be anticipated to endanger public health and welfare." Standards of Performance for New Stationary Sources and Guidelines for Control of Existing Sources: Municipal Solid Waste Landfills, 61 Fed. Reg. 9,905, 9,906 (Mar. 12, 1996). *See also* Priority List and Additions to the List of Categories of Stationary Sources; Final Rule, 44 Fed. Reg. 49,222, 49,226 (Aug. 21, 1979) (listing "crude oil and natural gas production" as a source category under Section 111). Once EPA has promulgated new standards, states are allowed to develop and submit to EPA for approval a procedure for implementing and enforcing the standards of performance for new, modified, or reconstructed stationary sources. 42 U.S.C. § 7411(b)(6). If the state procedure is approved, EPA will delegate implementation and enforcement authority to the state. *Id.*

[15] 42 U.S.C. § 7411(b)(1)(A).

[16] *Id.* § 7411(a).

[17] Id. For additional information regarding how the courts have interpreted BSER, see CRS Report R43699, *Key Historical Court Decisions Shaping EPA's Program Under the Clean Air Act*, by Linda Tsang and Alexandra M. Wyatt.

[18] 42 U.S.C. § 7411(d)(1). EPA used its authority under Section 111(d) to issue the Clean Power Plan (CPP) to regulate GHG emissions from existing power plants. *See generally* Carbon Pollution Emission Guidelines for Existing Stationary Sources: Electric Utility Generating Units, Final Rule, 80 Fed. Reg. 64,661 (Oct. 23, 2015).

[19] 42 U.S.C. § 7411(d)(1). Section 111(d) directs EPA to establish state plan procedures similar to Section 110 of the CAA, which requires states to develop and revise implementation plans to achieve EPA's national ambient air quality standards (NAAQS) and subsequent changes to those standards. *Id.*; *see also id.* § 7410.

establishing standards of performance for both new, modified, reconstructed sources and existing sources.[20]

Section 111(d)(1)(A) limits EPA's authority to issue Section 111(d) emission guidelines for existing sources. Based on its interpretation of Section 111 under the Obama Administration, for any air pollutant to which a Section 111(b) NSPS applied for new, modified, or reconstructed sources, EPA was required to regulate the same pollutant under Section 111(d) for existing sources *unless* that air pollutant was already regulated as (1) a "criteria pollutant" under the National Ambient Air Quality Standards (NAAQS) program[21] or (2) a "hazardous air pollutant" (HAP) under the National Emission Standards for Hazardous Air Pollutants (NESHAP) program.[22] For example, because GHGs such as methane are not considered "criteria pollutants" under the NAAQS or HAPs under the CAA, EPA reasoned that it had authority under Section 111(d) to address methane emissions from existing MSW landfills once it had issued NSPSs for methane emission standards for new, modified, and reconstructed sources in those categories.[23] However, EPA's interpretation and scope of its authority under Section 111(d) are currently subject to judicial challenge.[24]

Prior to 2016, EPA had issued Section 111(d) emission guidelines to address the following:

- GHG emissions from existing fossil fuel-fired power plants (known as the Clean Power Plan, or CPP);[25]
- VOC and methane emissions from MSW landfills;[26]
- organics, metals, and nitrogen oxides from municipal waste combustor units;[27]
- acid mist from sulfuric acid production units;[28]
- air pollutants from hospital/medical/infectious waste incinerators;[29]
- fluoride emissions from phosphate fertilizer plants;[30]
- reduced sulfur emissions from kraft pulp mills;[31] and

---

[20] *Compare id.* § 7411(b)(1)(B) (requiring EPA to establish "standards of performance"), *with id.* § 7411(d)(1) (requiring EPA to establish a procedure "under which each State shall submit ... a plan which ... establishes standards of performance for any existing source"). The definition of "standards of performance" is provided under 42 U.S.C. § 7411(a)(1), and applies to the provisions under Section 111.

[21] 42 U.S.C. §§ 7408-7410.

[22] *Id.* § 7412.

[23] See e.g., 2016 Landfill Emission Guidelines, supra note 140, at 59,277-78.

[24] For discussion of litigation related to Section 111(d)'s limits on EPA's authority, see CRS Report R44480, *Clean Power Plan: Legal Background and Pending Litigation in West Virginia v. EPA*, by Linda Tsang and Alexandra M. Wyatt.

[25] 40 C.F.R. pt. 60, subpt. UUUU. For more information about the CPP, see CRS Report R44341, *EPA's Clean Power Plan for Existing Power Plants: Frequently Asked Questions*, by James E. McCarthy et al., and CRS Report R44480, *Clean Power Plan: Legal Background and Pending Litigation in West Virginia v. EPA*, by Linda Tsang and Alexandra M. Wyatt.

[26] 40 C.F.R. pt. 60, subpt. Cf.

[27] *Id.* pt. 60, subpt. Cb.

[28] *Id.* pt. 60, subpt. Cd.

[29] *Id.* pt. 60, subpt. Ce. CAA Section 129 requires EPA to issue Section 111(d) emission guidelines for air pollution from hospital/medical/infectious waste incinerators. 42 U.S.C. § 7429(b).

[30] Phosphate Fertilizer Plants: Final Guideline Document Availability, 42 Fed. Reg. 12,022 (Mar. 1, 1977).

- fluoride emissions from primary aluminum plants.[32]

Reconsideration of Regulations Under CAA Section 307

Under CAA Section 307(d)(7)(B), EPA must convene a reconsideration proceeding if the petitioner meets certain conditions.[33] The petitioner must demonstrate that the objection to the final rule could not have been raised during the public comment period for the proposed rule, and the EPA Administrator must conclude that the "objection is of central relevance to the outcome of the rule."[34] In granting reconsideration, EPA must provide the "same procedural rights as would have been afforded had the information been available at the time the rule was proposed," such as public notice and comment.[35] If EPA denies the petition for reconsideration, the petitioner may seek judicial review of the denial.[36]

Further, under Section 307(d)(7)(B), EPA's reconsideration of a rule does not automatically postpone the effectiveness of the rule unless EPA or a court stays the rule for up to three months during the reconsideration.[37] In other words, if EPA determines that it must convene a reconsideration proceeding, the agency or a court may grant a stay of the effectiveness of the rule for a limited period of three months.[38]

As discussed below, EPA has granted reconsideration for specific requirements in the oil and natural gas sector and landfill methane rules and has issued three-month stays of the effective dates of those requirements for the rules.[39]

---

[31] Kraft Pulp Mills; Final Guideline Document; Availability, 44 Fed. Reg. 29,828 (May 22, 1979).

[32] Primary Aluminum Plants; Availability of Final Guideline Document, 45 Fed. Reg. 26,294 (April 17, 1980). In addition, EPA's 2005 Clean Air Mercury Rule (CAMR) delisted coal-fired power plants from CAA Section 112 and, instead, established a cap-and-trade system for mercury under Section 111(d). 70 Fed. Reg. 28,606 (May 18, 2005). The D.C. Circuit vacated CAMR in 2008 on grounds unrelated to the guidelines' substantive requirements. *See* New Jersey v. EPA, 517 F.3d 574, 581-84 (D.C. Cir. 2008) (holding that EPA's delisting of the source category from Section 112 was unlawful and that EPA was obligated to promulgate standards for mercury and other hazardous air pollutants under Section 112).

[33] 42 U.S.C. § 7607(d)(7)(B).

[34] *See id.* ("If the person raising an objection can demonstrate to the [EPA] Administrator that it was impracticable to raise such objection within such time or if the grounds for such objection arose after the period for public comment (but within the time specified for judicial review) and if such objection is of central relevance to the outcome of the rule, the Administrator shall convene a proceeding for reconsideration of the rule and provide the same procedural rights as would have been afforded had the information been available at the time the rule was proposed.").

[35] *Id.*

[36] *Id.*

[37] 42 U.S.C. § 7607(d)(7)(B).

[38] *Id.*

[39] *See infra* "Reconsideration of the 2016 Oil and Gas NSPSs" and "Reconsideration of 2016 MSW Landfill NSPSs and Emission Guidelines."

REGULATIONS TARGETING METHANE EMISSIONS FROM OIL AND GAS OPERATIONS[40]

Updating and Revising 2012 Emission Standards

The crude oil and natural gas sector is one of the highest-emitting industrial sectors of methane and VOCs in the United States.[41] In 2012, EPA issued a final rule setting standards of performance to limit emissions of VOCs from new, modified, and reconstructed sources in the oil and gas industry pursuant to CAA Section 111(b) (2012 Oil and Gas NSPSs).[42] In an effort to achieve the Obama Administration's goal of reducing methane emissions from the oil and gas sector and to respond to petitions for reconsideration of the 2012 NSPS to address methane emissions,[43] EPA proposed a rule revising and updating the 2012 NSPS in August 2015.[44] After receiving over 900,000 comments, the agency issued the final rule on June 3, 2016 (2016 Oil and Gas NSPSs).[45]

EPA's final rule amended the 2012 Oil and Gas NSPSs, expanding the emission sources covered by the rule and establishing performance standards for GHGs (i.e., methane limitations) from a number of oil and gas emission sources.[46] The final rule established, among other things,

- methane and VOC standards for emission sources and equipment not regulated under the 2012 Oil and Gas NSPSs, including hydraulically fractured oil well completions; pneumatic pumps; and fugitive emissions[47] from well sites and compressor stations; and

---

[40] In addition to the EPA's regulation of methane emissions from oil and gas sources operated by private entities, the Department of the Interior's (DOI's) Bureau of Land Management (BLM) issued a rule in 2016 to replace existing guidance and regulations issued 30 years prior under the Mineral Leasing Act of 1920 that would have reduced methane emissions from oil and natural gas production activities on public lands. Waste Prevention, Production Subject to Royalties, and Resource Conservation, 81 Fed. Reg. 83,008 (Nov. 18, 2016). For information on the BLM methane rule, see CRS Legal Sidebar WSLG1806, *UPDATE: BLM Venting and Flaring Rule Survives (For Now)*, by Linda Tsang, and CRS Report R42986, *Methane and Other Air Pollution Issues in Natural Gas Systems*, by Richard K. Lattanzio.

[41] For further information regarding air pollutants emitted from the oil and gas sector, see CRS Report R42986, *Methane and Other Air Pollution Issues in Natural Gas Systems*, by Richard K. Lattanzio.

[42] Oil and Natural Gas Sector: New Source Performance Standards and National Emission Standards for Hazardous Air Pollutants Reviews; Final Rule, 77 Fed. Reg. 49,542 (Aug. 16, 2012) (codified at 40 C.F.R. pt. 60, subpt. OOOO).

[43] *See e.g.*, Clean Air Council et al., Petition for Reconsideration, In the Matter of: Final Rule Published at 77 FR 49490 (Aug. 16, 2012), titled "Oil and Gas Sector: New Source Performance Standards and National Emission Standards for Hazardous Air Pollutants Reviews; Final Rule," Docket ID No. EPA–HQ–OAR–2010–0505, RIN 2060–AP76 (2012).

[44] Oil and Natural Gas Sector: Emission Standards for New and Modified Sources; Proposed Rule, 80 Fed. Reg. 56,593 (Sept. 18, 2015).

[45] *2016 Oil & Gas NSPSs*, *supra* note 9, at 35,840, 35,848.

[46] *Id.* at 35,825.

[47] EPA defines fugitive emissions as "methane losses can occur from leaks ... in all parts of the [natural gas] infrastructure, from connections between pipes and vessels, to valves and equipment." *Id.*

- methane emission standards for hydraulically fractured gas well completions and equipment leaks at natural gas processing plants that are currently regulated under the 2012 Oil and Gas NSPSs for VOC, but not for methane emissions.[48]

The final rule took effect on August 2, 2016, and applied only to equipment, processes, and activities in the production, processing, transmission, and storage phases of oil and natural gas systems constructed or modified after July 17, 2014; it did not apply to distribution entities.[49] EPA noted that while the new methane standards cover many sources, those sources already complying with the 2012 Oil and Gas NSPSs would not likely be required to install additional emissions controls, as VOC controls also curb methane emissions.[50] EPA estimated that the standards for new and modified sources are expected to reduce 510,000 short tons of methane in 2025, with estimated costs of $530 million and "climate benefits" of $690 million by 2025.[51]

Information Collection for Existing Oil and Natural Gas Sources

As discussed above, the release of the final 2016 Oil and Gas NSPSs under CAA Section 111(b) triggered EPA's duty to issue Section 111(d) emission guidelines for existing oil and natural gas sources covered under the 111(b) rule.[52] In 2016, EPA sent an Information Collection Request (ICR) to oil and natural gas companies seeking information on their existing oil and gas sources as a first step to regulating their methane emissions.[53] Effective March 2, 2017, EPA withdrew the ICR to assess the need for this information and to reduce the burden on businesses during this assessment.[54] The ICR withdrawal was issued shortly after nine state attorneys general and two governors submitted a letter to EPA, asking that the ICR be suspended and withdrawn.[55] EPA has not announced when or if it will reissue the ICR. Several states, cities, and environmental groups have sent EPA notices of intent to sue the agency, alleging that EPA failed to perform "its nondiscretionary duty" to issue emission guidelines limiting methane and VOC emissions from existing sources in the oil and natural gas sector.[56]

---

[48] *Id.* at 35,825-26.

[49] *Id.* at 35,843-48.

[50] EPA, EPA'S ACTIONS TO REDUCE METHANE EMISSIONS FROM THE OIL AND NATURAL GAS INDUSTRY: FINAL RULES AND DRAFT INFORMATION COLLECTION REQUEST 2 (2016), https://www.epa.gov/sites/production/files/2016-09/documents/nsps-overview-fs.pdf.

[51] *2016 Oil & Gas NSPSs*, *supra* note 9, at 35,827.

[52] *See supra* "Regulations Under CAA Section 111."

[53] Proposed Information Collection Request; Comment Request; Information Collection Effort for Oil and Gas Facilities, 81 Fed. Reg. 35,763 (June 3, 2016).

[54] Notice Regarding Withdrawal of Obligation To Submit Information, 82 Fed. Reg. 12,817 (Mar. 7, 2017).

[55] Letter from Ken Paxton, Att'y. Gen., Texas, et al., to Scott Pruitt, Admin., EPA (Mar. 1, 2017), https://www.epa.gov/sites/production/files/2017-03/documents/letter_from_attorneys_general_and_governors.pdf.

[56] Letter from Peter Zalzal, Envtl. Def. Fund, et al., to E. Scott Pruitt, Admin., EPA (Aug. 28, 2017), https://www.epa.gov/sites/production/files/2017-09/documents/edf_noi_08282017. pdf; Letter from Eric T.

Reconsideration of the 2016 Oil and Gas NSPSs

In April and June 2017, EPA granted administrative petitions for reconsideration of several aspects of the 2016 Oil and Gas NSPSs pursuant to CAA Section 307(d)(7)(B).[57] On April 18, 2017, EPA granted petitions to reconsider (1) the application of fugitive emissions requirements to low production well sites, and (2) the process and criteria used for requesting and approving the use of alternative means of emission limitations (AMEL)—an equivalent and enforceable state emissions limit—to comply with the fugitive emissions requirements.[58] EPA determined that the final rule subjected low production wells to fugitive emissions requirements based on information and rationale that differed significantly from the proposal, making it "impracticable" for petitioners to object to these requirements during the public comment period.[59] Similarly, EPA determined that petitioners did not have the opportunity to object or comment on the AMEL process and criteria because these provisions were introduced for the first time in the final rule.[60] Because these fugitive emission requirements relate directly to whether and how they apply to certain sources, EPA concluded that they are of "central relevance" to the outcome of the rule and granted reconsideration.[61]

In June 2017, EPA identified two other issues concerning the 2016 rule that, in the agency's view, met the Section 307(d)(7)(B) criteria for reconsideration.[62] Petitioners sought reconsideration of provisions that require a professional engineer (PE) to certify (1) the proper design of closed vent systems[63] and (2) "technical infeasibility" determinations to exempt a pneumatic pump at a well site from emission reduction requirements.[64] EPA acknowledged that it failed to present costs associated with the PE certification requirements in the proposal and determined that these compliance requirements are of central relevance to the outcome of the rule.[65] Thus, EPA has convened an administrative reconsideration proceeding to address these issues in addition to the fugitive emissions requirements.[66]

As part of the reconsideration proceeding, EPA plans to prepare a notice of proposed rulemaking that will provide the petitioners and the public an opportunity to comment on

---

Schneiderman, Att'y Gen., et al., to Scott Pruitt, Admin., EPA (June 29, 2017), https://www.epa.gov/sites/production/files/2017-07/documents/states_ noi_06292017.pdf.

[57] Oil and Natural Gas Sector: Emission Standards for New, Reconstructed, and Modified Sources; Grant of Reconsideration and Partial Stay, 82 Fed. Reg. 25,730, 25,731-32 (June 5, 2017). See *supra* "Reconsideration of Regulations Under CAA Section 307" for discussion of Section 307(d)(7)(B) requirements for reconsideration.

[58] 82 Fed. Reg. at 25,730.

[59] *Id.*

[60] *Id.*

[61] *Id.*

[62] *Id.* at 25,731-32.

[63] *See id.* at 25,732 (arguing that EPA failed to present costs associated with requiring a PE to certify the design and capacity of closed vent systems used to comply with the emission standards in the 2016 Oil and Gas NSPS).

[64] See id. (arguing that EPA did not propose that a PE must certify technical infeasibility determinations to qualify for an exemption for certain pneumatic pumps).

[65] Id.

[66] Id.

the rule requirements and associated issues identified above.[67] EPA also intends to look broadly at the entire 2016 Oil and Gas NSPSs, potentially indicating that EPA may revise other aspects of the emission standards. If EPA decides to revise the rule, CAA Section 307(d) requires that the agency follow the notice-and-comment rulemaking procedures that were used to promulgate the 2016 rule.[68]

EPA' s Stays for the 2016 Oil and Gas NSPSs and Resulting Litigation

Three-Month Stay of Certain Requirements

As part of the reconsideration process, on June 5, 2017, EPA stayed parts of the 2016 Oil and Gas NSPSs: the fugitive emissions requirements,[69] including the deadline for the initial monitoring survey, and other provisions related to the issues being reconsidered, until August 31, 2017.[70] EPA issued the three-month stay pursuant to its authority under CAA Section 307(d)(7)(B) without public notice and comment.[71] As discussed above, EPA or a court may stay the effectiveness of a rule for up to three months during the reconsideration once EPA grants reconsideration.[72] Environmental groups filed a petition to review and an emergency motion to stop or vacate the three-month stay, arguing that EPA did not have authority to issue the stay.[73]

On July 3, 2017, the D.C. Circuit, in a 2-1 split decision, held that the "90-day stay was unauthorized under section 307(d)(7)(B) and thus unreasonable," vacating the stay.[74] In its decision, the majority of the court first addressed EPA's claim that the court did not have jurisdiction to review the claim because the stay is a not a "final action."[75] Under the CAA, judicial review is limited to "final action taken, by the [EPA] Administrator."[76] In rejecting the argument, the majority explained that the "imposition of the stay is an entirely different matter" from the EPA's decision to reconsider the final rule, which was not a reviewable final agency action.[77] The majority reasoned that "EPA's stay . . . is essentially an order delaying the rule's effective date, and this court has held that such

[67] Id.

[68] 42 U.S.C. § 7607(d).

[69] The 2016 Oil and Gas NSPSs required regulated entities to conduct an "initial monitoring survey" to identify fugitive emissions at new well sites and compressor stations by June 3, 2017. 40 C.F.R. § 60.5397a(f); 2016 Oil & Gas NSPSs, supra note 9, at 35,824.

[70] 82 Fed. Reg. at 25,732-33.

[71] Id.

[72] 42 U.S.C. § 7607(d)(7)(B). See *supra* "Reconsideration of Regulations Under CAA Section 307" for discussion of Section 307(d)(7)(B) stay requirements.

[73] Petition for Review, Clean Air Council v. Pruitt, No. 17-1145 (D.C. Cir. June 5, 2017); Petitioners' Emergency Motion for a Stay or in the Alternative, Summary Vacatur, No. 17-1145 (D.C. Cir. June 5, 2017).

[74] Clean Air Council v. Pruitt, 862 F.3d 1, 8 (D.C. Cir. 2017) (per curiam).

[75] *Id.* at 6.

[76] 42 U.S.C. § 7607(b)(1).

[77] *Clean Air Council*, 862 F.3d at 6.

orders are tantamount to amending or revoking a rule."[78] The stay, the majority concluded, was a reviewable "final agency position" on the effective date of the rule that "affects regulated parties' 'rights or obligations.'"[79] In her dissent, Judge Janice Rogers Brown disagreed, arguing that the stay was not a final agency action because the stay (1) "facilitates" the reconsideration of the requirements and does not "resolve them"; and (2) does not compel compliance or impose "legal or practical requirements on anyone."[80]

After concluding that it had jurisdiction to review the stay, the majority of the court considered whether EPA had authority to issue the stay.[81] The majority rejected EPA's argument that the agency had "inherent" authority to issue the stay, explaining that EPA "may act only pursuant to authority delegated to them by Congress."[82] Further, the majority noted that EPA did not rely on its inherent authority, but purported to act pursuant to Section 307(d)(7)(B), a statute that, in the majority's view, "clearly delineates" that EPA is authorized to issue a three-month stay only if the agency was *required* to reconsider the rule.[83] Therefore, the majority reasoned that it must review EPA's reconsideration decision because "[u]nder CAA section 307(d)(7)(B), . . . the stay EPA imposed is lawful only if reconsideration was mandatory."[84]

The court concluded that the reconsideration was not mandatory because the petitioners did not object to an issue that was impracticable to raise during the public comment period.[85] The court determined that the industry groups petitioning for an administrative reconsideration had "ample opportunity to comment on all four issues" under reconsideration as evidenced by EPA directly addressing those comments in the final rule.[86] Because it was not "impracticable" to raise those issues during the comment period, the majority of the court held that EPA was not required under Section 307(d)(7)(B) to grant reconsideration and therefore not authorized to issue the three-month stay.[87] As a result, the majority concluded that EPA's decision to impose the stay was "arbitrary" and "capricious," and the court granted the motion to vacate the stay.[88] Once the stay was vacated, the formerly stayed requirements went back into effect.[89]

---

[78] *Id.*

[79] *Id.* at 7 (internal citations omitted).

[80] *Id.* at 14-17 (Brown, J., dissenting).

[81] *Id.* at 8 (majority opinion). In her dissent, Judge Brown did not reach the issue of whether EPA had authority to issue the stay. *Id.* at 18 (Brown, J., dissenting).

[82] *Id.* at 9 (majority opinion).

[83] *Id.*

[84] *Id.* at 9.

[85] *Id.* at 14.

[86] *Id.*

[87] *Id.*

[88] *Id.* Ten days after its decision, the court granted EPA's request to recall the mandate, and allowed the stay to remain in effect for two weeks (until July 17, 2017) in order to give EPA time to determine whether it would appeal the decision to vacate the stay. Order, Clean Air Council v. Pruitt, No. 17-1145 (D.C. Cir. July 13, 2017). The court issued the mandate on July 31, 2017. Per Curiam Order, Clean Air Council v. Pruitt, No. 17-1145 (D.C. Cir. July 31, 2017). In an 8-3 vote, a majority of the court denied the petition for rehearing en banc filed by intervenor states and industry groups. Per Curiam Order, Clean Air Council v. Pruitt, No. 17-1145 (D.C. Cir. Aug. 10, 2017).

[89] *See e.g.*, Action on Smoking & Health v. Civil Aeronautics Bd., 713 F.2d 795, 797 (1983) ("To 'vacate,' as the parties should well know, means 'to annul; to cancel or rescind; to declare, to make, or to render, void; to defeat; to deprive of force; to make of no authority or validity; to set aside.' Thus, by vacating or rescinding

Judicial review of EPA's attempts to stay rules already in effect could have broader effects on the Trump Administration's efforts to similarly stay or postpone requirements in other rules that are under reconsideration.[90] The majority's decision to vacate the three-month stay of the 2016 Oil and Gas NSPSs may affect future cases that challenge EPA's authority under CAA Section 307(d)(7)(B) to issue a three-month stay when reconsidering a rule.[91] In holding that EPA does not have inherent authority to issue temporary stays, the majority of the court limits EPA's authority to issue a three-month stay under Section 307(d)(7)(B) to circumstances when the agency is required to reconsider a rule.[92]

Further, some federal agencies that have attempted to use the Administrative Procedure Act (APA) to stay rules in effect have been struck down by the courts. For example, a federal district court granted summary judgment to plaintiffs that alleged that the Bureau of Land Management violated the APA when the Bureau temporarily postponed the compliance dates for certain sections of the rule regulating methane emissions from oil and natural gas production activities on public lands after the rule's effective date had already passed.[93]

Proposed Two-Year Stay

Before the D.C. Circuit issued its decision regarding EPA's authority to issue the three-month stay of the 2016 Oil and Gas NSPSs, EPA also proposed to further extend the three-month stay of the fugitive emissions requirements, well site pneumatic standards, and closed vent certification requirements for two years—until September 12, 2019.[94] In the proposal, EPA stated that the longer stay is "necessary" to provide sufficient time to complete its reconsideration of these requirements through the public notice-and-comment process.[95] However, the proposal did not identify the legal authority that permits EPA to issue the stay; nor did the agency rely on Section 307(d)(7)(B) for

---

the recissions [sic] proposed by ER-1245, the judgment of this court had the effect of reinstating the rules previously in force.... ").

[90] *See infra* "Stay of the MSW Landfill NSPSs and Emission Guidelines."

[91] *Id.*

[92] *Clean Air Council*, 862 F.3d at 9.

[93] California v. Bureau of Land Mgmt., No. 17-cv-03804-EDL, 2017 U.S. Dist. LEXIS 176620, at *13-14 (N.D. Cal. Oct. 4, 2017).

[94] Oil and Natural Gas Sector: Emission Standards for New, Reconstructed, and Modified Sources: Stay of Certain Requirements, 82 Fed. Reg. 27,645 (June 16, 2017) [hereinafter *Proposed 2-year Stay of the 2016 Oil & Gas NSPSs*]. Further, the agency also proposed a three-month stay to cover the gap period from when the two-year delay is finalized to its effective date pursuant to the Congressional Review Act. Oil and Natural Gas Sector: Emission Standards for New, Reconstructed, and Modified Sources: Three Month Stay of Certain Requirements, Proposed Rule, 82 Fed. Reg. 27,641 (June 16, 2017). In the proposal, EPA explains that if the two-year stay is finalized, it would likely be considered a "major rule" under the CRA. *Id.* at 27,642. Under the CRA, a major rule cannot take effect until 60 days after publication in the *Federal Register* or after Congress receives the rule report, whichever is later. 5 U.S.C. § 801(a)(3). Based on its expectation that courts are not likely to consider a three-month stay to be a major rule under the CRA, EPA anticipates that the three-month stay, if finalized, would take effect immediately upon its publication in the *Federal Register* and would stay the requirements for the period between the two-year stay's filing and effective dates. 82 Fed. Reg. at 27,642-43.

[95] *Proposed 2-year Stay of the 2016 Oil & Gas NSPSs*, 82 Fed. Reg. at 27,648.

the longer stay. The public comment period for the proposed stay closed on August 9, 2017.[96]

After reviewing comments submitted by stakeholders, EPA, on November 8, 2017, issued a notice of data availability (NODA) for the proposed stay.[97] The NODA seeks comment on additional information, ideas, and issues stakeholders raised regarding EPA's legal authority to issue the stay and challenges in implementing the stayed requirements.[98] For example, EPA is seeking comment on its legal authority to stay or extend the compliance periods for the requirements under reconsideration.[99] EPA argues that these proposed actions to stay or extend the compliance deadlines are "lawful exercises" of the agency's "inherent" authority to reconsider, revise, replace, or repeal past decisions and its general rulemaking authority under Section 301(a) of the CAA,[100] which authorizes EPA "to prescribe such regulations as are necessary to carry out [its] functions" under the Act.[101] The public comment period for the NODAs closed December 8, 2017.[102]

Petitions for Judicial Review

Although EPA is reconsidering the 2016 Oil and Gas NSPSs, the underlying rule is also being challenged in court. Prior to the start of the Trump Administration, North Dakota,[103] Texas (including the Railroad Commission of Texas and the Texas Commission on Environmental Quality as named parties),[104] a coalition of 13 states (including several state agencies),[105] and various gas associations filed petitions for review of the 2016 Oil and Gas NSPSs.[106] Eleven states, the city of Chicago, and various environmental advocacy groups intervened on behalf of EPA to support the final rule.[107]

---

[96] *Id.*

[97] Oil and Natural Gas Sector: Emission Standards for New, Reconstructed, and Modified Sources: Stay of Certain Requirements, Proposed rule; notice of data availability, 82 Fed. Reg. 51,788 (Nov. 8, 2017). EPA also issued a notice of data availability for the proposed three-month stay. Oil and Natural Gas Sector: Emission Standards for New, Reconstructed, and Modified Sources: Three Month Stay of Certain Requirements, Proposed rule; notice of data availability, 82 Fed. Reg. 51,794 (Nov. 8, 2017).

[98] *Proposed 2-year Stay of the 2016 Oil & Gas NSPSs*, 82 Fed. Reg. at 27,648.

[99] 82 Fed. Reg. at 51,790.

[100] *Id*

[101] 42 U.S.C. § 7601(a).

[102] 82 Fed. Reg. at 51,788; 82 Fed. Reg. at 51,794.

[103] Petition for Review, North Dakota v. EPA, No. 16-1242 (D.C. Cir. July 15, 2016).

[104] Petition for Review, Texas et al. v. EPA, No. 16-1257 (D.C. Cir. July 28, 2016).

[105] Petition for Review, West Virginia et al. v. EPA, No. 16-1264 (D.C. Cir. Aug. 2, 2016). Besides West Virginia, the state coalition includes Alabama, Arizona, Kansas, Kentucky, Louisiana, Attorney General Bill Schuette for the people of Michigan, Montana, Ohio, Oklahoma, South Carolina, Wisconsin, Kentucky Energy and Environment Cabinet, and the North Carolina Department of Environmental Quality (NCDEQ). *Id.* On March 20, 2017, the court granted NCDEQ's motion to withdraw as a petitioner. *Id.*

[106] Petition for Review, W. Energy Alliance v. EPA, No. 16-1266 (D.C. Cir. Aug. 2, 2016); Petition for Review, Indep. Petroleum Ass'n of Am. et al. v. EPA, No. 16-1262 (D.C. Cir. Aug. 2, 2016); Petition for Review, Interstate Nat. Gas Ass'n, No. 16-1263 (D.C. Cir. Aug. 2, 2016); Petition for Review, GPA Midstream Gas Ass'n, No. 16-1267 (D.C. Cir. Aug. 2, 2016); Petition for Review, Texas Oil & Gas Ass'n, No. 16-1269 (D.C. Cir. Aug. 2, 2016); Petition for Review, Am. Petroleum Inst., No. 16-1270 (D.C. Cir. Aug. 2, 2016).

[107] 107 *See* Clerk's Order, North Dakota v. EPA, No. 16-1242 (D.C. Cir. Dec. 19, 2016) (granting motion to intervene as respondents for the Clean Air Council, Earthworks, Environmental Defense Fund, Environmental

(See *Figure 1.*) All petitions have been consolidated with the lead case, *North Dakota v. EPA*.[108] In May 2017, the court granted EPA's request to hold the case in abeyance while the agency reviews the rule.[109]

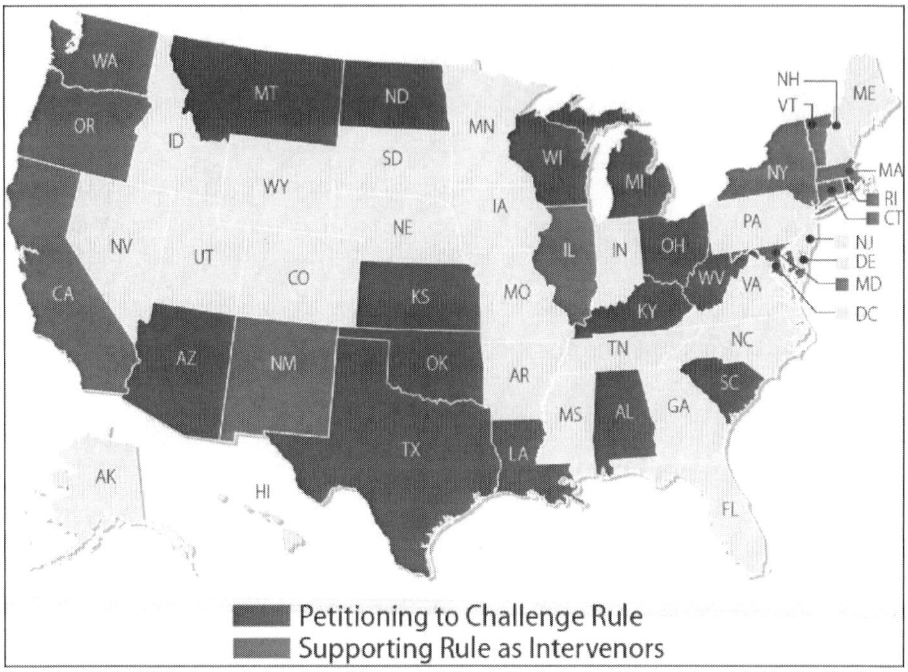

Source: Prepared by CRS from litigation filings in North Dakota v. EPA, No. 16-1242 (D.C. Cir.).

Notes: The Kentucky Energy and Environment Cabinet is a petitioner, in addition to the Commonwealth of Kentucky. The city of Chicago is an intervenor in support of EPA, in addition to the state of Illinois.

Figure 1. States Participating in 2016 Oil and Gas NSPSs Litigation. Consolidated Petitions: North Dakota v. EPA, D.C. Cir. No. 16-1242.

Litigation Issues

Before the litigation was paused, the petitioners filed their preliminary statements of the legal issues[110] that will be raised in the litigation.[111] The final statement of issues will be presented in the petitioners' briefs,[112] which have yet to be filed. This chapter does not

Integrity Project, Natural Resources Defense Council, and Sierra Club; the States of California, Connecticut, Illinois, Maryland, New Mexico, New York, Oregon, Rhode Island, Vermont, and Washington; the Commonwealth of Massachusetts; and the City of Chicago).

[108] *See* Clerk's Order, North Dakota v. EPA, No. 16-1242 (D.C. Cir. Aug. 8, 2016).

[109] Per Curiam Order, North Dakota v. EPA, No. 16-1242 (D.C. Cir. May 18, 2017).

[110] *See id.*; D.C. CIR. R. 15(c).

[111] *See e.g.*, Petitioner N.D.'s Statement of Issues to be Raised, North Dakota v. EPA, No. 16-1242 (D.C. Cir. Aug. 22, 2016) [hereinafter *N.D. Statement of Issues*]; Petitioner W. Energy Alliance's Non-Binding Statement of Issues to be Raised, North Dakota v. EPA, No. 16-1242 (D.C. Cir. September 1, 2016) [hereinafter *WEA Statement of Issues*]; Nonbinding Statement of Issues of the Am. Petroleum Inst., North Dakota v. EPA, No. 16-1242 (D.C. Cir. Sept. 6, 2016) [hereinafter *API Statement of Issues*].

[112] D.C. CIR. R. 28(a)(5).

aim to provide a comprehensive preview of the potential legal arguments that may be presented to the court for or against EPA's final rule. The sections below highlight some potential issues that may be raised in litigation, including a challenge to EPA's authority to issue the 2016 Oil and Gas NSPSs.

Endangerment Finding Under CAA Section 111(b)

CAA Section 111 requires EPA to first establish a list of source categories to be regulated, and then establish emission standards for new sources in that source category.[113] Section 111(b)(1)(A) requires that a source category be included on the list if, "in [the EPA Administrator's] judgment it causes, or contributes significantly to, air pollution which may reasonably be anticipated to endanger public health or welfare."[114] EPA commonly refers to this determination as an "endangerment finding."[115] In 1979, EPA listed "crude oil and natural gas production" as a source category under Section 111 based on its "endangerment finding."[116]

Several of the petitioners' statements of issues claimed that EPA failed to make the required "endangerment finding" for methane from the oil and gas sector under CAA Section 111.[117] One petitioner argued that EPA is required under Section 111(b) to make "endangerment findings" for particular pollutants such as methane that it seeks to regulate for listed source categories.[118]

In the final rule, EPA relied primarily on its previous 1979 "endangerment finding" for the "crude oil and natural gas production" source category as its authority to issue new and updated NSPSs for methane and VOC emissions.[119] In the response to comments section of the final rule, EPA reasoned that the agency did not need to make a separate "endangerment finding" for methane because the "2009 Endangerment Finding [for motor vehicles] that defines the aggregate group of six well-mixed [GHGs] as the air pollution addresses emissions of any individual component of that aggregate group and, therefore, supports the rational basis for this final rule."[120] EPA reiterated that the GHG is the "regulated pollutant," and the standards are expressed in form of limitations on methane emissions.[121] The final rule also included a new "endangerment finding" based on previous and new scientific analyses, and stated that "to the extent such a finding were necessary, pursuant to section 111(b)(1)(A), the Administrator hereby determines that, in her judgment, this source category, as defined above, contributes significantly to air

---

[113] 42 U.S.C. § 7411(b)(1).

[114] *Id.* § 7411(b)(1)(A).

[115] *2016 Oil & Gas NSPSs*, *supra* note 9, at 35,828.

[116] Priority List and Additions to the List of Categories of Stationary Sources; Final Rule, 44 Fed. Reg. 49,222, 49,226 (Aug. 21, 1979).

[117] *See N.D. Statement of Issues*, *supra* note 111; *WEA Statement of Issues*, *supra* note 111; *API Statement of Issues, supra* note 111; Comments from Kathleen M. Sgamma, V.P. of Govt. & Public Aff., W. Energy Alliance, to Gina McCarthy, Admin., EPA (Dec. 4, 2015).

[118] *API Statement of Issues,* at 2; Comments from Howard Feldman, Sen. Dir., Am. Petroleum Inst., to Gina McCarthy, Admin., EPA (Dec. 4, 2015), at 5-6.

[119] *2016 Oil & Gas NSPSs*, *supra* note 9, at 35,833.

[120] *Id.*

[121] *Id.* at 35,877.

pollution which may reasonably be anticipated to endanger public health or welfare."[122] Petitioners also seek to litigate whether the new "endangerment finding" is an arbitrary and capricious determination, prohibited by the CAA.[123]

Scope of the Oil and Gas Source Category

Several petitioners also argued that EPA has "unlawfully" expanded the listed "crude oil and natural gas production" source category to additional types of emission sources not previously regulated, such as hydraulically fractured oil well completions; pneumatic pumps; and fugitive emissions from well sites and compressor stations.[124] As discussed above, CAA Section 111 requires EPA to establish a list of source categories to be regulated if EPA determines that the source category "causes, or contributes significantly to, air pollution which may reasonably be anticipated to endanger public health or welfare."[125] One petitioner argued in its comments on the proposed rule that EPA's 1979 original source category determination for "crude oil and natural gas production" cannot "be read to include the numerous smaller emissions points by this proposal. The 1979 listing was focused on major-emitting operations and cannot be reasonably construed as encompassing small, discrete sources that are separate and apart from a large facility, like a processing plant."[126]

In the final rule, EPA claimed that the rule did not expand the oil and gas source category. The agency reasoned that 1979 source category listing of "crude oil and natural gas production" broadly covered the natural gas industry including production, processing, transmission, and storage of oil and natural gas.[127] EPA pointed to the 1979 source category listing analysis that indicated that the agency previously evaluated emissions from various segments of the natural gas industry, such as production and processing.[128] The agency explained that it provided notice in the rulemaking that smaller sources emitting less than 100 tons of emissions could be subject to NSPSs even though the 1979 listing of the "crude oil and natural gas production" was classified under the "major source categories."[129] EPA also highlighted the 1984 proposed NSPSs to limit VOC emissions from specific equipment in the natural gas production industry to support the broad scope of the oil and natural gas source category.[130] The 1984 proposal defined the crude oil and natural gas production industry to include "the operations of exploring

---

[122] *Id.* at 35,833.

[123] *WEA Statement of Issues, supra* note 111, at 2; *API Statement of Issues, supra* note 111 at 2.

[124] JEA Statement of Issues, supra note 111, at 2; API Statement of Issues, supra note 111, at 2; N.D. Statement of Issues, supra note 111, at 2.

[125] 42 U.S.C. § 7411(b)(1)(A).

[126] *API Comments, supra* note 111, at 4.

[127] *2016 Oil & Gas NSPSs, supra* note 9, at 35,832.

[128] *Id.*

[129] *Id.* When the "crude oil and natural gas production" source category was listed in 1979, EPA defined a "major source category" as "those categories for which an average size plant has the potential to emit 100 tons or more per year of any one pollutant." Priority List and Additions to the List of Categories of Stationary Sources, 44 Fed. Reg. 49,222 (Aug. 21, 1979). EPA used the "major source" categories to rank and prioritize setting standards for the listed sources. *Id.*

[130] 2016 Oil & Gas NSPSs, supra note 9, at 35,832.

for crude oil and natural gas products, drilling for these products, removing them from beneath the earth's surface, and processing these products from oil and gas fields for distribution to petroleum refineries and gas pipelines."[131]

However, EPA appeared to acknowledge the concerns and perceived ambiguity regarding the scope of the oil and gas category source. In the final rule, EPA stated that

> [T]o the extent that there is any ambiguity in the prior listing, the EPA hereby finalizes, as an alternative, its proposed revision of the category listing to broadly include the oil and natural gas industry. As revised, the listed oil and natural gas source category includes oil and natural gas production, processing, transmission, and storage. In support, the EPA has included in this action the requisite finding under section 111(b)(1)(A) that, in the Administrator's judgment, this source category ... contributes significantly to air pollution which may reasonably be anticipated to endanger public health or welfare.[132]

If the litigation moves forward, the court will most likely address whether this alternative category revision is necessary or sufficient under Section 111.

Next Steps

On May 18, 2017, the court granted EPA's request to hold the cases in abeyance as EPA reviews and reconsiders the rule.[133] Until a stay is finalized, the 2016 Oil and Gas NSPSs remain in effect. EPA has not issued any further notices regarding the reconsideration proceeding for the rule. See *Table 1* below for the current status of the oil and gas sector methane rules.

Regulations Targeting Methane Emissions from Municipal Landfills

Landfills emit gas that contains methane, carbon dioxide, and more than 100 different nonmethane organic compounds (NMOCs), such as vinyl chloride, toluene, and benzene.[134] In 1996, EPA listed municipal solid waste (MSW) landfills as a CAA Section 111 category of stationary sources that causes or contributes significantly to air pollution that may reasonably be anticipated to endanger public health or welfare.[135] In addition to listing MSW landfills as a source category, EPA, at that same time, issued NSPSs for

---

[131] Standards of Performance for New Stationary Sources; Onshore Natural Gas Processing Plants In the Natural Gas Production Industry; Equipment Leaks of VOC; Proposed Rule and Notice of Public Hearing, 49 Fed. Reg. 2,636, 2,637 (Jan. 20, 1984).

[132] *2016 Oil & Gas NSPSs*, *supra* note 9, at 35,833.

[133] Per Curiam Order, North Dakota v. EPA, No. 16-1242 (D.C. Cir. May 18, 2017).

[134] Standards of Performance for New Stationary Sources and Guidelines for Control of Existing Sources: Municipal Solid Waste Landfills, 61 Fed. Reg. 9,905, 9,906 (Mar. 12, 1996).

[135] 61 Fed. Reg. at 9,919 (codified at 40 C.F.R. § 60.16).

Table 1. EPA's Oil and Gas Sector Methane Rules

Regulatory Action	Status of the Regulatory Action (as of January 2018)	Status of Litigation on Regulatory Action (as of January 2018)
2016 Oil and Gas NSPSs 81 Fed. Reg. 35,824 (June 3, 2016)	In effect, but under reconsideration	Held in abeyance *North Dakota v. EPA*, No. 16-1242, (D.C. Cir. May 18, 2017)
Three-Month Stay: 2016 Oil and Gas NSPSs 82 Fed. Reg. 25,730 (June 5, 2017)	Vacated by D.C. Circuit (July 31, 2017)	D.C. Circuit ruled, 2-1, to vacate *Clean Air Council v. Pruitt*, 862 F.3d 1 (D.C. Cir. 2017)
Proposed Two-Year Stay: 2016 Oil and Gas NSPSsa 82 Fed. Reg. 27,645 (June 16, 2017)	Public comment period closed on August 9, 2017	N/A
Notice of Data Availability for the Proposed Two-Year Stay 82 Fed. Reg. 51,788 (Nov. 8, 2017)	Public comment period closed on December 8, 2017	N/A
2016 Proposed Information Collection Request for Existing Oil and Gas Facilities 81 Fed. Reg. 35,763 (June 3, 2016)	Withdrawn (March 2, 2017)	N/A

Source: Prepared by CRS from regulatory and litigation filings.

Notes: This table does not include all CAA regulations that affect the oil and gas sector.

[a] EPA also proposed a three-month stay to cover the gap period from when the two-year delay is finalized to its effective date pursuant to the Congressional Review Act. Oil and Natural Gas Sector: Emission Standards for New, Reconstructed, and Modified Sources: Three Month Stay of Certain Requirements, Proposed Rule, 82 Fed. Reg. 27,641 (June 16, 2017).

new, modified, and reconstructed MSW landfills pursuant to Section 111(b) and emission guidelines for existing landfills pursuant to Section 111(d) to address NMOCs and methane emissions.[138] In response to a consent decree resolving a lawsuit[139] and the Obama Administration's Methane Strategy, in 2014 EPA proposed updates to the 1996 MSW landfill NSPSs and sought public comments on whether EPA should update the companion emission guidelines for existing landfills.[144] Over 20 years after the original NSPSs and emission guidelines were issued, EPA published on August 29, 2016, its updated NSPSs to reduce landfill gas emissions from MSW landfills built, modified, or

[138] 61 Fed. Reg. at 9,905 (Mar. 12, 1996).

[139] Environmental Defense Fund filed a lawsuit, arguing that EPA failed to review the 1996 NSPS by the statutorily required deadline, which is every eight years pursuant to CAA Section 111(b). Envt'l Def. Fund v. EPA. No. 1:11-cv04492 (S.D.N.Y. June 30. 2011). Under a consent decree resolving that lawsuit, EPA agreed to review the 1996 landfill NSPS and take final action on whether to revise them. Consent Decree, Envt'l Def. Fund v. EPA. No. 1:11-cv-04492 (S.D.N.Y. Oct. 1, 2012).

[144] Standards of Performance for Municipal Solid Waste Landfills; Proposed Rule, 79 Fed. Reg. 41,795 (July 17, 2014); Emission Guidelines and Compliance Times for Municipal Solid Waste Landfills; Advanced Notice of Proposed Rulemaking, 79 Fed. Reg. 41,771 (July 17, 2014). Subsequently, EPA proposed revisions to the emission guidelines. Emission Guidelines and Compliance Times for Municipal Solid Waste Landfills; Proposed Rule; 80 Fed. Reg. 52,100 (Aug. 27, 2015).

reconstructed after July 17, 2014.[145] The agency also revised the 1996 emission guidelines for existing landfills operating prior to that date.[146]

The updated NSPSs aimed to further decrease methane emissions from MSW landfills. The revised NSPSs maintained the current design capacity applicability threshold for new or modified landfills at 2.5 million metric tons (MMT) of design capacity (or 2.5 million cubic meters of waste).[147] The updated NSPSs lowered the emission threshold that triggers requirements to capture landfill gas emissions (including methane) from the current 50 metric tons of NMOCs per year to 34 metric tons.[148] Under the rule, EPA would have required that a gas collection control system be installed and operational within 30 months after landfill gas emissions reach 34 metric tons of NMOCs or more per year.[149] The rule also updated monitoring requirements; expanded approved uses for treated landfill gas; and included other clarifications and updated definitions.[150] According to an agency fact sheet, 115 new, modified, or reconstructed landfills would be subject to the emission control requirements of the revised standards by 2025.[151]

Similar to the revised NSPSs, the revised emission guidelines for *existing* landfills built prior to July 17, 2014, would have required active landfills that emit more than 34 metric tons of NMOCs annually to install landfill gas collection and control systems.[152] Closed landfills would remain subject to the 1996 threshold of 50 metric tons per year when determining when controls must be installed or can be removed.[153]

As a result of the lower threshold for active landfills, EPA estimated that the emission control requirements will apply to 731 existing open and closed landfills, as compared to 638 facilities currently subject to the 1996 emission control requirements.[154] Both rules became effective on October 28, 2016 and remain in effect unless EPA finalizes a stay or revises the rules.[155]

---

145 Standards of Performance for Municipal Solid Waste Landfills, Final Rule, 81 Fed. Reg. 59,332 (Aug. 29, 2016) (codified at 40 C.F.R. pt. 60, subpt. XXX) [hereinafter *2016 Landfill NSPSs*]. The final rule became effective on October 28, 2016. *Id.*

146 Emission Guidelines and Compliance Times for Municipal Solid Waste Landfills; Final Rule, 81 Fed. Reg. 59,276 (Aug. 29, 2016) (to be codified at 40 C.F.R. pt. 60, subpt. WWW) [hereinafter *2016 Landfill Emission Guidelines*]. The final rule became effective on October 28, 2016. *Id.*

147 *2016 Landfill NSPSs*, *supra* note 139, at 59,333-334.

148 *Id.* at 59,334.

149 *Id.*

150 *Id.* at 59,342-345.

151 EPA, FINAL UPDATES TO PERFORMANCE STANDARDS FOR NEW, MODIFIED, AND RECONSTRUCTED LANDFILLS, AND UPDATES TO EMISSION GUIDELINES FOR EXISTING LANDFILLS: FACT SHEET 3, https://www.epa.gov/sites/production/ files/2016-09/documents/landfills-final-nsps-eg-factsheet.pdf [hereinafter "EPA Fact Sheet"].

152 *2016 Landfill Emission Guidelines*, *supra* note 140, at 59,278.

153 *Id.* at 59,279.

154 EPA Fact Sheet, at 2.

155 *Id.*; *2016 Landfill NSPSs*, *supra* note 139, at 59,322.

Reconsideration of 2016 MSW Landfill NSPSs and Emission Guidelines

In May 2017, EPA granted administrative petitions for reconsideration of several requirements in the revised and updated MSW landfill NSPSs and emission guidelines based on the criteria under CAA Section 307(d)(7)(B).[156] Under the Trump Administration, EPA found that the petitioners "demonstrated that it was impractical to raise the objection during the period for public comment" on a number of issues,[157] including Tier 4 surface emissions monitoring.[158] The 2016 MSW landfill NSPSs and emission guidelines added Tier 4 surface emissions monitoring procedures that would allow certain landfills that would have exceeded the applicability threshold for emission control requirements using other approved modeling procedures (i.e., Tiers 1 and 2) to use, as an alternative, surface monitoring to determine applicability.[159] A landfill that demonstrates that surface emissions are below 500 parts per million for four consecutive quarters using Tier 4 monitoring procedures would not be required to install emissions control technology even if Tiers 1 or 2 modeling calculations indicated that the landfill exceeded the applicability threshold.[160]

EPA highlighted the petitioners' objection to the Tier 4 surface emission monitoring as meeting the Section 307(d)(7)(B) criteria for reconsideration. EPA explained that the final rule imposed restrictions on the use of the optional Tier 4 surface emissions monitoring method that were not included in the proposed rule and were "added without the benefit of public comment."[161] EPA also determined that the objection to the Tier 4 monitoring limitations were of central relevance to the rule because they reduced "flexibility" to use the optional monitoring method.[162] Based on these determinations, EPA has convened an administrative reconsideration proceeding to address the issues and objections raised by the petitioners.[163] As part of the reconsideration proceeding, EPA plans to prepare a notice of proposed rulemaking that will provide the petitioners and the public an opportunity to comment on the rule requirements and associated issues identified above.[164]

---

[156] Stay of Standards of Performance for Municipal Solid Waste Landfills and Emission Guidelines and Compliance Times for Municipal Solid Waste Landfills, 82 Fed. Reg. 24,878 (May 31, 2017). [hereinafter *Stay of 2016 MSW Landfill Rules*].

[157] Other issues under reconsideration include annual liquids reporting; corrective action timeline procedures; overlapping applicability with other rules; the definition of cover penetration; and design plan approval. 82 Fed. Reg. at 24,878-79. In granting reconsideration of these issues, EPA did not provide details regarding the objections that the petitioners raised regarding these issues except for the Tier 4 surface emission monitoring. *Id.*

[158] *2016 Landfill Emission Guidelines*, *supra* note 140, at 59,287.

[159] Id.; 2016 Landfill NSPSs, supra note 139, at 59,322.

[160] Id.

[161] Stay of 2016 MSW Landfill Rules, at 24,879.

[162] Id.

[163] Id.

[164] Id.

Stay of the MSW Landfill NSPSs and Emission Guidelines

In addition to granting reconsideration of the issues discussed above, EPA stayed for three months the effectiveness of both MSW landfills NSPSs and emission guidelines, in their entirety, starting from May 31 until August 29, 2017, pursuant to its authority under Section 307(d)(7)(B).[165] EPA reasoned that staying all requirements of the rules was necessary because the Tier 4 monitoring provisions in the two rules are "integral to how the rules function as a whole."[166] Because this three-month stay expired on August 29, 2017, the 2016 rules are currently in effect during the reconsideration process. At this time, EPA has not formally proposed a longer stay of the rules or initiated the public comment period for issues under reconsideration.

Similar to their challenge to the 2016 Oil and Gas NSPSs stay, environmental groups filed a petition to review and a motion to vacate the three-month stay of the MSW landfills NSPSs and emission guidelines in the D.C. Circuit.[167] Although the court denied the petitioners' motion to vacate the stay,[168] the court is moving forward with briefing of the case and has asked the parties to address standing and mootness now that the three-month stay has expired.[169] The final briefs in the case are due on February 26, 2018.[170]

The petitioners' litigation issues mirror those raised in the challenge to the 2016 Oil and Gas NSPSs stay, arguing that EPA exceeded its authority under Section 307(d)(7)(B) because the agency failed to meet preconditions for mandatory reconsideration.[171] The petitioners also claim that the stay is arbitrary and capricious because the stay of both regulations in their entirety is "overbroad" relative to the "narrow" issues under reconsiderations.[172]

Petitions for Review of the 2016 MSW Landfill Emission Guidelines

Although EPA is reconsidering the 2016 MSW landfills emission guidelines, the guidelines themselves are being challenged in court.[173] Prior to the start of the Trump Administration, industry trade associations and waste management and recycling companies challenged EPA's 2016 revised emission guidelines for existing MSW

[165] Id.
[166] Id.
[167] Petition for Review, Nat. Res. Def. Council v. Pruitt, No. 17-1157 (D.C. Cir. June 15, 2017); Motion for Summary Vacatur, No. 17-1157 (D.C. Cir. Aug. 4, 2017).
[168] Per Curiam Order, Nat. Res. Def. Council v. Pruitt, No. 17-1157 (D.C. Cir. Sept. 28, 2017).
[169] Clerk's Order, Nat. Res. Def. Council v. Pruitt, No. 17-1157 (D.C. Cir. Oct. 4, 2017).
[170] Clerk's Order, Nat. Res. Def. Council v. Pruitt, No. 17-1157 (D.C. Cir. Jan. 12, 2018).
[171] Petitioners' Non-Binding Statement of Issues, Nat. Res. Def. Council v. Pruitt, No. 17-1157 (D.C. Cir. July 20, 2017).
[172] Id.
[173] The 2016 MSW landfill NSPSs were not challenged in court.

landfills in the D.C. Circuit.[174] On June 14, 2017, the court granted EPA's request to hold the case in abeyance while the agency reviews and reconsiders the rule.[175]

Litigation Issues

Although the litigation is on hold, the petitioners have submitted preliminary statements of issues that challenge EPA's authority to revise emission guidelines. Among the various legal arguments raised, petitioners make a unique and potentially far-reaching challenge to EPA's authority to revise emission guidelines issued under Section 111(d) of the CAA.

EPA's Authority to Revise Section 111(d) Emission Guidelines

Petitioners challenging the landfill emission guidelines question whether EPA has the authority to revise the 1996 emission guidelines for existing MSW landfills under CAA Section 111(d).[176] The core of the legal argument is whether Section 111(d) implicitly allows EPA to revise and update emission standards. Pursuant to CAA Section 111(b), EPA is required to "at least every 8 years review and, if appropriate, revise" NSPSs for new, modified, and reconstructed sources.[177] However, Section 111(d) does not include a similar mandated review period for emission guidelines for *existing* sources.[178] EPA's implementing regulations for Section 111(d) emission guidelines only address revisions to the agency's determination regarding adverse public health effects that may alter compliance times under the guidelines and require states to make corresponding revisions to their state plans.[179]

Under the Obama Administration, EPA asserted in the 2016 landfill emission guidelines' preamble that the agency "is not statutorily obligated to conduct a review of the Emission Guidelines, but has the discretionary authority to do so when circumstances indicate that it is appropriate."[180] EPA reasoned that this is a permissible interpretation of the CAA, and "is consistent with the gap filling nature of section 111(d)" to regulate air pollutants and sources not addressed in other sections of the CAA.[181] EPA argued that it would be "illogical" for the agency to be precluded from requiring existing sources to

[174] Nat'l Waste & Recycling Ass'n v. EPA, No. 16-1371 (D.C. Cir. Oct. 27, 2016); Utility Air Regulatory Group v. EPA, No. 16-1374 (D.C. Cir. Oct. 28, 2016).

[175] Clerk's Order, Nat'l Waste & Recycling Ass'n v. EPA, No. 16-1371 (D.C. Cir. June 14, 2017).

[176] Nonbinding Statement of Issues of Petition Utility Air Reg. Group, No. 16-1374 (consolidated with No. 16-1371) (D.C. Cir. Dec. 1, 2016). *See e.g.*, Comments from Allison D. Wood, Counsel, Utility Air Regulatory Group (UARG), to Gina McCarthy, Admin., EPA (Oct. 26, 2015), at 5-8 [hereinafter *UARG Comments*].

[177] 42 U.S.C. § 7411(b)(1)(B).

[178] *See id.* § 7411(d).

[179] *See* 40 C.F.R. § 60.22(d) (requiring EPA to "publish notice of the determination in the Federal Register, revise the guideline document as necessary ... , and propose and promulgate emission guidelines and compliance times" if the agency concludes that on the basis of new information that a prior determination regarding adverse health effects of a designated pollutant is incorrect or no longer correct).

[180] 2016 Landfill Emission Guidelines, supra note 140, at 59,277.

[181] Id.

update air pollution controls when the BSER will change over time as these sources continue to operate for decades.[182]

In support of this position, EPA highlighted jurisprudence that has recognized the power of regulatory agencies to reassess and revise promulgated rules even without an express statutory mandate to make such a revision.[183] The Supreme Court, for example, has stated that "[r]egulatory agencies do not establish rules of conduct to last forever; they are supposed, within the limits of the law and of fair and prudent administration, to adapt their rules and practices to the Nation's needs in a volatile, changing economy."[184] EPA, under the Trump Administration, is making a similar argument to support delaying and extending the compliance timelines for the 2016 Oil and Gas NSPSs.[185] The agency contends that it has the "inherent authority to reconsider past decisions and to revise, replace, or repeal a decision to the extent permitted by law and supported by a reasoned explanation."[186]

EPA's authority to revise Section 111(d) emission guidelines has not been questioned previously because this was the agency's first attempt to do so. Some argue in comments on the proposed emission guidelines that Congress's explicit mandate authorizing EPA to review and potentially revise Section 111(b) NSPSs for new, modified, and reconstructed every eight years demonstrates that Congress knew how to provide for such review and revision and intentionally did not provide a similar provision with respect to Section 111(d) emission guidelines.[187] One commentator has pointed out that the Supreme Court has held that "where Congress includes particular language in one section of a statute but omits it in another section of the same Act, it is generally presumed that Congress acts intentionally and purposely in the disparate inclusion or exclusion."[188] In contrast, EPA argued in the revised emission guidelines that "[h]ad Congress intended to preclude the EPA from updating the emission guidelines to reflect changes, it would surely have specifically said so, something it did not do."[189]

The legislative history of CAA Section 111 does not discuss the eight-year review for NSPSs under Section 111(b) or review of emission guidelines under Section 111(d). The history does provide some discussion regarding the need for EPA to look to currently available and evolving techniques in determining what constitutes the BSER that has been adequately demonstrated for standards of performance and emission guidelines. For example, when amending the CAA in 1970, the Senate Public Works Committee stated that "[t]he performance standards should be met through application of the latest available emission control technology or through other means of preventing or

[182] Id. at 59,278.

[183] See id. at 59,277 n.5 (citing American Trucking Ass'n v. Atchison, Topeka & Santa Fe Ry., 387 U.S. 397, 416 (1967)).

[184] American Trucking Ass'n, 387 U.S. at 416.

[185] See supra "Proposed Two-Year Stay."

[186] Oil and Natural Gas Sector: Emission Standards for New, Reconstructed, and Modified Sources: Stay of Certain Requirements, Proposed rule; notice of data availability, 82 Fed. Reg. 51,788, 51,790 (Nov. 8, 2017).

[187] UARG Comments, supra note 170, at 2-4.

[188] See id. at 3 (quoting Russello v. United States, 464 U.S. 16, 23 (1983)).

[189] 2016 Landfill Emission Guidelines, supra note 140, at 59,278.

controlling air pollution."[190] The committee also stated that "standards of performance are not static," and that they "should provide an incentive for industries to work toward constant improvement in techniques for preventing and controlling emissions from stationary sources."[191] In a report regarding the 1977 CAA amendments, the Senate Environment and Public Works Committee reiterated that "[i]n passing the Clean Air Amendments of 1970, the Congress for the first time imposed a requirement for specified levels of control technology.... This requirement sought to assure the use of available technology and to stimulate the development of new technology."[192] This history appears to suggest that EPA can broadly review and revise the "standards of performance," a term of art that applies to both Section 111(b) NSPSs and Section 111(d) emission guidelines,[193] to ensure that the standards reflect the improvement of systems of emission reduction.

In addition, a reference to Section 110 procedures in Section 111(d) raises questions as to whether Congress intended for EPA to revise emission guidelines and state plans in a similar manner as Section 110. Specifically, Section 111(d) requires EPA to "prescribe regulations which shall establish a procedure similar to that provided by section 110 under which each State shall submit to the [EPA] Administrator a plan which ... establishes standards of performance for any existing source for any air pollutant."[194] In turn, CAA Section 110 requires states to adopt and submit to EPA for approval state implementation plans (SIPs) that provide for the "implementation, maintenance, and enforcement" of NAAQS promulgated by EPA under CAA Section 109.[195] EPA is required to review the NAAQS every five years.[196] If EPA decides to revise the NAAQS, states are required to revise their SIPs to implement and enforce the revised NAAQS.[197]

It could be argued that this regulatory scheme is similar to Section 111(d) because emission guidelines are implemented through state plans. Once EPA sets guidelines for standards of performance for existing sources under Section 111(d), states are required to submit to EPA for approval a plan to implement and enforce the standards of performance for existing sources.[198] Therefore, it could be argued that in referencing Section 110 in Section 111(d), Congress may have intended to allow EPA to revise Section 111(d) guidelines and state plans in the same manner that NAAQS and SIPs are allowed to be revised in Section 109 and Section 110, respectively.

However, Section 110 and 111(d) are not completely analogous. In both cases, state plan revisions are necessary when the standards are revised because the state is the primary implementation and enforcement authority; but unlike Section 111(d), Section 110

[190] S. REP. 91-1196, at 16 (1970).
[191] *Id.* at 17.
[192] S. REP. 95-127, at 17 (1977).
[193] *See* 42 U.S.C. § 7411(a) (providing term definitions "[f]or purposes of this section [111]," including "standard of performance").
[194] 42 U.S.C. § 7411(d)(1).
[195] *Id.* § 7410(a)(1).
[196] *Id.* § 7409(d)(1).
[197] *Id.* § 7410(a)(1).
[198] *Id.* § 7411(d)(1).

revisions are triggered through a statutorily mandated five-year review under Section 109.[199] In contrast, Congress did not mandate a review and revision cycle for Section 111(d) emission guidelines for existing sources.

If the case moves forward, a ruling upholding EPA's implicit authority to review and revise Section 111(d) guidelines could support future EPA review and revision of other Section 111(d) emission guidelines to require additional emission reductions for existing sources based on a new analysis of the BSER.[200]

Next Steps

The litigation challenging the substantive requirements of the MSW landfill emission guidelines is currently on hold pending EPA's review and reconsideration[201] See *Table 2* below for the current status of the landfill methane rules.

Table 2. EPA'S MSW Landfill Methane Rules

Regulatory Action	Status of the Regulatory Action (as of January 2018)	Status of Litigation on Regulatory Action (as of January 2018)
2016 MSW Landfill NSPSs 40 C.F.R. pt. 60, subpt. XXX	In effect for MSW landfills constructed, modified, or reconstructed after July 17, 2014, but under reconsideration Stayed from May 31, 2017, until Aug. 29, 2017	N/A
2016 MSW Landfill Emission Guidelines 40 C.F.R. pt. 60, subpt. Cf	In effect for MSW landfills constructed, modified, or reconstructed before July 17, 2014, but under reconsideration Stayed from May 31, 2017, until August 29, 2017	Held in abeyance *Nat'l Waste & Recycling Ass'n v. EPA,* No. 16-1371 (D.C. Cir. June 14, 2017)
Three-Month Stay: 2016 MSW Landfill Rules 82 Fed. Reg. 24,878 (May 31, 2017)	Expired on August 29, 2017	Litigation moving forward Briefs due in February 2018

Source: Prepared by CRS from regulatory and litigation filings.
Note: This table does not include all CAA regulations that affect MSW landfills.

199 *Id.* § 7409(d)(1).
200 For list of CAA Section 111(d) emission guidelines, see "Regulations Under CAA Section 111" section.
201 Clerk's Order, North Dakota v. EPA, Nat'l Waste & Recycling Ass'n v. EPA, No. 16-1371 (D.C. Cir. June 14, 2017).

In: Environmental Protection Agency (EPA)
Editor: Coby Schaefer
ISBN: 978-1-53614-933-3

Chapter 8

EPA'S AND BLM'S METHANE RULES*

Richard K. Lattanzio

In 2016, the U.S. Environmental Protection Agency (EPA) and the Bureau of Land Management (BLM) promulgated separate regulations intended to control "methane emissions" at crude oil and natural gas production facilities. Some stakeholders have argued that the EPA and BLM rules are duplicative and outside of the agencies' statutory authorities. On January 30, 2017, a joint resolution of disapproval under the Congressional Review Act was introduced to overturn the BLM rule. H.J. Res. 36 would vacate the rule and preempt the agency from promulgating similar proposals in the future. The measure passed the House by a 221 to 191 vote and now moves to the Senate.

Below is a brief discussion of the two rules, their similarities, and their differences.

- EPA's "Oil and Natural Gas Sector: Emission Standards for New, Reconstructed, and Modified Sources, Final Rule" (81 *Federal Register* 35824, June 3, 2016) was promulgated under the authority of the Clean Air Act, as amended (CAA; 42 U.S.C. §7401 et seq.). CAA Section 111 requires EPA to establish a list of source categories to be regulated and emission standards for those source categories. Specifically, CAA Section 111(b)(1)(A) requires that a source category be included on the list if, "in [the EPA Administrator's] judgment it causes, or contributes significantly to, air pollution which may reasonably be anticipated to endanger public health or welfare." Once a source category is listed, CAA Section 111(b)(1)(B) requires that the EPA propose and then promulgate "standards of performance" for new sources in the source category. CAA Section

* This is an edited, reformatted and augmented version of Congressional Research Service, Publication No. IN10645, dated February 3, 2017.

111(a)(1) defines "a standard of performance" as "a standard for emissions of air pollutants which reflects the degree of emission limitation achievable through the application of the best system of emission reduction which (taking into account the cost of achieving such reduction and any non-air quality health and environmental impact and energy requirement) the Administrator determines has been adequately demonstrated."

- BLM's "Waste Prevention, Production Subject to Royalties, and Resource Conservation, Final Rule" (81 *Federal Register* 83008, November 18, 2016) was promulgated under the authority of the Mineral Leasing Act of 1920, as amended (MLA; 30 U.S.C. §181 et seq.). Section 225 of the MLA requires BLM to ensure that lessees "use all reasonable precautions to prevent waste of oil or gas developed in the land," and, under Section 187, that leases include "a provision that such rules ... for the prevention of undue waste as may be prescribed by [the] Secretary shall be observed."

Since natural gas that is vented, flared, or leaked from crude oil and natural gas production, processing, and transmission activities can be defined as both "air pollution which may reasonably be anticipated to endanger public health or welfare" and "undue waste," EPA and BLM have promulgated rules to control for it under their respective statutory authorities.

EPA's rule targets natural gas emissions for the control of methane pollution (the primary component of natural gas and a potent greenhouse gas). EPA's rule sets standards of performance for methane emissions from several types of new, reconstructed, or modified facilities in the oil and gas production sector. EPA estimates that the rule would reduce methane emissions by 510,000 tons in 2025 and yield climate benefits of $690 million in 2025 (in 2012 dollars), outweighing the estimated costs of $530 million.

BLM's rule targets natural gas emissions as a potential waste of public resources and loss of royalty revenue. BLM's rule requires operators of crude oil and natural gas facilities on federal and Indian lands to take various actions to reduce the waste of gas, establishes clear criteria for when flared gas will qualify as waste and therefore be subject to royalties, and clarifies which on-site uses of gas are exempt from royalties. BLM estimates that the rule would avoid an estimated 175,000-180,000 tons of methane emissions per year and yield total benefits from $209 million to $403 million per year, outweighing the estimated costs of $110-$275 million per year. BLM estimates annual royalties to the federal government, tribal governments, states, and private landowners to increase by $3-$10 million per year.

Some stakeholders have argued that EPA's and BLM's rules are duplicative. Several similarities exist between the two, including the following:

- Many emission sources are covered under both rules (e.g., well completions, compressors, pneumatic controllers and pumps, storage tanks, and equipment leaks).
- BLM's rule sets controls for these sources to align with the EPA's final requirements.
- BLM's rule states that "this final rule seeks to minimize regulatory overlap. Thus, if EPA and/or States or tribes have adopted requirements that are at least as effective as and would potentially overlap with the provisions of this rule, the final rule provides a means for operators to comply with the EPA, State, local or tribal requirements in lieu of the BLM requirements."

Others point to differences between the two rules, including the following:

- EPA's rule covers facilities constructed, modified, or reconstructed after September 18, 2015; BLM's rule covers all facilities, both new and existing.
- EPA's rule covers selected facilities in the "crude oil and natural gas production" and "natural gas transmission and storage" sectors on all lands, both public and private; BLM's rule covers all oil and gas facilities on federal or tribal lands.
- EPA's rule provides for "combustion control devices" (i.e., flaring) as the best system of emission reduction in specific cases; BLM's rule requires operators to reduce flaring by capturing it for sale or use during production. The required capture percentage increases over time, and the base level of "allowable" flaring ramps down over time.
- EPA's rule does not have controls for well maintenance and liquids unloading activities; BLM's rule does. EPA's rule does not speak to royalty payments; BLM's rule revises 43 C.F.R. §3103.3, which governs royalty rates applicable to onshore oil and gas leases and updates the preexisting royalty provisions in BLM's 1980 Notice to Lessees, NTL-4A, to more clearly and specifically define when a loss of gas is considered "unavoidable" and royalty-free, and when it is considered "avoidable" and subject to royalties.

In: Environmental Protection Agency (EPA)
Editor: Coby Schaefer
ISBN: 978-1-53614-933-3

Chapter 9

EPA's Wood Stove/Wood Heater Regulations: Frequently Asked Questions*

James E. McCarthy and Kate C. Shouse

Summary

On March 7, 2018, the House passed H.R. 1917, a bill that would delay for three years the implementation of more stringent emission standards for new residential wood heaters. The emission standards were promulgated by the U.S. Environmental Protection Agency (EPA) in 2015, and are becoming effective through a two-step process. Step 1 standards took effect on May 15, 2015; unless delayed, more stringent Step 2 standards will become effective on May 15, 2020. EPA's action revises standards for wood stoves and pellet stoves that were set in 1988, and establishes standards for other types of wood heaters, principally forced air furnaces and hydronic heaters, for the first time.

According to EPA, smoke from wood heaters and fireplaces contributes "hundreds of thousands of tons" of fine particles to the air each year, nationally accounting for nearly 25% of all area source air toxics cancer risks and 15% of non-cancer respiratory effects. In many areas, in wintertime, wood heaters are the largest source of particulate air pollution. Until the implementation of this rule, however, many heater types were not subject to any federal emission standard.

The rule will only gradually reduce this pollution, because it only applies to new heaters (not those already in use) and it gives the industry a five-year grace period (until 2020) before its most stringent (Step 2) standards take effect. Nevertheless, EPA estimated that implementing the rule will eliminate 360 to 810 premature deaths annually in the 2015-2020 period as well as reduce hospital admissions and lost work days due to respiratory illness. EPA quantified these benefits at $3.4 billion to $7.6 billion per year

* This is an edited, reformatted and augmented version of Congressional Research Service Publication No. R43489, dated March 12, 2018.

during the 2015-2020 period, more than 70 times the agency's estimated annualized compliance cost to manufacturers, $46 million.

Trade associations representing the affected industries and affected companies have mixed views of the 2015 standards. While generally supporting revision of the 1988 standards and the inclusion of additional heater types, some have expressed concern that the standards will impose too great a cost. Homeowners may continue to use current, highly polluting equipment, rather than buy more expensive replacements. In its analysis of the final rule, EPA estimated the vast majority of cost (88%) would be borne by the hydronic heater and forced air furnace segments of the industry. Specifically, EPA estimated compliance costs for hydronic heaters and forced-air furnaces will be about 17% of total sales revenues for the two product types. EPA did not estimate the effect of these costs on prices or output, but the agency's analysis does conclude that high unit compliance costs could lead to "potential nontrivial increases in market price to wood-burning appliance consumers and potential decreases in output."

In addition to cost issues, many commenters on the *proposed* rule expressed concerns regarding the process to be used in certifying compliance and the short period of time in which units available at the time of the rule's promulgation could be tested and certified. EPA addressed many of these comments in the final rule, making changes intended to ease the burden of certifying compliance. Industry representatives continue to express concern that there will not be enough time to develop new models, certify compliance, and meet lead-time requirements of product retailers for the Step 2 (2020) compliance deadline.

This chapter presents some of the most frequent questions raised concerning the rule, so as to provide basic background information about the final wood heater rule, its potential impacts, and stakeholder reactions to it.

INTRODUCTION

On March 16, 2015, the Environmental Protection Agency (EPA) published final emission standards for new residential wood heaters, including wood stoves, pellet stoves, hydronic heaters, and forced air furnaces.[1] Proposed in 2014,[2] the rule resulted from a regular review process that Section 111 of the Clean Air Act requires for all New Source Performance Standards (NSPS). Wood stoves and pellet stoves have been subject to NSPS since 1988. The act requires a review of NSPS and, if appropriate, revision of the standards, at least every eight years. The 2015 rule marks the first revision, however, since EPA promulgated the wood stove NSPS in 1988.

Release of the proposed and final versions of the rule generated a substantial amount of interest, particularly in areas where wood is used as a heating fuel, and some Members of Congress from those areas have written EPA to express concerns regarding the

---

[1] U.S. Environmental Protection Agency (EPA), "Standards of Performance for New Residential Wood Heaters, New Residential Hydronic Heaters and Forced-Air Furnaces," 80 *Federal Register* 13672, March 16, 2015. Hereinafter, "Final Rule."

[2] EPA took public comment on the proposal until May 5, 2014. For the proposal, see U.S. EPA, "Standards of Performance for New Residential Wood Heaters, New Residential Hydronic Heaters and Forced-Air Furnaces, and New Residential Masonry Heaters; Proposed Rule," 79 *Federal Register* 6329, February 3, 2014. Hereinafter, "Proposed Rule."

proposed rule's possible impacts.[3] On March 7, 2018, the House passed H.R. 1917, which would delay implementation of the standards for three years.[4]

This chapter addresses questions posed by various requesters in order to provide basic information about EPA's action, its potential impacts, and industry and other reactions to the rule. The following 10 questions are among those frequently asked about the wood heater rule.

TO WHAT SOURCES OF EMISSIONS DOES THE FINAL RULE APPLY?

The final rule amended the NSPS for Residential Wood Heaters, which applied to wood stoves and pellet stoves, and, for the first time, set standards for new residential hydronic heaters, forced air furnaces, and single burn rate stoves.[5]

In addition to these five types of wood heaters, EPA had proposed NSPS for new residential masonry heaters that burn wood, but did not take final action on them in the 2015 final rule. The agency stated that it wanted "to allow additional time for the Masonry Heater Association (MHA) to finish their efforts to develop revised test methods and alternative compliance calculation procedures."[6] As of publication of this chapter, EPA has not taken action on masonry heaters.

The final rule does not apply to outdoor fireplaces, pizza ovens, barbecues or chimineas (freestanding outdoor chimneys), and it does not apply to new or existing heaters that are fueled solely by oil, gas, or coal.

---

[3] A letter to EPA Administrator McCarthy from Senators Collins and King of Maine is available at https://www.king.senate.gov/newsroom/press-releases/senators-collins-king-express-concern-with-epas-proposedwood-stove-rules. A letter to the EPA Administrator from Senator Thune of South Dakota is available at http://www.thune.senate.gov/public/index.cfm/press-releases?ID=008e36fa-35c2-42f0-a50c-6165fb33d982.

[4] The provisions affecting wood heaters were reported by the House Energy and Commerce Committee as H.R. 453 (H.Rept. 115-508). The Rules Committee combined that bill with H.R. 1917 when it established a rule for floor consideration of the latter, on March 6, 2018.

[5] Hydronic heaters, which are typically located outside the buildings they heat in small sheds with short smokestacks, are appliances that burn wood to heat a liquid (water or a water-antifreeze mixture) that is piped to provide heat and hot water to occupied buildings, such as homes. Hydronic heaters may also be located indoors. The forced air furnaces that would be subject to these regulations are similar to natural gas-fired or oil-fired forced air furnaces used in millions of American homes, except that they are designed to burn wood (Proposed Rule, p. 6360). According to EPA, "Industry information suggests that there are three times more sales of wood-fired, forced-air furnaces each year compared to wood-fired hydronic heaters. These units are relatively easy to retrofit into existing structures, and their sales price is substantially less than hydronic heaters but greater than gas or oil furnaces" (Proposed Rule, p. 6360). A single burn rate wood heater is "a wood heater that is not equipped with or installed with a burn control device to allow the operator to vary burn rate conditions. Burn rate control devices include stack dampers that control the outflow of flue gases from the heater to the chimney ... and air control slides, gates or any other type of mechanisms that control combustion air flow into the heater" (Final Rule, p. 13704).

[6] U.S. EPA, *Regulatory Impact Analysis (RIA) for Residential Wood Heaters NSPS Revision*, Final Report, February 2015, p. 1-1, at https://www.epa.gov/sites/production/files/2015-02/documents/20150204-residential-wood-heatersria.pdf. Hereinafter, "Final Rule RIA." Masonry heater operation, according to EPA, involves one or two short high burn rate cycles where hot gases are generated during combustion of a fuel load in the firebox. The hot gases then pass through channels, saturating the masonry mass with heat. The masonry mass then radiates heat into the area around the masonry heater for 12 to 24 hours.

In addition, EPA did not include standards for indoor fireplaces in the regulation, because the agency determined that, "Fireplaces are not effective heaters."[7] The agency sought additional data and comments on this issue in the preamble to the proposed rule, but the final rule continued to exempt fireplaces from emission standards. Fireplaces are included in an agency voluntary program that encourages manufacturers to make cleaner-burning fireplaces and retrofits available for consumers,[8] and some industry representatives had asked EPA to include fireplaces in the final rule.[9]

WHEN DOES THE FINAL RULE GO INTO EFFECT?

The final rule's standards take effect in two steps, which are phased in over a five-year period. The "Step 1" standards took effect on May 15, 2015, for all covered heaters except forced air furnaces. Forced air furnaces had until May 2016 or May 2017 (depending on whether they are small or large) to comply with Step 1. Unless the implementation schedule is modified, the more stringent "Step 2" standards will become effective for all covered heaters on May 15, 2020.

For most new wood stoves or pellet stoves, there will be relatively little change in emission standards until the Step 2 deadline in 2020. Other types of wood heaters (notably, hydronic heaters and wood-fired forced air furnaces) were not subject to federal emission standards prior to the 2015 rule. As with wood stoves, far more stringent standards will affect new hydronic heaters and forced air furnaces beginning with models sold in 2020.

DOES THE FINAL RULE AFFECT WOOD STOVES AND OTHER WOOD HEATERS ALREADY PURCHASED OR IN USE PRIOR TO ITS EFFECTIVE DATE?

No. The standards only apply to new wood stoves and heaters sold after the final rule's effective date. See previous question for effective dates for each compliance phase of the rule.

The absence of standards for most wood heaters currently in use limits the effectiveness of EPA and state efforts to reduce wood smoke emissions. According to the

[7] U.S. EPA, "Overview of Proposed Updates to Air Emissions Requirements for New Residential Wood Heaters," p. 2, at http://www2.epa.gov/sites/production/files/2013-12/documents/proposed_wood_heater_nsps_overview_fact_sheet_1.pdf.

[8] For information, see "Burn Wise," U.S. EPA's Voluntary Program for Wood-Burning Fireplaces, July 11, 2013, at http://www.epa.gov/burnwise/pdfs/fireplaceqa.pdf.

[9] See, for example, statement of Mark Klein, President and Chief Executive Officer of Innovative Hearth Products, at U.S. EPA Public Hearing on the Proposed New Source Performance Standards for Wood Heaters, Boston, MA, February 26, 2014, transcript, p. 39, at http://www.regulations.gov/#!documentDetail;D=EPA-HQ-OAR-2009-0734- 1037.

U.S. Energy Information Administration, 11.5 million homes use wood heaters as their primary or secondary heating source.[10] An estimated 6 million of the heaters in operation predate EPA's 1988 standards.[11] EPA projected annual sales of wood heaters were to be about 270,000 in 2015,[12] less than 2.4% of the number of units in use. New units do not necessarily replace existing units, thus it may be decades before older, higher polluting equipment is retired. Industry, EPA, and a number of states have initiated wood stove change-out programs, with subsidies for purchases that replace equipment that predates EPA standards, and at least one manufacturer proposed banning the sale of replacement parts for non-EPA-compliant models.[13] Wood smoke impacts, particularly those from heaters in use prior to 2015, are likely to remain an issue of concern for some time.

WHAT DOES THE FINAL RULE REQUIRE?

The final rule's standards take effect in two steps.[14] Step 1 took effect on the effective date of the rule, May 15, 2015. Starting on that date, new wood heaters were required to meet standards that EPA described at proposal as "emission levels that almost all models can readily achieve now using today's designs and technology."[15] Many in industry disputed this characterization, but in its RIA for the final rule, EPA stated that the Step 1 limit for wood stoves and pellet stoves "is the 1995 Washington State standard for non-catalytic stoves."[16] The agency reported that 130 wood and pellet stoves (out of 145 models then on the market) already met the standard, using industry supplied data.[17] Of hydronic heaters, the agency has stated: "The Step 1 limit is identical to the EPA 'Phase 2' voluntary program limit and is therefore already met by all 50 of the 50 currently Phase 2 qualified hydronic heater models built by U.S. manufacturers participating in the voluntary program."[18] The emissions standard for forced air furnaces was based on testing done in March 2010 for the Canadian standard for these units.[19]

---

[10] U.S. Energy Information Administration, *Today in Energy*, March 17, 2014, at https://www.eia.gov/todayinenergy/ detail.php?id=15431.

[11] See statement of Jack Goldman, President and CEO of the Hearth, Patio, and Barbecue Association, at U.S. EPA Public Hearing on the Proposed New Source Performance Standards for Wood Heaters, Boston, MA, February 26, 2014, transcript, p. 17, at http://www.regulations.gov/#!documentDetail;D=EPA-HQ-OAR-2009-0734-1037. The six (or occasionally seven) million figure also appears in statements from other stakeholders made at the public hearing at pp. 94, 96, 200, 221, 265, 272, and 325 of the transcript.

[12] Final Rule RIA, Table 4-4.

[13] Statement of Brett Watson, President of Jotul North America, at U.S. EPA Public Hearing on the Proposed New Source Performance Standards for Wood Heaters, Boston, MA, February 26, 2014, transcript, p. 263, at http://www.regulations.gov/#!documentDetail;D=EPA-HQ-OAR-2009-0734-1037.

[14] In addition to the two-step five-year phase-in approach taken in the final rule, in its proposal the agency asked for comment on an alternative approach that would have had three steps. Under this approach, the final standards would have taken effect in 2023 rather than 2020, with an intermediate set of standards in 2018. EPA did not choose this approach for the final standards.

[15] Proposed Rule, p. 6353.

[16] Final Rule RIA, p. 4-5.

[17] Final Rule RIA, p. 4-5.

[18] Final Rule RIA, p. 4-5.

[19] Final Rule RIA, p. 4-6.

Step 2, which requires a 56% reduction in particulate matter (PM) emissions from the Step 1 standards for wood stoves and pellet stoves, a roughly 70% reduction for hydronic heaters, and an 84% reduction in emissions from forced air furnaces, takes effect five years after the rule's effective date, May 15, 2020.[20] In proposing the standards, EPA stated that the "... second step represents stronger emission levels achievable for all appliance types at reasonable cost, but allows appropriate lead times for manufacturers to redesign their model lines to accommodate the improved technology across multiple model lines and test, field evaluate, and certify the new model lines."[21] The agency's RIA also provided data on numerous heaters that it stated currently achieved emissions at or below the required Step 2 level.[22]

WHAT WILL THE RULE COST?

In 2015, EPA estimated that the annualized cost of the rule in the 2015-2020 timeframe will be $45.7 million,[23] and that it will affect an average of 296,000 heaters sold annually during this period.[24] As shown below in Table 1, EPA's estimate of the cost impact of the rule varies by industry segment. EPA expected the vast majority of the cost (88%) to be borne by the hydronic heater and forced air furnace segments of the industry, neither of which was previously subject to emission limits. In percentage terms, EPA estimated that increased costs will range from a low of 1.1% of sales for pellet stoves to a high of 17.1% of sales for hydronic heaters.

The total estimated costs are relatively small compared to other Obama-era EPA rules that received considerable attention in Congress. For example, the Mercury and Air Toxics Standards (MATS), which apply to electric generating units, were estimated to cost $9.6 billion annually[25]— more than 200 times the estimated cost of the wood heater rule. However, given that the revenues of the electric utilities, to which the MATS rule applies, are about 750 times those of the wood heater industry, the estimated costs of this rule would impose a greater percentage burden on this small segment of U.S. industry than MATS would impose on electric utilities.[26] Most of the manufacturers of wood

---

[20] The rule allows somewhat higher emissions, generally, for wood stoves, pellet stoves, and hydronic heaters if they are certified using cord wood. See Final Rule, pp. 13678, 13681.

[21] Proposed Rule, p. 6353.

[22] Proposed Rule RIA, pp. 9-15 to 9-17. EPA also presented this data in the Final Rule RIA, pp. 4-5 to 4-6.

[23] The $45.7 million estimate was based on a 7% discount rate. The agency also produced a cost estimate using a 3% discount rate (which resulted in lower annualized cost), but most of the agency's explanatory materials use the $45.7 million estimate. See Final Rule RIA, p. 1-4. For detailed cost estimates, summarized in this paragraph, see Section 5 of the final rule RIA, https://www.epa.gov/sites/production/files/2015-02/documents/20150204-residential-woodheaters-ria.pdf.

[24] Final Rule RIA, Table 4-4.

[25] See U.S. Environmental Protection Agency, *Regulatory Impact Analysis for the Final Mercury and Air Toxics Standards*, EPA-452-R-11-011, December 2011, pp. ES-2, https://www.epa.gov/sites/production/files/2015-11/ documents/matsriafinal.pdf.

[26] Revenue from sales of electricity to ultimate customers was $388,089 million in 2017 according to the U.S. Energy Information Administration, *Electric Power Monthly*, February 27, 2018, Table 5-2. Annual revenue

heaters are small businesses—EPA concluded that approximately 90% of U.S. companies that manufacture residential wood heaters meet the Small Business Administration's small entity definition.[27] Small businesses do not have large revenues over which to spread research and development costs. As a result, stakeholders raised concerns that stringent standards would drive them out of business.[28]

In the final rule RIA, EPA stated that, lacking estimates of the price elasticity of both supply and demand, it did not estimate wood heater price or output changes that might result from the rule. On the other hand:

> While we are unable to estimate price and output changes, we note that the high increases in unit costs for some affected appliances could lead to potential nontrivial increases in market price to wood-burning appliance consumers and potential decreases in output for such appliances if supply elasticities are determined to be low. This could be true for hydronic heaters and forced air furnaces, but is unlikely for wood stoves and pellet stoves.[29]

Table 1. Annualized Compliance Cost and Average Annual Product Sales, 2015-2020, by Product Type

Product Type	Annualized Compliance Cost ($ million)	Average Annual Product Sales, 2015-2020 ($ million)	Compliance Cost/ Product Sales
Pellet stoves	$1.5	$132.2	1.1%
Wood stoves	$3.0	$123.1	2.4%
Single burn rate stoves	$0.9	$11.6	7.8%
Forced-air furnaces	$15.4	$91.8	16.8%
Hydronic heating systems	$24.9	$145.7	17.1%

Source: CRS analysis of EPA's Final Rule RIA, Table 5-5.

HAS THERE BEEN CONTROVERSY REGARDING THE AGENCY'S TESTING AND CERTIFICATION REQUIREMENTS?

Yes. Many concerns and comments in response to EPA's *proposed* rule had to do with the testing that would be required for certification of wood heating equipment under the proposal. In order to demonstrate compliance with the proposed standards, EPA

from sales of wood heaters, as shown in Table 1 of this report was estimated by EPA at approximately $500 million.

[27] Final Rule RIA, p. 6-13.

[28] U.S. Environmental Protection Agency, Response to Comment on Proposed Rule, 'Standards of Performance for New Residential Wood Heaters, New Residential Hydronic Heaters and Forced-Air Furnaces, and New Residential Masonry Heaters', EPA-HQ-OAR-2009-0734-1775, February 2015, https://www.regulations.gov. See pp. 455-456.

[29] Final Rule RIA, p. 5-23.

would rely on testing and certification provided by independent labs—just as it did with the 1988 standards for wood and pellet stoves. The labs rely on an EPA protocol that specifies the type and amount of wood, moisture content, startup procedures, duration of the test, etc., to test whether or not a specific model meets the NSPS standards.

As noted above (in "What does the final rule require?"), EPA has stated that a large number of wood heaters (more than 85% of wood stove models available when the rule was developed, for example)[30] already met the Step 1 standards at the time of their promulgation. Controversy over the proposed rule centered on EPA's desire to alter the test method to better account for real world conditions. In particular, the agency proposed the use of cordwood (split wood) in place of "cribs" (lumber assembled in a standardized configuration) in the certification process. Recertifying under a new testing protocol would have imposed costs on already certified equipment and could have resulted in many of the heaters not being certified.

EPA resolved this concern for Step 1 compliance by grandfathering heaters certified at emission levels below the new NSPS under the 1988 test methods. Under the final rule, heaters/stoves with EPA certifications under the 1988 emission standards that showed compliance with the Step 1 emission levels were automatically deemed as certified to meet the Step 1 emission limits until the Step 2 compliance date (in 2020). Hydronic heaters, which had not previously been subject to emission standards, were allowed to show compliance with the Step 1 standards if they qualified under a pre-existing EPA voluntary program or were certified by a New York State program. More than 90 heater models met the Step 1 standards (and some met the Step 2 standards) under these provisions.

These provisions addressed most of the industry's concerns about the Step 1 compliance process, but Step 2 compliance remains an issue. At House and Senate hearings held in late 2017, witnesses presented different opinions about the extent to which new wood heaters can demonstrate compliance with Step 2 standards. For example, an environmental organization stated that there is "a long list of devices that already meet" the Step 2 standards.[31] An industry association disagreed with that characterization, stating that few devices can meet the testing criteria for Step 2, which includes a carbon monoxide measurement.[32] According to this association, "it is not feasible" for one type of heater—warm air furnaces—to meet the Step 2 emission

---

[30] Final Rule RIA, p. 2-3.

[31] Statement of John D. Walke, Natural Resources Defense Council, U.S. Congress, hearing before the Subcommittee on Clean Air and Nuclear Safety, Senate Committee on Environment and Public Works, *Legislative Hearing on S. 1857, S. 203, S. 839, and S. 1934,* 115th Cong., 1st sess., November 14, 2017, Serial No. 115-127 (Washington: GPO, 2017), pp. 64-84. On p. 81, the witness cited "U.S. EPA. Compliance, List of EPA Certified Wood Stoves, October 2017 *available at* https://www.epa.gov/compliance/list-epa-certified-wood-stoves."

[32] U.S. Congress, hearing before the Subcommittee on Clean Air and Nuclear Safety, Senate Committee on Environment and Public Works, *Legislative Hearing on S. 1857, S. 203, S. 839, and S. 1934,* 115th Cong., 1st sess., November 14, 2017, Serial No. 115-127 (Washington: GPO, 2017). See attached letter from Jack Goldman, Hearth, Patio, and Barbecue Association, p. 122.

standards.[33] As of March 2018, the EPA identified 66 of 559 certified model lines as "meet[ing] the 2020 particulate matter emission standards"; the list showed a carbon monoxide measurement for 38 of these model lines.[34]

In addition, witnesses from heater manufacturing companies and the nation's largest certification lab expressed concern that there would not be enough time to certify products for Step 2 compliance before the 2020 deadline, in part, they alleged, because retailers—not wanting to be left with non-compliant inventory when the Step 2 deadline hits—will refuse to accept anything but Step 2 compliant products beginning in 2018. As a result, many in the industry have rallied behind legislation (H.R. 453[35] and S. 1857) that would move the Step 2 deadline to 2023.[36]

EPA did not testify at either 2017 hearing. In a written statement submitted for the record of the House Committee hearing, EPA Administrator Pruitt stated that the Administration had no official position on H.R. 453, but "I am very supportive of the committee's efforts to provide additional flexibilities, extensions, and clarifications for industries complying with various Clean Air Act regulations."[37] The agency had stated, in proposing and promulgating the rule, that the five-year phased approach was developed to provide the affected industry sufficient time to develop, test, and certify new products. The period of time EPA has allowed for the development of Step 2 compliant products (five years) is longer than the time allowed for compliance with most Clean Air Act emission standards.

DOES THE RULE HAVE ANY SPECIFIC PROVISIONS FOR SMALL BUSINESSES?

In general, the emission standards apply equally to manufacturers of all sizes with the same compliance deadlines for all. More than 90% of the manufacturers of wood heaters qualify as small businesses.[38]

---

[33] Hearth, Patio & Barbecue Association, "EPA's NSPS," https://www.hpba.org/Advocacy/EPAs-NSPS.

[34] According to EPA, "model line" refers to all wood heaters offered for sale by a single manufacturer that are similar in burning technology but have a different cosmetic or aesthetic look. List accessed on March 8, 2018. U.S. EPA, *List of EPA Certified Wood Stoves*, https://www.epa.gov/compliance/list-epa-certified-wood-stoves.

[35] As noted earlier, the House passed H.R. 453 on March 7, 2018, as part of H.R. 1917.

[36] Statement of Frank Moore, Hardy Manufacturing Company, Inc., U.S. House, Committee on Energy and Commerce, Subcommittee on Environment hearing, *Big Relief for Small Business: Legislation Reducing Regulatory Burdens on Small Manufacturers and Other Job Creators*, September 13, 2017, and attached letter from Rachel Feinstein, Hearth, Patio, and Barbecue Association. Also, statement of Paul Williams, United States Stove Company, in U.S. Senate, Committee on Environment and Public Works, Subcommittee on Clean Air and Nuclear Safety hearing, *Legislative Hearing on S. 1857, S. 203, S. 839, and S. 1934,* November 14, 2017, and Attachment 2, letter from OMNI Test Laboratories.

[37] Statement for the Record by Administrator Scott Pruitt, U.S. Environmental Protection Agency, in U.S. House, Committee on Energy and Commerce, Subcommittee on Environment, *Big Relief for Small Business: Legislation Reducing Regulatory Burdens on Small Manufacturers and Other Job Creators*, September 13, 2017, p. 148.

[38] Final Rule RIA, p. 6-13.

Nevertheless, EPA incorporated numerous suggestions in the rule in response to the concerns raised by small businesses. As part of the regulatory process, in 2010, EPA created a Small Business Advocacy Review panel under the Regulatory Flexibility Act, as amended by the Small Business Regulatory Enforcement Fairness Act.[39] The panel included representatives of the Office of Advocacy of the Small Business Administration (SBA), the Office of Information and Regulatory Affairs at the Office of Management and Budget, and 26 small businesses or organizations from the wood heating industry.

The panel produced a number of recommendations, which are summarized in the preamble to the proposed wood heater rule. EPA stated in the preamble that it collected additional information and refined its economic and technical analyses based on the panel's input. The agency stated that its stepped compliance approach is a response to the panel's recommendation for delayed compliance dates for low volume producers.[40] The panel recommended that EPA allow the use of International Standards Organization (ISO)-accredited labs and certifying bodies to expand the options for product testing and certification, which the final rule does. The panel recommended that EPA allow manufacturers to test a representative unit for a model line to determine compliance with the standards, rather than each individual unit, which EPA agreed to. The panel also recommended that EPA not move forward with proposed emission limits for 12 types of heaters, 8 of which EPA agreed to.[41]

These responses seem to have addressed many small business concerns. The preamble to the final rule notes that SBA's Chief Counsel for Advocacy did not file any comments to the proposed rule.[42]

WHAT ARE THE ESTIMATED BENEFITS OF THE RULE?

EPA has found that wood smoke—a mixture of fine particles, carbon monoxide, volatile organic compounds, and toxic air pollutants—causes respiratory illness and even premature death in some people exposed to it. In 2015, the agency estimated that the reductions in fine particulate (PM2.5) emissions that will result from implementation of the final standards will avoid 360 to 810 premature deaths per year in the years 2015-2020, as well as reduce nonfatal heart attacks, emergency room visits for asthma, and lost work days. Using its standard methods for monetizing health benefits, EPA estimated that reduced health impacts resulting from the proposal would provide a benefit of $3.4

[39] 5 U.S.C. §609.
[40] Proposed Rule, p. 6370, and Final Rule, p. 13673.
[41] Proposed Rule, pp. 6370-6371.
[42] Final Rule, p. 13698.

billion to $7.6 billion per year over the period 2015-2020, or $74 to $165 in benefits for every dollar spent to comply.[43]

EPA estimated that 98% of the monetized benefits of the rule would come from the reduction in premature mortality due to reduced $PM_{2.5}$ exposure.[44] To estimate the number of avoided premature deaths, the agency relied on two peer-reviewed epidemiological studies (the American Cancer Society study and the Harvard Six Cities study), which have formed the backbone of the agency's conclusions on the health effects of $PM_{2.5}$ for the past two decades.[45] In line with other executive branch agencies,[46] and as reviewed by its outside Science Advisory Board, EPA valued statistical lives saved at $8.5 million per life in 2013 dollars.[47]

The agency attributed 99% of the monetized benefits to the standards for hydronic heaters, forced air furnaces, and single burn-rate stoves (none of which were previously subject to regulation).[48]

The projected monetized benefits did not include the value of other expected emission reductions, including in volatile organic compounds, carbon monoxide, black carbon, and hazardous air pollutant emissions.[49] EPA also identified improved wood heater efficiency as a benefit of the rule—i.e., new wood heaters would use less wood to produce an equivalent amount of heat. EPA did not quantify this benefit, noting that it lacked robust data to do so.[50]

DO ANY OF THE RULE'S PROVISIONS APPLY TO THE OPERATORS OF RESIDENTIAL WOOD HEATERS?

Although the final rule mainly sets emission standards to be met by manufacturers, it also has some provisions applicable to the equipment's operator. Specifically, the final rule sets requirements for the type of fuel that may be used, and it prohibits the burning of garbage, yard waste, tires, plastic, waste petroleum products or paint, material containing asbestos, construction and demolition waste, paper products, railroad ties, pressure-treated wood, animal remains, unseasoned wood, and salt water driftwood in those heaters.[51] For all wood heaters, it is a violation of federal regulations to operate the

[43] These are the estimated benefits as expressed in 2013 dollars, using a 3% discount rate. EPA also estimated benefits using a 7% discount rate, producing a range of $3.1 billion to $6.9 billion in 2013 dollars. The benefits of the rule are discussed in Section 7 of the Final Rule RIA.

[44] Final Rule RIA, p. 7-6.

[45] Final Rule RIA, p. 7-5.

[46] For a discussion of the value of a statistical life used by various federal agencies, see CRS Report R41140, *How Agencies Monetize "Statistical Lives" Expected to Be Saved By Regulations*, by Curtis W. Copeland.

[47] Final Rule RIA, pp. 7-6 and 7-7.

[48] CRS calculation, based on data in Final Rule RIA, Figure 7-2, p. 7-13.

[49] Final Rule RIA, p. 7-1.

[50] Final Rule RIA. See pp. 5-26 and 5-27 for a discussion of energy impacts of the rule.

[51] Final Rule, p. 13717.

equipment in a manner inconsistent with operating instructions in the owner's manual.[52] For pellet-fueled appliances, the rule makes it clear that operation according to the owner's manual includes operation only with pellet fuels that have been specified in the manual and "graded under a licensing agreement with a third-party organization approved by the EPA. The Pellet Fuels Institute, ENplus, and CANplus are initially deemed to be approved third-party organizations for this purpose."[53]

What Have Stakeholders' Reactions Been to the 2015 Standards?

According to EPA, many of the 1,300 comment letters submitted in response to the 2014 proposal generally supported it and the importance of reducing particle pollution.[54] Commenters represented the wood heat industry, state and local governments, environmental and health advocacy groups, and private citizens.

There was a general consensus in the wood heat industry that the updated standards were welcome, in part to provide uniformity and predictability.[55] For example, in a May 8, 2017, letter to the leadership of the House of Representatives and the House Energy and Commerce Committee, the Hearth, Patio, and Barbecue Association (HPBA, which represents the makers of wood heaters) stated:

> The 2015 NSPS program reflects today's modern wood heating devices that provide important benefits to millions of Americans, especially those living in rural communities. The program fosters the market for wood, which is an important domestic source of heating fuel. It will save consumers money, many of whom are low income households, by lowering fuel costs through increased appliance efficiency. Replacing non-EPA-certified stoves with today's modern stoves will reduce health risks from exposure to wood smoke [footnote omitted], but this can only be done if products are clean burning, fuel efficient, and affordable. Finally, this program will ensure continued innovation in U.S. manufacturing that will help keep domestic companies competitive in the solid fuel industry.[56]

---

[52] Final Rule, p. 13712.

[53] Final Rule, p. 13705.

[54] Final Rule, p. 13685. See also U.S. Environmental Protection Agency, Response to Comment on Proposed Rule, 'Standards of Performance for New Residential Wood Heaters, New Residential Hydronic Heaters and Forced-Air Furnaces, and New Residential Masonry Heaters,' EPA-HQ-OAR-2009-0734-1775, February 2015, https://www.regulations.gov.

[55] U.S. Congress, House Committee on Energy and Commerce, Subcommittee on Environment, *Big Relief for Small Business: Legislation Reducing Regulatory Burdens on Small Manufacturers and Other Job Creators*, 115th Cong., 1st sess., September 13, 2017, Serial No. 115-54 (Washington: GPO, 2017), p. 12.

[56] The letter was jointly signed by the Hearth, Patio, and Barbecue Association and Northeast States for Coordinated Air Use Management and written to "express our joint concern with H.R. 694 that would rescind the 2015 New Source Performance Standards (NSPS) for residential wood heating devices promulgated by the U.S. Environmental Protection Agency (EPA)." See U.S. Senate, Committee on Environment and Public Works, Subcommittee on Clean Air and Nuclear Safety, *Legislative Hearing on S. 1857, S. 203, S. 839, and S. 1934,* November 14, 2017, pp. 125-126.

Many states echoed the concerns about uniform standards and endorsed the 2014 proposal to strengthen and expand the NSPS. The National Association of Clean Air Agencies (NACAA), which at the time represented air pollution control agencies in 42 states and the District of Columbia, identified residential wood heaters as a challenging source of emissions for states and localities to regulate. NACAA concluded that it

> cannot overstate the importance of federal standards for these sources. Residential wood combustion is extremely difficult to regulate at the state level because the devices are installed and operated in private homes and consumers are able to purchase wood heaters and stoves outside their own state (which could enable a consumer to purchase a device that does not meet the standards of his state of residence). In the absence of strong new federal standards, states and localities will have no choice but to pursue or build upon their own regulatory programs in order to attain and maintain [federal air quality standards] and/or meet other clean air goals.[57]

The prevalence of wood smoke and its health consequences, particularly in small communities, led some jurisdictions to ban the installation of new wood heaters or to establish burn bans on days when air quality is poor. Prior to the rule's development, several states adopted emission standards more stringent than EPA's 1988 NSPS. According to the Connecticut Department of Energy and Environmental Protection, local health districts in the state have issued cease and desist orders to individuals violating local bans on the use of hydronic heaters. In monitoring conducted over a two-year period, Connecticut found the monthly contribution of wood smoke to total concentrations of $PM_{2.5}$ to be as high as 41.3%, with daily contributions as high as 74.3%.[58] The existence of bans and the threat of future bans discourage sales of wood heating equipment; regardless, many in the industry want to see the pollution problem addressed.

The wood heater industry—as represented by HPBA and others—generally supported EPA's decision in 2015 to revise the NSPS and expand it to cover additional heater types. As noted earlier, however, HPBA and a number of heater manufacturers have expressed concern regarding the timing of the Step 2 standards. For example, one manufacturer testified at a 2017 Congressional hearing that the wood burning industry may not have enough time to meet the standard in 2020.[59] (See "Has there been controversy over the agency's testing and certification requirements?" above).

---

[57] Letter from George (Tad) S. Aburn, Jr., Co-Chair, NACAA Criteria Air Pollutants Committee, and Lynne A. Liddington, Co-Chair, NACAA Criteria Air Pollutants Committee, to EPA Docket (EPA-HQ-OAR-2009-0734), May 1, 2014, EPA-HQ-OAR-2009-0734-1417 at https://www.regulations.gov.

[58] Statement of Richard Pirolli, Bureau of Air Management, Connecticut Department of Energy and Environmental Protection, at U.S. EPA Public Hearing on the Proposed New Source Performance Standards for Wood Heaters, Boston, Massachusetts, February 26, 2014, transcript, p. 190, at http://www.regulations.gov/#!documentDetail;D=EPA-HQ-OAR-2009-0734-1037.

[59] U.S. Congress, House Committee on Energy and Commerce, Subcommittee on Environment, *Big Relief for Small Business: Legislation Reducing Regulatory Burdens on Small Manufacturers and Other Job Creators*, 115th Cong., 1st sess., September 13, 2017, Serial No. 115-54 (Washington: GPO, 2017), p. 12.

Environmental and health advocacy groups, state and local pollution control agencies, and private citizens were among the commenters that generally supported the 2015 NSPS due to concerns about the health effects of wood smoke.[60] Comments from other private citizens and some industry organizations, however, raised concerns that the proposed rule would increase the price of new wood heaters and therefore negatively affect the rural poor.[61]

More recently, some stakeholders reiterated concerns about the difficulties that some jurisdictions face in limiting emissions from residential wood heaters. The Sierra Club testified at a 2017 congressional hearing that areas such as Fairbanks, AL, and Salt Lake City, UT, "that are struggling with unhealthy levels of particulate pollution driven primarily or partially by wood smoke cannot afford" delays to cleaner technology.[62] EPA has reclassified these two nonattainment areas from moderate to serious nonattainment for fine particle pollution.[63]

According to EPA Region 10, "Fairbanks North Star Borough faces an especially difficult challenge of meeting existing pollution standards for a number of reasons including a high reliance on woodstoves and wood heaters to stay warm."[64]

The challenge, which the industry, EPA, and individual states all face, has been to find the sweet spot, where emissions are significantly reduced, without needlessly harming the industry or unwittingly providing incentives for equipment owners to continue using old, inefficient, highly polluting equipment. This has not been an easy task.

Given the large number of wood heating devices already in use—many of which are not subject to any emission standard—the NSPS, on their own, will have relatively small impacts on total wood heater emissions for many years, unless other programs address the retirement or change-out of old, uncontrolled wood heaters. As a result, controversy over wood heater emissions is likely to remain an issue.

---

[60] U.S. Environmental Protection Agency, Response to Comment on Proposed Rule, 'Standards of Performance for New Residential Wood Heaters, New Residential Hydronic Heaters and Forced-Air Furnaces, and New Residential Masonry Heaters,' EPA-HQ-OAR-2009-0734-1775, February 2015, https://www.regulations.gov, p. 8.

[61] Ibid., p. 452.

[62] U.S. Congress, House Committee on Energy and Commerce, Subcommittee on Environment, *Big Relief for Small Business: Legislation Reducing Regulatory Burdens on Small Manufacturers and Other Job Creators*, 115th Cong., 1st sess., September 13, 2017, Serial No. 115-54 (Washington: GPO, 2017), p. 57.

[63] U.S. Environmental Protection Agency, "Determinations of Attainment by the Attainment Date, Determinations of Failure to Attain by the Attainment Date and Reclassification for Certain Nonattainment Areas for the 2006 24-Hour Fine Particulate Matter National Ambient Air Quality Standards," 82 *Federal Register* 21711, May 10, 2017.

[64] U.S. Environmental Protection Agency, "EPA, Alaska to Work Together on Improving Fairbanks Air Quality," press release, May 1, 2017, https://www.epa.gov/newsreleases/epa-alaska-work-together-improving-fairbanks-air-quality.

In: Environmental Protection Agency (EPA)
Editor: Coby Schaefer
ISBN: 978-1-53614-933-3

Chapter 10

EPA's Role in Emergency Planning and Notification at Chemical Facilities*

Richard K. Lattanzio and David M. Bearden

Summary

Chemicals and the facilities that manufacture, store, distribute, and use them are essential to the U.S. economy. However, incidents occasioned by natural disasters, unintentional events, or security threats show that the handling and storage of chemicals are not without risk. Federal agencies implement a number of programs to help prevent chemical facility accidents, reduce risks of terrorist attacks on chemical facilities, protect chemical facility workers, collect and share relevant information with the public and decisionmakers, and prepare communities and local, tribal, and state first-responders to respond to potential large-scale accidents.

This chapter reviews the U.S. Environmental Protection Agency's (EPA's) authorities regarding risk management, emergency planning, and release notification, among others, at chemical facilities. In doing so, it describes the statutory authorities—and makes note of some of the more prominent, subsequent regulations—as provided by the following:

- *Facility risk management planning requirements under Section 112(r)(7) of the Clean Air Act (CAA).* EPA's Risk Management Program (RMP) is aimed at reducing chemical risk at the local level. EPA regulations require owners and operators of a facility that manufactures, uses, stores, or otherwise handles certain listed flammable and toxic substances to develop a risk management program that includes hazard assessment (including an evaluation of worst-case and alternative accidental release scenarios), prevention mechanisms, and emergency response measures.

---

* This is an edited, reformatted and augmented version of Congressional Research Service Publication No. R44952, dated September 18, 2017.

- *Emergency planning notification requirements under the Emergency Planning and Community Right-to-Know Act of 1986 (EPCRA).* The requirements are designed to promote emergency planning and preparedness at the state, local, and tribal levels. EPCRA helps ensure local communities and first responders have needed information on potential chemical hazards within their communities in order to develop community emergency response plans.
- *Emergency release notification requirements under the Comprehensive Environmental Response, Compensation, and Liability Act of 1980 (CERCLA).* CERCLA obligates a facility to report certain releases of hazardous substances to the National Response Center to inform decisions about federal involvement in responding to the incident to coordinate with state and local officials. The requirements also establish liability for response costs and natural resource damages.
- *Duties of the Chemical Safety and Hazard Investigation Board, known as the Chemical Safety Board (CSB), under Section 112(r)(6) of the CAA.* The purpose of the CSB is to investigate accidents to determine the conditions and circumstances that led up to the event and to identify the cause or causes so that similar events might be prevented.
- *Toxic release inventory reporting requirements.* EPCRA authorizes EPA to establish and maintain a Toxic Release Inventory (TRI) of facilities that manufacture, import, process, or use certain types of toxic chemicals by providing public disclosure of the locations of such facilities.

INTRODUCTION

Chemicals and the facilities that manufacture, store, distribute, and use them are essential to the U.S. economy. However, incidents occasioned by natural disasters (e.g., hurricanes, earthquakes, floods), unintentional events (e.g., fire, accidents), or security threats (e.g., terrorism) show that the handling and storage of chemicals are not without risk. Incidents such as the 2017 fire at the Arkema chemical plant in Crosby, Texas, the 2013 explosion at the West Fertilizer Company in West, Texas, and the 1984 release of methyl isocyanate at the Union Carbide plant in Bhopal, India, have motivated many in federal, state, and local governments to back efforts to reduce the risk of chemical accidents in the United States.

Federal agencies implement a number of programs to help prevent chemical facility accidents, reduce risks of terrorist attacks on chemical facilities, protect chemical facility workers, collect and share relevant information with the public and decisionmakers, and prepare communities and local, tribal, and state first-responders to respond to potential large-scale accidents. State, local, and tribal authorities also have critical responsibilities in managing risks from chemical facility accidents through setting and enforcing requirements for zoning, siting, and emergency response and planning.[1]

[1] As described in The White House, Office of the Press Secretary, "FACT SHEET: Executive Order on Improving Chemical Facility Safety and Security," August 1, 2013.

This chapter reviews the U.S. Environmental Protection Agency's (EPA's) authorities regarding risk management, emergency planning, and release notification, among others, at chemical facilities. In doing so, it describes the statutory authorities—and makes note of some of the more prominent, subsequent regulations—as provided by the following:

- facility risk management planning requirements under Section 112(r)(7) of the Clean Air Act (CAA);
- emergency planning notification requirements under the Emergency Planning and Community Right-to-Know Act of 1986 (EPCRA);
- emergency release notification requirements under the Comprehensive Environmental Response, Compensation, and Liability Act of 1980 (CERCLA);
- duties of the Chemical Safety and Hazard Investigation Board, known as the Chemical Safety Board (CSB), under Section 112(r)(6) of the CAA; and
- toxic release inventory reporting requirements under EPCRA.

This chapter does not address other federal agencies' authorities regarding safety, planning, and notification, such as the Department of Labor, Occupational Safety and Health Administration (OSHA) requirements covering occupational safety and health risks to workers, the Department of Homeland Security (DHS) requirements covering homeland security risks from security threats to facilities, or the Department of Transportation (DOT) requirements covering risks from the transportation of hazardous materials. Further, it does not address the federal response framework for releases of oil and chemicals into the environment.[2]

FACILITY RISK MANAGEMENT PLANNING REQUIREMENTS

In the CAA Amendments of 1990, P.L. 101-549, Congress enacted Section 112(r)(1), also known as the General Duty Clause (GDC). It applies to any facility where extremely hazardous substances are present.[3] GDC is a performance-based authority recognizing that owners and operators have a general duty and responsibility to prevent and mitigate the consequences of chemical accidents.

---

[2] For a discussion of the federal role in responding to releases of hazardous substances, see CRS Report R43251, *Oil and Chemical Spills: Federal Emergency Response Framework*, by David M. Bearden and Jonathan L. Ramseur. For additional reports on these topics, see CRS Report R43360, *Chemical, Hazardous Substances, and Petroleum Spills: CRS Experts*, coordinated by James D. Werner.

[3] "Extremely hazardous substances" are defined as any chemicals listed in 40 C.F.R. §68, "or any other extremely hazardous substance."

Further, Section 112(r)(7) authorizes EPA to require the establishment of approved risk management planning requirements for stationary sources that manage certain types of toxic or flammable substances, if the quantity of the substance exceeds a threshold established in regulation for that substance.[4] Section 112(r)(7)(B) requires EPA to promulgate regulations and guidance for the development of facility Risk Management Plans (RMPs) to prevent and detect accidental releases into the ambient air from these stationary sources.[5] Facility owners and operators also must incorporate measures into their plans to respond to an accidental release of such substances, if such release were to occur. For these purposes, Section 112(r)(2)(A) defines an "accidental release" to be an "unanticipated emission of a regulated substance or other extremely hazardous substance into the ambient air from a stationary source."[6]

Pursuant to Section 112(r)(7)(B), the major elements of an RMP must include

- an assessment of the potential severity of the hazards if an accidental release into the ambient air were to occur (including off-site consequence analysis of worst-case scenarios);
- a facility-specific program to prevent accidental releases (including safety precautions and employee training); and
- a facility-specific response program to take actions that may be necessary to protect human health and the environment in the event of an accidental release (including procedures for notifying federal, state, and local agencies responsible for responding to an incident).

Section 112(r)(7)(E) makes the operation of a stationary source subject to RMP requirements unlawful if the facility owner or operator does not prepare and implement an RMP in accordance with applicable regulatory requirements.[7]

---

[4] Both the U.S. Environmental Protection Agency's (EPA's) 40 C.F.R. §68 Risk Management Plan (RMP) regulations and the Department of Labor, Occupational Safety and Health Administration's (OSHA's) 29 C.F.R. §1910.119 Process Safety Management (PSM) standard were authorized in the CAA Amendments of 1990. This was in response to a number of catastrophic chemical accidents occurring worldwide that had resulted in public and worker fatalities and injuries, environmental damage, and other community impacts. Most notably, the sudden, accidental release in December 1984 of methyl isocyanate in an industrial incident at the Union Carbide plant in Bhopal, India, and the attendant loss of thousands of lives and widespread injuries motivated many in Congress to support legislation to reduce the risk of chemical accidents in the United States.

[5] 42 U.S.C. §7412(r)(7)(B). EPA promulgated regulations to designate regulated substances and threshold quantities under CAA Section 112(r)(7) in 1994, and promulgated the regulations to establish RMP requirements for covered stationary sources in 1996. These regulations are codified together at 40 C.F.R. §68, including a list of 140 regulated substances (77 toxic and 63 flammable chemicals), and their threshold quantities (40 C.F.R. §68.130). CAA Section 112(r)(7)(D) requires EPA to coordinate the development of its regulations and guidance for RMPs with requirements for "comparable" purposes established by OSHA and DOT. Information on these regulations and accompanying agency guidance are available on the EPA website at https://www.epa.gov/rmp.

[6] 42 U.S.C. §7412(r)(2)(A).

[7] 42 U.S.C. §7412(r)(7)(E).

Facility owners or operators must submit their RMPs to EPA, the Chemical Safety Board (CSB) established pursuant to Section 112(r)(6), and state and local emergency response officials.[8] Section 114(c) also requires RMPs to be made available to the public, with the exception of certain confidential business or trade information that an RMP may contain.[9] For facility security purposes, Section 112(r)(7)(H) also limits the public availability of off-site consequence analysis of worst case scenarios presented in an RMP.[10] These plans must be revised and resubmitted to EPA every five years. In overseeing regulated facilities, EPA may also require facilities to revise their RMPs if necessary to ensure compliance with federal requirements.

EPA has delegated RMP oversight responsibility to some states and localities. Where the RMP has been delegated to a state, the state may have additional requirements for the federally listed chemicals and/or additional listed chemicals.

Revisions under the Obama Administration

Under the Obama Administration, EPA revised the RMP requirements in response to Executive Order 13650, *Improving Chemical Facility Safety and Security,* issued by President Obama on August 1, 2013.[11] EPA promulgated the final rule revisions on January 13, 2017, highlighting the fertilizer facility incident in 2013 in West, Texas, among "catastrophic chemical facility incidents" that were the primary impetuses for Executive Order 13650.[12]

The revisions include

- additional analysis of safer technology and alternatives as part of the process hazard analysis for some sources;
- third-party audits and incident investigation root cause analysis for some sources;
- enhancements to the emergency preparedness requirements; and
- increased public availability of chemical hazard information to assist local emergency authorities in planning for and responding to accidents and to improve public awareness of chemical hazards at regulated sources.

---

[8] "Plans prepared pursuant to this subparagraph shall also be submitted to the Chemical Safety and Hazard Investigation Board, to the State in which the stationary source is located, and to any local agency or entity having responsibility for planning for or responding to accidental releases which may occur at such source, and shall be available to the public under section 7414(c) of this title." 42 U.S.C. §7412(r)(7)(B)(iii).

[9] 42 U.S.C. §7414(c).

[10] 42 U.S.C. §7412(r)(7)(H).

[11] Executive Order 13650, "Improving Chemical Facility Safety and Security," 78 *Federal Register* 48029-48033, August 7, 2013.

[12] EPA, "Accidental Release Prevention Requirements: Risk Management Programs Under the Clean Air Act," 82 *Federal Register* 4594-4705, January 13, 2017.

Revisions under the Trump Administration

Under the Trump Administration, the EPA Administrator published a final rule on June 14, 2017, to delay the effective date of the RMP rule amendments for 20 months until February 19, 2019.[13] The action was taken under CAA section 307(d)(7)(B).[14] The rule states that the action "allows the Agency time to consider petitions for reconsideration of the Risk Management Program Amendments and take further regulatory action, as appropriate, which could include proposing and finalizing a rule to revise or rescind these amendments." EPA summarized stakeholders' submitted reasons to delay the effective date of the amendments as follows:[15]

- lack of sufficient notice for comments or the addition of new provisions to the final rule that were not in the 2016 RMP proposed amendments,
- safety and security concerns related to implementation of the final rule,
- cost burden to regulated facilities and emergency response organizations,
- insufficient coordination with OSHA by EPA, and
- insufficient coordination with stakeholders or consideration of stakeholder comments.

EMERGENCY PLANNING NOTIFICATION REQUIREMENTS

EPCRA was enacted in 1986 as Title III of the Superfund Amendments and Reauthorization Act of 1986 (P.L. 99-499). EPCRA requires facilities to report the presence of hazardous chemicals or extremely hazardous substances to state and local emergency response officials, if the quantity present would exceed certain thresholds.[16] This information is intended to assist state and local officials in developing their own emergency response plans in the event of an incident at a facility. The universe of facilities subject to reporting under EPCRA is larger than facilities subject to RMP requirements under the CAA, because EPCRA applies to a broader body of chemicals.

Section 311 of EPCRA[17] specifies the applicability of reporting requirements under that statute to hazardous chemicals that require the preparation of a material safety data

---

[13] EPA, "Accidental Release Prevention Requirements: Risk Management Programs Under the Clean Air Act; Further Delay of Effective Date," 82 *Federal Register* 27133-27144, June 14, 2017.

[14] 42 U.S.C. §7607(d)(7)(B).

[15] EPA, "Response to Comments on the 2017 Proposed Rule Further Delaying the Effective Date of EPA's Risk Management Program Amendments (April 3, 2017; 82 FR 16146)." Office of Emergency Management, USEPA, Washington, DC. June 8, 2017, https://www. regulations.gov/document?D=EPA-HQ-OEM-2015-0725-0881.

[16] 42 U.S.C. §§11001-11050. For a broader discussion of the authorities of EPCRA than presented in this memorandum, see CRS Report RL32683, The Emergency Planning and Community Right-to-Know Act (EPCRA): A Summary, by David M. Bearden.

[17] 42 U.S.C. §11021.

sheet (MSDS) pursuant to the Occupational Safety and Health Act of 1970 (OSH Act).[18] These hazardous chemicals encompass a broad array of substances commonly found in industrial, commercial, or other workplace settings. Considering this breadth, there is not a singular consolidated list. They are defined by certain characteristics and properties specified in federal regulations promulgated under the OSH Act.[19] Certain uses of hazardous chemicals are excluded from reporting requirements, as outlined in Section 311(e) of EPCRA.[20]

Section 312 of EPCRA requires the owner or operator of a facility storing a hazardous chemical in a quantity equal to or exceeding 10,000 pounds to report the presence of the chemical to the State Emergency Response Commission (SERC), the appropriate Local Emergency Planning Committee (LEPC), and the local fire department with jurisdiction over the facility.[21] States established SERCs and LEPCs pursuant to Section 301 of EPCRA.[22]

The general threshold of 10,000 pounds for the reporting of hazardous chemicals applies to "Tier I" reporting under which the facility owner or operator is required to report this information at its own initiative to the SERC, the appropriate LEPC, and the local fire department. There is a "zero" threshold (i.e., no minimum quantity) for "Tier II" reporting for which a SERC, LEPC, or local fire department may require additional information from the facility owner or operator about the presence of a hazardous chemical present at the facility.

If a hazardous chemical also is designated under EPCRA as an extremely hazardous substance, the reporting threshold generally is 500 pounds, and may be less under separate reporting requirements. Section 302(a) of EPCRA directs EPA to determine which chemicals warrant designation as extremely hazardous substances and to establish separate reporting thresholds for them.[23] In making these determinations, EPA must take into account the "toxicity, reactivity, volatility, dispersability, combustability, or flammability" of a substance. Section 302(c) of EPCRA requires the owner or operator of a facility storing an extremely hazardous substance exceeding the applicable threshold to report the presence of the substance to the SERC and the LEPC (but not the local fire

---

[18] 29 U.S.C. §§651-678.

[19] 29 C.F.R. §1910.1200(c).

[20] 42 U.S.C. §11021(e). Exclusions include (1) any food, food additive, color additive, drug, or cosmetic regulated by the Food and Drug Administration; (2) any substance present as a solid in any manufactured item to the extent exposure to the substance does not occur under normal conditions of use; (3) any substance to the extent it is used for personal, family, or household purposes, or is present in the same form and concentration as a product packaged for distribution and use by the general public; (4) any substance to the extent it is used in a research laboratory or a hospital or other medical facility under the direct supervision of a technically qualified individual; and (5) any substance to the extent it is used in routine agricultural operations or is a fertilizer held for sale by a retailer to the ultimate customer.

[21] 42 U.S.C. §11022.

[22] 42 U.S.C. §11001.

[23] 42 U.S.C. §11002(a).

department).[24] Section 303(d) also requires facilities subject to the reporting of an extremely hazardous substance to designate a representative to serve on the LEPC.[25]

Section 324 of EPCRA requires information reported by facilities to SERCs, LEPCs, and local fire departments to be made available to the public, with the exception of facility information that may be subject to protection as confidential business or trade information.[26] Section 322 of EPCRA specifies the types of confidential business or trade information that a facility owner or operator may choose to protect from public disclosure.[27] For security purposes, a facility owner or operator also may opt not to publicly disclose the exact location of a specific chemical within a facility boundary.

EMERGENCY RELEASE NOTIFICATION REQUIREMENTS

Requirements to report releases of a hazardous substance into the environment are enumerated in both EPCRA and CERCLA.[28] EPCRA requires reporting of certain releases to the SERC and the appropriate LEPC to make state and local officials aware of the release so as to inform emergency response actions that may be appropriate within their respective jurisdictions. CERCLA requires a facility to report certain releases to the National Response Center to inform decisions about federal involvement in responding to the incident to coordinate with state and local officials.[29] Section 107 of CERCLA also establishes liability for response costs and natural resource damages.[30] Similar to emergency planning notification requirements discussed above, whether the owner or operator of a facility would be required under EPCRA or CERCLA to report an actual release into the environment would depend primarily on the quantity of the release.

Section 103 of CERCLA[31] generally requires persons who release hazardous substances into the environment to notify the federal National Response Center of the incident as soon as the person has knowledge of the release, if the quantity of the release is equal to or exceeds the threshold for that substance. Section 102 directed EPA to

---

[24] 42 U.S.C. §11002(c).

[25] 42 U.S.C. §11003(d).

[26] 42 U.S.C. §11044.

[27] 42 U.S.C. §11042. For exclusions due to trade purposes, each of the following must be shown: (1) such person has not disclosed the information to any other person, other than a member of a local emergency planning committee, an officer or employee of the United States or a State or local government, an employee of such person, or a person who is bound by a confidentiality agreement, and such person has taken reasonable measures to protect the confidentiality of such information and intends to continue to take such measures; (2) the information is not required to be disclosed, or otherwise made available, to the public under any other Federal or State law; (3) disclosure of the information is likely to cause substantial harm to the competitive position of such person; and (4) the chemical identity is not readily discoverable through reverse engineering.

[28] 42 U.S.C. §§9601-9675. For a broader discussion of the authorities of CERCLA than presented in this report, see CRS Report R41039, *Comprehensive Environmental Response, Compensation, and Liability Act: A Summary of Superfund Cleanup Authorities and Related Provisions of the Act*, by David M. Bearden.

[29] For a discussion of the federal role in responding to releases of hazardous substances, see CRS Report R43251, *Oil and Chemical Spills: Federal Emergency Response Framework*, by David M. Bearden and Jonathan L. Ramseur.

[30] 42 U.S.C. §9607.

[31] 42 U.S.C. §9603.

designate specific chemicals as hazardous substances for the purpose of CERCLA and to establish thresholds for reporting releases into the environment.[32] A list of designated hazardous substances and the reporting threshold (i.e., reportable quantity) for each substance is specified in federal regulation.[33]

In certain circumstances, a release may not be subject to reporting under CERCLA, even if the release otherwise would exceed the reportable quantity. Section 103 excludes federally permitted releases of hazardous substances from reporting requirements under the statute. Section 101(10) defines the term "federally permitted release" to include releases of hazardous substances authorized in permits issued under certain other federal environmental laws cited in that definition.[34] Section 107(j) of CERCLA also exempts federally permitted releases from liability under the statute.[35] Federally permitted releases are exempt from reporting requirements and liability under CERCLA based on the premise that permit requirements would address potential risks, and that the exemption may avoid potential conflicts between one federal law allowing a release and another imposing liability for the same action.

Section 304 of EPCRA generally requires the owner or operator of a facility from which an extremely hazardous substance is released into the environment to report the release to the SERC and the appropriate LEPC, if the volume of the release is a reportable quantity.[36] Parallel with CERCLA, Section 304 of EPCRA however federally permitted releases from these reporting requirements. Most extremely hazardous substances designated under EPCRA also are designated as hazardous substances under CERCLA, but some are not.

CHEMICAL SAFETY BOARD

As amended in 1990, Section 112(r)(6) of the CAA authorized the establishment of the Chemical Safety and Hazard Investigation Board, often referred to as the Chemical Safety Board (CSB) for short.[37] The principal mission of the CSB is to

> investigate (or cause to be investigated), determine and report to the public in writing the facts, conditions, and circumstances and the cause or probable cause of any accidental release resulting in a fatality, serious injury or substantial property damages.[38]

---

[32] 42 U.S.C. §9602.
[33] 40 C.F.R. Part 302.
[34] 42 U.S.C. §9601(10). Permits issued under the authorities of these federal environmental laws include permits issued by states with delegated federal authority.
[35] 42 U.S.C. §9607(j).
[36] 42 U.S.C. §11004.
[37] 42 U.S.C. §7412.
[38] 42 U.S.C. §7412(6).

Based on the findings of its investigations, the CSB is authorized to recommend measures that may reduce the likelihood or consequences of accidental releases in the future, and to propose "corrective steps" to mitigate the safety risks of chemical production, processing, handling, and storage. The CSB is not a regulatory agency, however, and is not authorized to enforce or compel compliance with such recommendations or corrective steps. Other agencies with regulatory authority may develop enforceable requirements based on CSB recommendations, such as EPA for accidental release prevention requirements under Section 112(r)(7) of the CAA, or OSHA for worker protection requirements under the OSH Act.[39]

TOXIC RELEASE INVENTORY REPORTING REQUIREMENTS

In addition to emergency planning and release notification requirements, Section 313 of EPCRA authorized EPA to establish and maintain a Toxic Release Inventory (TRI) of facilities that manufacture, import, process, or use certain types of toxic chemicals.[40] These facilities are diverse in terms of their industrial and commercial operations. TRI does not necessarily identify actual releases into the environment that may present a particular level of risk to human health or the environment, nor does the TRI track facilities that have violated any particular environmental requirements under either federal or state law. The TRI only provides public disclosure of the locations of certain facilities at which toxic chemicals are present in various quantities, consistent with the "community right-to-know" objective of EPCRA.

Section 313 requires the owner or operator of a facility to submit an annual report to EPA, and a state official designated by the governor of the state in which the facility is located, identifying the quantities of toxic chemicals manufactured, imported, processed, or otherwise used at that facility the previous year, if the quantity would exceed the threshold required for reporting. The specific chemicals subject to these reporting requirements are designated in federal regulations promulgated by EPA, pursuant to Section 313.[41] Section 313(d) outlines the designation criteria, including chemical toxicity, potential adverse impacts on human health if exposure were to occur, and certain types of illnesses or health conditions that may be associated with potential exposure to the chemical.[42] Section 313(f) establishes a general reporting threshold of 10,000 pounds for toxic chemicals used at a facility during a calendar year, and 25,000 pounds for toxic chemicals manufactured, imported, or processed at a facility during a calendar year.[43]

---

[39] The CSB website (http://www.csb.gov/) offers additional information on the mission, history, and current membership of the Board; and maintains a collection of completed and ongoing chemical accident investigations conducted by the Board.

[40] 42 U.S.C. §11023.

[41] 40 C.F.R. Part 372.

[42] 42 U.S.C. §11023(d).

[43] 42 U.S.C. §11023(f).

INDEX

A

Administrative Procedure Act, 65, 74, 194
adverse effects, 170
advocacy, 121, 132, 136, 195, 224, 226
aesthetic, 221
agencies, ix, xii, xiii, xiv, xv, 8, 12, 13, 33, 41, 42, 47, 50, 53, 55, 58, 59, 60, 66, 69, 70, 73, 74, 76, 78, 80, 82, 83, 84, 85, 87, 89, 93, 96, 99, 111, 132, 143, 144, 153, 158, 164, 179, 180, 182, 183, 185, 194, 195, 205, 209, 223, 225, 226, 227, 228, 229, 230, 236
agency actions, 65
agriculture, 36, 42, 66, 73, 87
air emissions, 50, 67
Air Force, 143
air pollutants, 2, 21, 25, 67, 162, 166, 187, 189, 204, 210, 222
air quality, 6, 7, 10, 23, 150, 157, 158, 162, 167, 170, 172, 173, 174, 177, 186, 210, 225
air toxics, xiv, 24, 27, 213
airports, 38
Alaska, 17, 32, 130, 159, 160, 161, 165, 226
algae, 35
ambient air, 19, 20, 22, 149, 153, 175, 186, 230
American Recovery and Reinvestment Act of 2009, 93
ammonium, 185
ANS, x, 119, 120, 121, 123, 124, 125, 126, 131, 132, 133, 134, 136
APA, 194
appropriations, ix, 3, 7, 50, 54, 77, 89, 92, 93, 94, 115
Appropriations Act, 77, 89, 92
Appropriations Committee, xii, 115, 146, 177
aquatic life, 128
aquatic nuisance species (ANS), x, 119, 120
aquatic systems, 132
arsenic, 40, 43
asbestos, 76, 97, 223
asthma attacks, 166
at-risk populations, 151, 155
Attorney General, viii, 50, 195
audit, 55, 87, 231
auto manufacturers, xiii, 13, 179
automobiles, 150
autonomy, viii, 49, 52
average costs, 111
aviation industry, 19

B

background information, vii, xv, 2, 130, 214
bacteria, 127, 129
barges, x, 119, 120, 123, 125, 128, 130
benefits, vii, viii, xii, xiv, 1, 2, 5, 8, 12, 14, 16, 17, 18, 20, 21, 22, 23, 24, 25, 26, 27, 31, 32, 33, 43, 44, 53, 55, 70, 83, 87, 95, 124, 125, 131, 142, 145, 146, 147, 164, 166, 167, 190, 210, 213, 222, 223, 224
benzene, 199
biodiesel, 45
biodiversity, 131, 132
biofuel, 28, 29, 45
biomass, 28, 29, 157
blend wall, 27
boilers, 4, 25, 43, 44

bronchitis, 147, 166
Bureau of Land Management (BLM), v, xiii, 163, 189, 194, 209, 210, 211
Bush, President George Walker, 4

C

cables, 129
California, xii, xiii, 13, 15, 20, 112, 133, 134, 145, 147, 160, 164, 165, 166, 167, 173, 175, 179, 180, 181, 182, 194, 196
Canadian shippers, xi, 120
cancer, xiv, 27, 213
carbides, 185
carbon, xiii, 5, 11, 13, 16, 18, 19, 22, 31, 67, 149, 183, 184, 185, 199, 220, 222, 223
carbon dioxide, xiii, 11, 13, 18, 31, 183, 184, 185, 199
carbon emissions, 19
carbon monoxide, 22, 67, 149, 185, 220, 222, 223
case law, 82
CERCLA, xv, 50, 61, 67, 70, 79, 98, 103, 104, 228, 229, 234, 235
Chamber of Commerce, 3
chemical, x, xv, 119, 120, 124, 163, 171, 175, 227, 228, 229, 230, 231, 233, 234, 236
chemical reactions, 171, 175
chemicals, xv, xvi, 76, 99, 130, 227, 228, 229, 230, 231, 232, 233, 235, 236
chlorine, 128, 130
clean air, 19, 150, 225
Clean Air Act (CAA), vii, xii, xiii, xv, xvi, 1, 2, 4, 6, 10, 11, 13, 14, 16, 17, 19, 20, 21, 22, 23, 25, 26, 27, 28, 30, 44, 50, 57, 61, 64, 67, 68, 79, 82, 96, 97, 98, 101, 103, 104, 146, 147, 149, 150, 151, 153, 155, 156, 158, 162, 164, 165, 167, 170, 171, 172, 173, 176, 180, 181, 183, 185, 186, 187, 188, 189, 190, 191, 192, 193, 194, 195, 197, 198, 199, 200, 202, 204, 205, 206, 207, 209, 214, 221, 227, 228, 229, 230, 231, 232, 235, 236
Clean Water Act (CWA), ix, x, xi, 3, 6, 10, 33, 34, 35, 37, 38, 39, 40, 41, 42, 50, 52, 57, 61, 63, 64, 67, 68, 73, 76, 79, 82, 96, 97, 98, 99, 101, 103, 104, 105, 106, 107, 108, 109, 111, 112, 113, 114, 115, 116, 119, 120, 121, 122, 123, 125, 134, 135, 136, 138, 139, 141, 143
cleanup, 61, 69, 82
climate change, 11, 19
Clinton, President William Jefferson, 4
coal, 9, 16, 17, 22, 23, 24, 33, 34, 35, 43, 163, 168, 188, 215
Coast Guard, xi, 60, 76, 120, 122, 126, 127, 128, 129, 130, 131, 132, 133, 134, 137, 139, 140, 142, 143
Code of Federal Regulations, 4
combustion controls, 23
commercial fishing vessels, x, xi, 119, 120, 122, 123, 129, 138, 140
competitive advantage, 81
compliance, viii, ix, xiv, 4, 7, 21, 22, 23, 24, 25, 27, 30, 33, 39, 40, 41, 45, 49, 50, 51, 52, 53, 55, 56, 57, 58, 59, 60, 61, 62, 63, 64, 65, 66, 67, 68, 69, 70, 71, 72, 73, 74, 77, 78, 79, 80, 81, 82, 83, 84, 85, 86, 87, 88, 89, 90, 91, 92, 93, 94, 95, 96, 98, 99, 100, 101, 103, 104, 105, 106, 107, 108, 109, 110, 112, 114, 124, 126, 129, 131, 133, 136, 137, 138, 140, 142, 143, 148, 150, 170, 191, 193, 194, 195, 200, 201, 202, 204, 205, 214, 215, 216, 219, 220, 221, 222, 231, 236
Comprehensive Environmental Response, Compensation, and Liability Act (CERCLA), xv, 10, 50, 104, 228, 229, 234
Comprehensive Environmental Response, Compensation, and Liability Act of 1980, xv, 228, 229
Conference Report, 77
congressional hearings, 3, 55, 84, 137
Consolidated Appropriations Act, ix, 50, 55, 77, 89
consumer choice, 181
consumers, xiv, 29, 214, 216, 219, 224, 225
consumption, 15, 29
contamination, 37, 45, 50, 70, 130
control measures, 36, 150, 151, 170, 173
cooperative agreements, 85, 170
corn starch, 28
corporate average fuel economy, 12
cost, vii, ix, xii, xiv, 1, 3, 5, 11, 12, 14, 15, 16, 17, 20, 21, 22, 23, 24, 25, 26, 27, 31, 32, 33, 38, 40, 82, 88, 105, 106, 107, 112, 131, 133, 142, 145, 147, 164, 165, 166, 167, 168, 169, 170, 171, 186, 210, 214, 218, 232
cost saving, 88
cost-benefit analysis, 24, 170
costs of compliance, 136
counsel, ix, 50, 81
Court of Appeals, xiii, 12, 21, 22, 23, 24, 31, 32, 33, 44, 121, 137, 156, 176, 177, 184, 185
crimes, 76, 77, 78
criminal fines, ix, 50, 81, 103

criminal investigations, 79
criminal violations, ix, 50, 64, 71, 75, 76, 79
critical infrastructure, 91, 92
criticism, 3, 7, 8, 34, 36, 64, 86
crop yields, xii, 19, 145, 155
crude oil, xiii, 17, 186, 189, 197, 198, 209, 210, 211
cruise ships, x, 119, 120, 123, 130
CWA permits, ix, 35, 42, 105, 106, 108, 109, 111, 115, 121, 125, 134, 135
cycles, 215

D

data availability, 195, 205
database, 67, 76, 95, 96, 97, 99
deaths, 24, 147, 166, 222
Deepwater Horizon, ix, 50, 72, 81, 103
demographic characteristics, 86
Department of Defense, 55, 68, 143
Department of Energy, 28, 55, 225
Department of Homeland Security (DHS), 60, 127, 229
Department of Justice, viii, 50, 54, 59, 110
Department of Labor, 229, 230
Department of the Interior, 33, 58, 115, 189
Department of Transportation, 11, 12, 14, 15, 60, 229
detergents, 141
discharges, x, xi, 9, 35, 36, 38, 39, 40, 42, 50, 56, 67, 107, 119, 120, 121, 122, 123, 124, 125, 126, 128, 129, 130, 131, 132, 134, 135, 137, 138, 139, 140, 141, 142, 143
disclosure, xvi, 87, 228, 234, 236
disinfection, 128
distribution, 18, 30, 45, 84, 89, 190, 199, 233
District of Columbia, xii, xiii, 25, 110, 111, 135, 145, 148, 154, 158, 160, 161, 174, 177, 184, 185, 225
diversity, 66, 67, 93
DOJ, viii, 50, 54, 57, 58, 59, 60, 75, 76, 81, 82, 89, 90
DOT, 60, 76, 229, 230
draft, 13, 21, 59, 86, 116, 121, 124, 125, 133, 134, 141, 151, 154, 156, 164
drinking water, 10, 45, 65, 66, 73, 76, 112, 113, 116

E

earthquakes, 228
economic activity, vii, 1
economic downturn, 29
economic impacts, vii, 1, 120, 131, 142, 168
economic values, 156
effluent, xi, 2, 38, 40, 110, 119, 123, 124, 126, 127, 128, 129, 130, 132, 135, 137, 141
electricity, 3, 168, 218
ELGs, 38, 40
emergency, xv, 31, 45, 63, 83, 130, 166, 192, 222, 227, 228, 229, 231, 232, 234, 236
emergency planning, xv, 227, 228, 229, 234, 236
Emergency Planning and Community Right-to-Know Act (EPCRA), xv, xvi, 80, 98, 101, 103, 104, 228, 229, 232, 233, 234, 235, 236
emergency preparedness, 231
emergency response, xv, 227, 228, 231, 232, 234
emission, xii, xiii, xiv, 2, 4, 9, 11, 12, 13, 14, 16, 17, 19, 20, 21, 23, 25, 26, 27, 31, 32, 33, 43, 44, 56, 67, 100, 146, 147, 150, 162, 165, 167, 168, 169, 170, 173, 174, 175, 183, 185, 186, 187, 189, 190, 191, 192, 197, 198, 200, 201, 202, 203, 204, 205, 206,207, 209, 211, 213, 214, 216, 217, 218, 220, 221, 222, 223, 225, 226, 230
emission standards, xiii, xiv, 2, 4, 11, 12, 14, 16, 17, 18, 19, 24, 25, 26, 27, 30, 31, 32, 33, 43, 44, 67, 183, 185, 187, 189, 190, 191, 192, 194, 195, 197, 200, 204, 205, 209, 213, 214, 216, 220, 221, 223, 225
emitters, 13, 66
endangered species, 131
Energy Independence and Security Act, 12, 14, 28
Energy Policy Act of 2005, 45
Energy Policy and Conservation Act, 180
energy recovery, 44
enforcement, viii, ix, x, 42, 49, 50, 51, 52, 53, 54, 55, 56, 57, 58, 59, 60, 61, 62, 63, 64, 65, 66, 67, 68, 69, 70, 71, 72, 73, 74, 75, 76, 77, 78, 79, 80, 81, 82, 83, 84, 86, 88, 89, 90, 93, 94, 95, 96, 97, 98, 99, 100, 103, 106, 107, 108, 109, 137, 139, 176, 177,186, 206
environmental audit, 87
environmental community, xiii, 179
environmental conditions, 63, 73, 85, 86
environmental groups, vii, x, xi, 1, 4, 32, 33, 34, 37, 39, 40, 42, 86, 106, 110, 114, 117, 120, 126, 134, 136, 137, 190, 192, 203
environmental impact, 84, 149, 186, 210
environmental issues, 56, 64, 86
environmental laws, ix, 2, 50, 51, 54, 56, 57, 58, 59, 60, 63, 65, 68, 69, 70, 72, 74, 88, 97, 235
environmental management, 63, 69, 88

environmental organizations, 46
environmental policy, 52
environmental protection, ix, 53, 64, 76, 88, 90, 105, 106
environmental quality, 50, 84, 94, 131, 142
environmental regulations, 10, 59, 98
environmental standards, 74
environmental statutes, vii, 1, 7, 54, 57, 65, 70, 75, 77, 81, 82, 93, 98
epidemiology, 151
equipment, viii, xiii, xiv, 7, 24, 25, 27, 28, 30, 32, 40, 49, 123, 129, 183, 185, 189, 190, 198, 211, 214, 217, 219, 220, 223, 225, 226
ethanol, 27, 28, 45
executive order, xiii, 35, 64, 69, 79, 84, 85, 164, 170, 183, 184, 185, 228, 231
exposure, 24, 85, 151, 153, 155, 157, 223, 224, 233, 236

F

federal assistance, 79
Federal Bureau of Investigation (FBI), 60
federal facilities, 55, 68, 69, 70, 82, 83, 87
federal government, xiii, 37, 41, 52, 54, 63, 65, 66, 69, 79, 80, 90, 93, 179, 210
federal law, 61, 63, 83, 235
federal pollution control laws, v, viii, 49, 50, 52, 53, 57, 60, 61, 64, 65, 69, 75, 80, 90, 94
Federal Register, 11, 12, 13, 14, 15, 16, 18, 19, 20, 21, 22, 23, 24, 25, 26, 27, 28, 29, 30, 31, 32, 33, 35, 37, 38, 39, 40, 41, 43, 44, 45, 46, 64, 80, 84, 87, 88, 121, 123, 124, 127, 143, 146, 148, 154, 159, 161, 162, 164, 171, 172, 174, 194, 204, 209, 210, 214, 226, 231, 232
federal regulations, 17, 22, 223, 233, 236
federalism, 52
field crops, 147
financial capability assessment, x, 106, 111, 112, 113, 116
financial support, 115
first responders, xv, 228
fiscal year, ix, 50, 51, 60, 71, 73, 76, 89, 92, 95, 100, 101, 102, 103, 104
fish, 34, 37, 39, 124, 125, 129, 141
Fish and Wildlife Service, 60, 76
fisheries, 131
fishing, x, xi, 119, 120, 122, 123, 125, 129, 138, 140, 142
flexibility, ix, x, 36, 53, 70, 105, 106, 108, 109, 110, 112, 114, 116, 130, 135, 202
fluctuations, 71
flue gas, 215
food additive, 233
Food and Drug Administration, 233
forced air furnaces, xiv, 26, 213, 214, 215, 216, 217, 218, 219, 223
forest fire, 158
formation, 98, 149, 162, 163, 171, 174, 177
freighters, x, 119, 120, 123
fuel consumption, 15, 181
fuel economy, xii, 11, 12, 14, 179, 180
fuel efficiency, 4, 14
fuel management, 141
fuel prices, 13
funding, viii, ix, 45, 49, 50, 51, 53, 54, 55, 63, 77, 79, 89, 90, 94, 106, 114, 115

G

GAO, 52, 53, 54, 62, 66, 88, 95
garbage, 129, 223
global warming, xiii, 11, 18, 183, 184
governments, 5, 53, 61, 64, 65, 96, 99, 169, 210
GPA, 195
greenhouse gas, xii, xiii, 2, 9, 11, 12, 13, 14, 15, 17, 18, 100, 179, 183, 184, 210
greenhouse gas (GHG), xii, xiii, 9, 11, 12, 13, 14, 15, 16, 18, 31, 100, 179, 180, 181, 183, 184, 185, 186, 187, 197
greenhouse gas (GHG) emissions, v, 2, 9, 11, 12, 13, 14, 15, 16, 18, 19, 180, 184, 186, 187
Gross Domestic Product (GDP), 4, 102, 104, 168
groundwater, 34, 38, 43, 45, 73
guidelines, xiii, 5, 16, 38, 40, 76, 81, 111, 113, 129, 183, 185, 186, 187, 188, 190, 200, 201, 202, 203, 204, 205, 206, 207
Gulf of Mexico, ix, 50, 72, 81, 103

H

HAP, 187
Hawaii, 17, 165
hazardous air pollutants, 18, 30, 67, 149, 188
hazardous materials, 76, 229
hazardous substance, xv, 228, 229, 230, 232, 233, 234, 235
hazardous waste, viii, 49, 50, 70, 73, 76, 97, 98, 99
hazards, xv, 46, 228, 230, 231

health and environmental effects, 153
health condition, 236
health effects, 146, 147, 153, 204, 223, 226
health risks, 85, 224, 229
heart attack, 19, 147, 166, 222
heart disease, 31
homeland security, 78, 229
House Committee on Transportation and Infrastructure, 110, 116
House of Representatives, 54, 224
household income, 112
human, x, 18, 43, 56, 73, 83, 84, 93, 109, 119, 120, 131, 151, 166, 230, 236
human activity, 84
human exposure, 151
human health, x, 43, 56, 73, 83, 93, 109, 119, 120, 131, 166, 230, 236
hurricanes, 228
hydrogen peroxide, 128
hydronic heaters, xiv, 26, 27, 213, 214, 215, 216, 217, 218, 219, 220, 223, 224, 225, 226

I

ICE, 60
Immigration and Customs Enforcement, 60
immunity, 82
IMO, 127, 128, 129, 132, 133, 134, 137, 139
imprisonment, 76
incarceration, 71, 79
industrial sectors, xiii, 183, 189
industry, xii, xiii, xiv, 4, 9, 10, 11, 16, 19, 20, 24, 25, 27, 29, 30, 31, 32, 33, 34, 38, 39, 40, 41, 68, 84, 86, 124, 134, 136, 140, 142, 145, 169, 179, 181, 189, 193, 198, 199, 203, 213, 214, 215, 216, 217, 218, 220, 221, 222, 224, 225, 226
inflation, ix, 50, 54, 80, 103, 104
infrastructure, viii, ix, x, 28, 49, 52, 55, 105, 106, 107, 108, 109, 110, 111, 113, 114, 116, 131, 189
insecticide, 35
inspections, 32, 61, 62, 65, 73, 98, 99, 101
inspectors, 59, 72, 78
interest groups, xii, 51, 146
Internal Revenue Service, 60
International Maritime Organization, 32, 127
international standards, 32
intrusions, 171, 172
invasions, 131

J

Jackson, Lisa, 3
Johnson, Stephen, 154
jurisdiction, 10, 42, 64, 97, 172, 192, 193, 233

L

labeling, 28
labor unions, xiii, 179
landfills, xiii, 43, 183, 185, 186, 187, 199, 200, 201, 202, 203, 204, 207
law enforcement, 76, 77, 78
laws, viii, ix, 7, 28, 49, 50, 51, 52, 53, 55, 56, 57, 58, 59, 60, 61, 63, 64, 65, 68, 69, 70, 72, 74, 76, 79, 80, 83, 88, 89, 90, 94, 97, 235
laws and regulations, viii, 28, 49, 61, 68, 69, 74, 88
lead, 46
leadership, 88, 224
legal issues, xiii, 184, 185, 196
legislation, ix, xi, 3, 7, 8, 21, 30, 35, 36, 41, 50, 54, 55, 61, 63, 67, 84, 94, 114, 115, 116, 119, 122, 138, 139, 140, 171, 221, 230
legislative proposals, 7, 107, 139
light trucks, xii, 12, 26, 179
light-duty vehicles, xii, 179
litigation, 55, 59, 60, 63, 90, 150, 187, 196, 199, 200, 203, 204, 207
local authorities, 61
local community, 88
local conditions, 135
local government, ix, x, 36, 51, 57, 61, 64, 82, 87, 95, 96, 105, 106, 108, 109, 111, 112, 116, 150, 175, 224, 228, 234
lubricants, 129, 131, 132
lung disease, xii, 145

M

MACT, 6, 24, 25, 30, 44, 67
majority, xiv, 3, 24, 57, 61, 66, 69, 71, 74, 81, 90, 103, 125, 192, 193, 194, 214, 218
management, xi, xv, 10, 35, 37, 38, 66, 77, 108, 113, 115, 119, 121, 123, 124, 126, 127, 131, 132, 133, 134, 137, 138, 139, 140, 141, 227
manufacturing companies, 221
Marine Corps, 143
marketplace, 13, 29, 180
median household income (MHI), x, 106, 112, 113

mercury, 24, 31, 33, 40, 188, 188
metals, x, 40, 68, 119, 120, 123, 124, 187
methane, v, xiii, 11, 17, 18, 171, 183, 184, 185, 186, 187, 188, 189, 190, 194, 197, 199, 200, 201, 207, 209, 210
methane emissions, xiii, 17, 18, 183, 184, 186, 187, 189, 190, 194, 197, 199, 200, 201, 209, 210
methodology, 5, 166, 176
metropolitan areas, 148
micrograms, 21, 128
microorganisms, 127
modifications, 9, 38, 80, 109, 110, 125, 130, 137, 139
moisture content, 220
molecular oxygen, 171
moratorium, x, xi, 119, 120, 121, 122, 123, 124, 138, 139, 140, 141
mosquitoes, 35
MSW, xiii, 183, 185, 186, 187, 188, 194, 199, 201, 202, 203, 204, 207
multiple factors, 29
municipal solid waste (MSW), xiii, 183, 185, 186, 187, 188, 194, 199, 200, 201, 202, 203, 204, 207

N

National Academy of Sciences, 132
National Ambient Air Quality Standards (NAAQS), xi, 10, 19, 20, 21, 22, 23, 67, 145, 146, 147, 148, 149, 150, 151, 152, 153, 154, 155, 158, 159, 160, 161, 162, 163, 164, 165, 167, 168, 169, 170, 171, 172, 173, 174, 175, 176, 177, 186, 187, 206, 226
National Association of Manufacturers, xii, 20, 145, 168
National Defense Authorization Act, 140
National Highway Traffic Safety Administration (NHTSA), xii, 12, 13, 14, 15, 179, 180, 181, 182
National Oceanic and Atmospheric Administration (NOAA), 60, 163
National Priorities List, 70
National Research Council, 34, 132, 133
national security, 68, 78, 143
natural disaster, xv, 227, 228
natural gas, xiii, 16, 17, 18, 30, 163, 183, 184, 185, 186, 188, 189, 190, 194, 197, 198, 199, 209, 210, 211, 215
natural resources, 34, 79, 90, 130
negative effects, xii, 19, 145
NEPA, 91
nitrates, 129
nitrogen, 22, 23, 33, 36, 67, 149, 162, 187
nitrogen dioxide, 22, 67
nitrous oxide, 11
noncompliance, viii, ix, 49, 61, 63, 70, 73, 74, 79, 81, 82, 87, 97, 105, 109, 124, 126, 130
NPL, 70
NRC, 132, 134
nutrient, 36, 37, 130
nutrients, x, 36, 40, 119, 120, 123, 124, 129, 141

O

Obama Administration, viii, xii, xiii, 2, 6, 8, 10, 20, 34, 41, 175, 176, 179, 182, 183, 184, 185, 187, 189, 200, 204, 231
Obama, President Barack, vii, ix, 1, 2, 26, 42, 50, 89, 184, 231
Occupational Safety and Health Act, 233
Office of Management and Budget (OMB), 5, 17, 35, 38, 102, 104, 140, 153, 154, 156, 170, 222
oil, x, xiii, 15, 18, 24, 30, 31, 40, 41, 42, 44, 72, 103, 119, 120, 123, 124, 125, 129, 130, 131, 141, 142, 163, 169, 183, 184, 185, 186, 188, 189, 190, 194, 197, 198, 199, 200, 210, 211, 215, 229
oil and grease, x, 119, 120, 123, 124
Oil Pollution Act of 1990, 40
oil production, 18
oil spill, 40, 42, 72
Omnibus Appropriations Act, 84
operating revenue, 131
operations, viii, 34, 37, 49, 52, 55, 66, 67, 69, 73, 79, 83, 122, 125, 163, 198, 233, 236
organic compounds, 199
organism, 133, 137
OSHA, 229, 230, 232, 236
oversight, vii, viii, 1, 3, 7, 44, 49, 52, 53, 61, 62, 67, 94, 98, 111, 114, 231
oxygen, 27, 37
ozone, v, xi, xii, 6, 10, 18, 19, 20, 21, 22, 23, 67, 128, 145, 146, 147, 148, 149, 150, 151, 152, 153, 154, 155, 156, 157, 158, 159, 160, 161, 162, 163, 164, 165, 166, 167, 168, 169, 171, 172, 173, 174, 175, 176, 177
ozone concentrations, xii, 20, 145, 147, 157, 163, 164, 174
ozone standards, xi, 20, 21, 145, 148, 149, 154, 158, 159, 161, 162, 165, 176

P

P.L. 99-499, 232
paints, 163
passenger cars, xii, 26, 179
passenger vessels, x, 119, 120, 141
pathogens, x, 119, 120, 123, 124
penalties, viii, ix, 49, 52, 55, 62, 69, 70, 71, 74, 75, 79, 80, 81, 82, 83, 84, 87, 94, 95, 97, 98, 99, 100, 101, 102, 103, 105, 107, 109, 156
performance measurement, 86
permit, x, xi, 7, 11, 13, 14, 34, 35, 36, 37, 42, 43, 56, 67, 69, 72, 74, 97, 98, 106, 108, 109, 111, 113, 114, 115, 119, 120, 121, 122, 123, 124, 125, 126, 128, 129, 130, 131, 132, 133, 134, 135, 136, 137, 138, 139, 140, 141, 142, 235
personal communication, 175
pesticide, 35, 61, 135, 163
petroleum, 4, 33, 41, 130, 163, 168, 169, 195, 196, 197, 199, 223, 229
pH, 129
phosphate, 187
phosphorus, 36
plant growth, 157
plants, 16, 17, 18, 24, 31, 39, 40, 147, 157, 175, 187, 188, 190
plastics, 4
policy issues, 35
policymakers, 107, 114
pollutants, viii, x, 2, 10, 14, 18, 19, 21, 24, 25, 26, 30, 31, 33, 38, 40, 49, 56, 67, 72, 119, 120, 121, 123, 124, 129, 131, 141, 142, 149, 150, 151, 153, 162, 163, 166, 169, 187, 188, 189, 197, 204, 210, 222, 225
pollution, vii, viii, ix, xii, xiv, 1, 2, 4, 11, 13, 16, 18, 19, 23, 24, 25, 27, 32, 33, 36, 37, 38, 41, 49, 50, 52, 53, 54, 55, 56, 57, 60, 61, 63, 64, 65, 66, 67, 68, 69, 70, 73, 74, 75, 76, 79, 80, 81, 82, 83, 86, 89, 90, 93, 94, 95, 97, 105, 109, 127, 129, 138,143, 145, 148, 150, 153, 162, 163, 164, 165, 167, 169, 171, 174, 175, 176, 186, 187, 197, 198, 199, 205, 206, 209, 210, 213, 224, 225, 226
pollution control, vii, viii, xii, 1, 2, 16, 23, 24, 25, 32, 36, 49, 50, 51, 52, 53, 54, 55, 56, 57, 60, 63, 64, 65, 66, 67, 68, 69, 70, 73, 74, 75, 76, 79, 81, 82, 86, 89, 90, 93, 94, 95, 106, 143, 145, 150, 151, 169, 205, 225, 226
pollution control statutes, vii, 1, 2, 52, 54, 57, 60, 63, 65, 66, 68, 81, 82, 86, 90, 94
polycyclic aromatic hydrocarbon, 129
power generation, 17
power plants, xii, 6, 9, 13, 16, 17, 22, 23, 24, 39, 40, 43, 146, 149, 150, 163, 165, 175, 186, 187, 188
premature death, xii, xiv, 5, 19, 21, 22, 24, 25, 27, 31, 145, 147, 166, 213, 222, 223
prevention, xv, 40, 45, 83, 210, 227, 236
price elasticity, 219
procedural right, 188
procedural rule, 81
profitability, 142
property taxes, 113
proposed regulations, 3
public assistance, 65, 112
public awareness, 231
public health, vii, ix, xii, 1, 5, 8, 11, 50, 63, 83, 85, 86, 105, 106, 108, 145, 146, 149, 150, 153, 155, 166, 173, 177, 186, 197, 198, 199, 204, 209, 210
public health advocates, vii, 1
public interest, 46, 84, 94
public resources, 210
public safety, 130
public welfare, 149, 155, 156, 157
pulp, 187

Q

quality assurance, 175
quality standards, 22, 36, 37, 153, 186

R

radiation, 67, 152, 158, 168
railroad ties, 223
ratepayers, 107
Reagan, Ronald, 64
real property, 90
recycling, 43, 44, 203
redevelopment, 39
reform, 8, 52, 170
regulations, vii, viii, xii, xiii, xv, 1, 2, 3, 4, 5, 6, 7, 8, 9, 11, 12, 17, 19, 24, 25, 26, 30, 31, 34, 37, 38, 39, 40, 41, 42, 43, 44, 45, 46, 53, 54, 58, 59, 66, 71, 75, 76, 85, 94, 112, 122, 126, 143, 146, 158, 164, 168, 172, 173, 175, 176, 177, 179, 183, 184, 185, 186, 189, 195, 200, 203, 204, 206, 209, 215, 216, 217, 219, 221, 224, 225, 226, 227, 229, 230, 232
regulatory affairs, 5
regulatory agencies, 7, 8, 56, 205

regulatory requirements, 17, 50, 56, 66, 72, 75, 102, 114, 121, 140, 143, 230
rehabilitation, 111
relaxation, 114
remote sensing, 73
renewable fuel, 4, 28, 29
Renewable Fuel Standard, 4, 28, 29
Renewable Fuel Standard (RFS), 28, 29, 30
Renewable Fuels Association, 28
Republican majorities, vii, 1, 7
research institutions, 153
Reservations, 64
Residential, 27, 46, 112, 214, 215, 216, 219, 223, 224, 225, 226
resources, viii, xiii, 29, 39, 49, 53, 59, 61, 62, 65, 69, 71, 77, 78, 79, 81, 82, 84, 85, 86, 88, 89, 94, 95, 96, 106, 170, 183, 184
restrictions, 202
risk management, xv, 227, 229, 230
rubber products, 4
runoff, 36, 37, 38, 39, 109, 110
rural areas, 22, 163, 172, 174, 175

S

saltwater, 126, 134, 140
Science Advisory Board (SAB), 8, 133, 137, 223
Secretary of Defense, 143
security threats, xv, 227, 228, 229
selenium, 40, 43
Senate, vii, xi, xiv, 1, 3, 7, 26, 42, 52, 55, 77, 89, 115, 116, 120, 138, 139, 140, 171, 176, 205, 209, 220, 221, 224
September 11, 96
service industries, 38
sewage, 107, 110, 121, 130
shellfish, 39
Sierra Club, 33, 46, 151, 196, 226
sludge, 10
small business, 66, 82, 142, 219, 221, 222
small communities, 225
small firms, 142
smog, 147
solvents, 163
soybeans, 157
spill prevention control and countermeasure (SPCC), 40, 41
stakeholder groups, 58
stakeholders, ix, xiii, xiv, 9, 15, 19, 44, 55, 85, 88, 105, 109, 111, 114, 116, 176, 179, 181, 182, 184, 195, 209, 210, 217, 219, 226, 232
State Implementation Plan, xii, 145, 150, 162, 176, 177
state planning, 53
state regulators, viii, 49, 112, 114
statutes, vii, 1, 2, 7, 10, 50, 51, 52, 53, 54, 55, 56, 57, 59, 60, 63, 64, 65, 66, 67, 68, 70, 73, 75, 76, 77, 80, 81, 82, 86, 90, 93, 94, 98
statutory authority, xiii, 184, 185
statutory provisions, 57, 63
storage, xv, 16, 18, 30, 40, 41, 82, 190, 198, 199, 211, 227, 228, 236
stormwater, ix, 36, 38, 39, 105, 106, 107, 108, 109, 110, 111, 112, 113, 115, 116, 134
stoves, xiv, 26, 213, 214, 215, 216, 217, 218, 219, 220, 221, 223, 224, 225
strategic planning, 58, 63, 67
sulfur, 11, 22, 23, 26, 32, 33, 67, 149, 187
sulfur dioxide, 22, 23, 33, 67, 149
sulfuric acid, 187
Superfund, 51, 57, 61, 70, 79, 90, 92, 97, 98, 104, 232, 234
Superfund Amendments and Reauthorization Act of 1986, 232
Supreme Court, 2, 9, 11, 12, 14, 17, 23, 24, 35, 42, 82, 168, 170, 176, 205

T

tankers, x, 119, 120, 123, 125, 130, 131
tanks, 45, 126, 128, 129, 134, 140, 141, 211
Task Force, 76, 170
technical assistance, 39, 85, 111, 115
temperature, 162, 163
Tennessee Valley Authority, 43
terrorist attack, xv, 227, 228
time frame, viii, 2, 74, 108, 167, 180
Title I, 59, 77, 84, 232
Title II, 59, 77, 84, 232
Title V, 13, 14, 77, 85, 89
toluene, 199
total energy, 184
tourism, 131
toxic chemical compounds, x, 119, 120, 124
toxic effect, 123
toxic substances, xv, 50, 85, 227
toxicity, 233, 236
toxicology, 151

transmission, 18, 30, 163, 190, 198, 199, 210, 211
transportation, 19, 28, 66, 87, 123, 125, 140, 229
tribal lands, 64, 211
Trump, Donald, President, xiii, 183, 184, 185

U

U.S. Army Corps of Engineers, 33
U.S. Department of Agriculture, 41
U.S. economy, xv, 227, 228
U.S. history, 72
U.S. Treasury, 79
U.S. waters, x, 34, 42, 119, 120, 121, 122, 127, 132, 135, 139
unemployment rate, 112
Union Carbide, 228, 230
unit cost, 219
United Nations, 19, 127, 184
United Nations Framework Convention on Climate Change, 184
urban areas, 21, 163, 174

V

vegetation, 149, 151, 155, 157, 171
vessel discharges, x, xi, 119, 120, 121, 129, 132, 135, 138
Vessel General Permit (VGP), v, x, xi, 119, 120, 121, 122, 123, 124, 125, 126, 127, 128, 129, 130, 131, 132, 133, 134, 135, 136, 137, 139, 141, 142, 144
vessels, x, xi, 119, 120, 121, 122, 123, 124, 125, 126, 127, 128, 129, 130, 131, 132, 133, 134, 135, 137, 138, 139, 140, 141, 142, 143, 189
vinyl chloride, 199
viruses, 127, 129
volatile organic compound (VOC), xiii, 18, 30, 162, 163, 168, 169, 183, 185, 187, 189, 190, 197, 198, 199, 222, 223
volatile organic compounds, 30, 162, 185, 222, 223
volatility, 233

W

waste, 2, 34, 43, 44, 61, 66, 67, 72, 76, 81, 97, 123, 124, 141, 187, 201, 203, 210, 223
waste disposal, 34
waste incineration, 44
waste incinerator, 44, 187
waste management, 2, 141, 203
wastewater, ix, 36, 38, 40, 66, 105, 106, 107, 108, 109, 110, 111, 113, 114, 115, 116, 125
water, viii, ix, x, xi, 6, 7, 9, 34, 35, 36, 37, 39, 40, 42, 45, 49, 50, 56, 58, 61, 67, 72, 97, 98, 99, 105, 106, 107, 108, 109, 111, 112, 113, 114, 115, 116, 120, 121, 122, 123, 124, 125, 126, 127, 128, 129, 130, 131, 132, 133, 134, 135, 136, 137, 138, 139, 140, 141, 142, 215, 223
water quality, ix, 35, 36, 37, 39, 42, 105, 106, 107, 108, 109, 111, 115, 128, 134, 135, 136
water quality standards, 35, 36, 37, 42, 128, 135
watershed, 37
waterways, 36, 41, 110, 141
welfare, xi, 11, 19, 20, 145, 147, 149, 150, 155, 156, 186, 197, 198, 199, 209, 210
wells, 18, 30, 45, 163, 164, 191
wetlands, 42, 61, 97
wildlife, 34, 90
wood, xiv, xv, 26, 27, 213, 214, 215, 216, 217, 218, 219, 220, 221, 222, 223, 224, 225, 226
wood heaters, xiv, 26, 27, 213, 214, 215, 216, 217, 219, 220, 221, 223, 224, 225, 226
workplace, 233
worst case scenarios, 142, 231

Key Reports from the U.S. Environmental Protection Agency Office of the Inspector General

Editor: Debroah Ledoux

Series: American Political, Economic, and Security Issues

Book Description: This book provides meaningful reports from the U.S. Environmental Protection Agency Office of the Inspector General. In addition to providing details of some key successful operations within the agency, the reports also provide details about operations within the agency that need updated measures and controls to be effective.

Hardcover ISBN: 978-1-53614-182-5
Retail Price: $270

Oversight and Controls of the U.S. Environmental Protection Agency

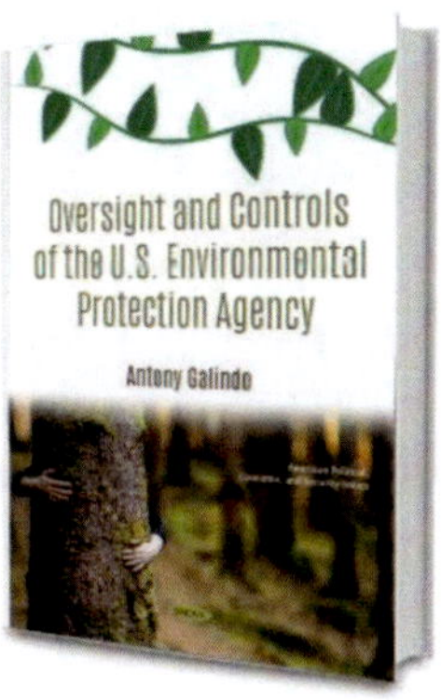

Author: Antony Galindo

Series: American Political, Economic, and Security Issues

Book Description: This book provides accounts of the oversight and operations of the U.S. Environmental Protection Agency as reported by the office of the Inspector General.

Hardcover ISBN: 978-1-53614-180-1
Retail Price: $230